Forum for Interdisciplinary Mathematics

The *Forum for Interdisciplinary Mathematics* is a Scopus-indexed book series. It publishes high-quality textbooks, monographs, contributed volumes and lecture notes in mathematics and interdisciplinary areas where mathematics plays a fundamental role, such as statistics, operations research, computer science, financial mathematics, industrial mathematics, and bio-mathematics. It reflects the increasing demand of researchers working at the interface between mathematics and other scientific disciplines.

Manoranjan Kumar Singh

Applied Fuzzy Mathematics

Manoranjan Kumar Singh
Department of Mathematics
Magadh University
Bodh-Gaya, Gaya, Bihar, India

ISSN 2364-6748 ISSN 2364-6756 (electronic)
Forum for Interdisciplinary Mathematics
ISBN 978-981-97-3259-3 ISBN 978-981-97-3257-9 (eBook)
https://doi.org/10.1007/978-981-97-3257-9

This Springer imprint is published by the registered company Springer Nature Singapore Pte Ltd.
The registered company address is: 152 Beach Road, #21-01/04 Gateway East, Singapore 189721, Singapore

To my grandparents, the late Dharamdeo Singh and late Jhalka Devi
To my parents, the late Kamala Devi and late Lalan Prasad Singh

Foreword

It is my great pleasure to write a few lines about the effort of Prof. Manoranjan Kumar Singh for his book, *Applied Fuzzy Mathematics*. The author has successfully managed to bring out a compact and complete textbook for undergraduate and graduate students, computer scientists, and professionals who need knowledge of the application of fuzzy mathematics. With the help of some instructive examples, the book illustrates the various shades of fuzzy mathematics used in the world of knowledge. The book is intended for readers who are particularly interested in fuzzy mathematics applicability to various fields of mathematics, technology, and daily life. While each chapter begins with the fundamental facts or principles of the underlying mathematics necessary for a foundational comprehension of the techniques described, the book is applications focused.

The major goal of the book is to make the topic easier to understand and more visually appealing, in which the author executes superbly. The book is easy to understand because each chapter begins with the fundamentals of the concerned chapter. Each chapter has supplemented with multiple-choice questions and true/false type statements that aid students in developing their critical thinking skills. Those studying medicine, engineering, and undergraduate and graduate students will all benefit from this book, as fuzzy logic is essential to computing with words (CWW) and vice-versa. The application aspects of fuzzy sets, fuzzy relation, and fuzzy logic in certain crucial areas of human existence and knowledge are well covered in the book. Fuzzy relation equations and neural networks are used to evaluate cardiovascular diabetes mellitus and the life insurance sector problems, and display the use of fuzzy set theory in computer science, sciences, and engineering, as well as industrial applications. The use of fuzzy sets in social sciences has also been explained lucidly. It also depicts a glimpse of the use of fuzzy graphs. Decision-making, anticipating the outcome of a cricket match, and applications in medical science have all been illustrated elegantly.

Ultimately, I wish the author all the success and congratulate him on his creative initiative.

April 2023

Prof. Santosh Kumar
Department of Mathematic, College
of Natural and Applied Sciences
University of Dar es Salaam
Tanzania

Preface

Applications of Fuzzy Mathematics

There is a common saying that beauty lies in the eyes of the beholder. The world's best and most beautiful things cannot be seen or even touched; they must be felt with the heart. This means that beauty doesn't exist independently but has been created by observers. This famous quote can help us remember that a beholder sees or experiences things, becoming aware of them. Fuzzy mathematics has made mathematicians and scientists feel about the subject and attracts the admirer due to its usefulness. It has an aura of its own. No one can put fuzzy mathematics within the bounds of words.

This book portrays different shades of colors of fuzzy mathematics in the realm of knowledge and some illustrative examples. The book is designed for those interested primarily in applications of fuzzy mathematics in mathematics, technology, and different spheres of life. Focusing on applications, each chapter begins with the basic facts or principles of the underlying mathematics required for a fundamental understanding of the methods discussed.

In Stanley P. Gudder's (1988) words, "a slight variation in the axioms at the foundation of a theory can result in massive change at the frontiers." Classical sets to fuzzy sets are a significant paradigm shift. This is due to a wide array of successful applications of fuzzy set theory ranging from consumer products to automotive applications. In this book, an attempt has been made to have such a depiction of applications of fuzzy set theory and fuzzy logic. Here fuzzy logic confers to a logical system that induces classical two-valued reason for argument under uncertainty. In the far-reaching sense, fuzzy logic confers to all theories that utilize fuzzy sets, which are classes with unsharp boundaries.

Chapter Organization

The book is structured into seven comprehensive chapters. Chapter 1 elaborates on the concepts of fuzzy relation equations with coefficients represented as fuzzy sets of real numbers [Buckley & Om, 1990]; that is, an equation is viewed as a fuzzy equation if its coefficients are fuzzy sets of real numbers. These fundamental ideas lay the groundwork for exploring complex mathematical relationships in the realm of fuzzy logic.

Chapter 2 introduces the precepts of the fuzzy graph and intuitionistic fuzzy graph theories by introducing some simple knowledge of classical graph theory—based on which fuzzy and intuitionistic fuzzy graphs are established. Here, the notions of adjacent graphs, weighted graphs, simple graphs, multiple graphs, pseudo graphs, etc., are given under fuzzy and intuitionistic fuzzy environments. Also, the degree of a node, the degree of an edge, regular graph, irregular graph, complete graph, the order of a graph, etc., are mentioned. Again, a fuzzy and intuitionistic fuzzy subgraph of a graph, spanning subgraph, induced subgraph, walk, path, trail, connected graph, and the distance between nodes are discussed. Additionally, binary trees, forests, deletion, cut vertex, bridge, eccentricity, center, radius, and the diameter of a graph are covered. Finally, the metric, union, and intersection of graphs, sum, and product of two graphs, along with illustrative examples, are discussed.

Chapter 3 describes the possibility theory and its intertwined relationship with the idea of fuzzy sets. The whereabouts of possibility theory surrounded by fuzzy measure theory are also dealt with within it. Several distinct classes of fuzzy measures exist, including plausibility or belief measures, possibility or necessity measures, and probability measures have been discussed. It also presents Dempster's rule of combination, a lattice of the possibility distribution, and possibility theory vis-a-vis probability theory. Finally, glimpses of Sugeno's and Choquet's integral have been dealt with.

Chapter 4 presents the compositional rule of inference, the main inference rule in approximate reasoning. This chapter overviews the fundamental aspects or features of two-valued logic. Under this, in brief, we have illustrated different connectives and truth tables for them along with NAND, NOR, and XOR. Further, it deals with functionally complete connectives, logical equivalence, the principle of duality, the principle of substitution and tautological implications, quantification, three-valued logic, and n-valued logics for an arbitrary number of truth values ($n \geq 2$). Further, fuzzy logics, fuzzy propositions, fuzzy quantifiers, linguistic hedges, inference from conditional fuzzy propositions, inference from qualified propositions, and inference from quantified propositions have been sketched. In brief, this chapter has dealt with essential aspects of fuzzy logic and fuzzy propositions.

Chapter 5 discusses the application of fuzzy set theory—to the realm of topological space, usually referred to as fuzzy topological spaces. C. L. Chang (1968) propounded the notion of fuzzy topological spaces. It covers the comparison of fuzzy topologies, the neighborhood of a point and fuzzy set, the interior and closure of a fuzzy set, fuzzy limit point, derived fuzzy set, basis and sub-basis of a fuzzy set, fuzzy set induced by mapping, fuzzy continuous function, fuzzy homomorphism, the cover of a fuzzy set and compact fuzzy space, along with valuable results over them.

Chapter 6 is dedicated to the fuzzification of algebraic notions: fuzzy groupoid, fuzzy group, fuzzy level subgroups, fuzzy normal subgroups, fuzzy cosets, and their algebraic properties, along with some quirky/eccentric illustrations. This chapter showed how Rosenfeld (1971) and Singh (1993, 2003, 2006, 2010) formulated the fundamental concepts of fuzzy subgroups; Das (1981), the conception of fuzzy level subgroups; and Bhattacharya (1985,1987), the perception of fuzzy normal subgroups and fuzzy cosets. It also shows that fuzzy group, fuzzy normal subgroup, and fuzzy coset are an analog of the concept of the group, normal subgroup, and coset of a group.

Chapter 7 is devoted to the applicational aspects of fuzzy sets, fuzzy relations, and fuzzy logic in some critical facets of human life and knowledge, as fuzzy logic plays a pivotal role in computing with words (CWW) and vice-versa. CWW is a methodology in which words are used instead of numbers for computing, and reasoning has been sketched elegantly. It consists of fuzzy set theory applications in computer science, science, and engineering, industrial applications, fuzzy relation equations, and neural networks in evaluating cardiovascular diabetes, mellitus, and life insurance. We have also dealt with the applications of fuzzy sets in psychology and economics. Further, it also covers applications of fuzzy graphs. Applications in medical science, forecasting the score of a cricket match, and decision-making have also been dealt with. It also examines applications of fuzzy groups and fuzzy topology in geographic information systems (GIS) and the area of innovative city projects and transshipment problems with CPM/PERT.

Gaya, India — Manoranjan Kumar Singh

Acknowledgements

The author expresses his sincere gratitude to all the people who have had a direct or indirect influence on the production of this book:

- Professor L. A. Zadeh propounded this ground-breaking objective. To Profs. A. Rosenfeld, C. L. Chang, R. Lowen, K. T. Atanassov, S. Stoeva, Smarandache, and many more mathematicians whose works and research papers have motivated me to write a book in this domain.
- To my teachers and motivators Profs. Dr. Sudama Singh, Dr. Bishwanath Singh, Dr. Rajeswar. Pd. Singh, and Dr. Surendra Prasad Singh from Magadh University Bodh Gaya and Gaya College, Gaya, Bihar, and Dr. R. P. Singh, of T. D. College, Jaunpur, Uttar Pradesh, India.
- To all my friends and well-wishers who knowingly or unknowingly have motivated me. To name a few: Dr. Gajendra Pratap Singh of the School of Computational and Integrative Science, Jawaharlal Nehru University, New Delhi; Dr. Sunil Singh Chandel, Ujjain (M.P.) and Dr. Jitendra Tiwari, S. N. D. T. University, Mumbai; to my students, Dr. Dheeraj Kumar, Government Tilak P. G. College, Katni, Madhya Pradesh; Dr. Sudipto Gayan, Faculty of Engineering and Technology, Siksha "O" Anusandhan, Bhubaneswar, Odisha, India; and other students of various batches. I thank them all for their pertinent suggestions and valuable input.
- My family members have been my most consistent source of support and inspiration. I owe them much more than any words can express for their sound counseling, cheerful support, love, and best wishes, which has been keeping my spirit up. Without their support and encouragement, this arduous task of writing a book on such a topic could never have seen the light of day. So, my heartiest thanks and gratitude go to my wife, Dr. Ajanta Singh, for helping me realize my strengths and weaknesses, and to my two sons, Aniket Ranjan and Brajesh Ranjan, and my daughter, Ananta Ranjan, for their love, affection and sparing time with me during this arduous mission. Owing to their patience and constant support throughout the drafting of this book, this book has become a reality.

- I feel blessed for my father-in-law, the late Rambilas Singh, and mother-in-law, the late Sumitra Devi; my maternal grand father & mother late Mangal Rai and late Resma Devi; my Vice-Chancellor, the late Ishwar Chandra Kumar of V. K. S. U. Ara, Bihar; my teacher, late Prof. Baij Nath Prasad and late Prof. N. C. Agrawal, whose love, blessings, support, and inspiration have made me what I am today. Lastly, to the one person who has been the source of my strength and courage and whose very existence is my own, my beloved wife, Dr. Ajanta Singh along with my children Engrs. Aniket Ranjan, Ananta Ranjan & Brajesh Ranjan. I pay my utmost regards to my mother Maa Mangla.
- Many thanks are also due to Shamim Ahmad, Executive Editor, Springer Nature India, New Delhi, for guiding me throughout this project. In the absence of his guidance, it would have been much harder to sail. My thanks are also due to Springer Nature for accepting the book for publication.
- The author is thankful to all the referees for their valuable suggestions for the improvements and suggestions needed in this book.

Contents

Chapter 1
Solving Fuzzy Linear Equations

This chapter addresses different types of fuzzy relation equations whose coefficients and unknowns are fuzzy sets of real numbers. We will consider the fuzzy equivalents of $a + x = c$, $ax = c$, and $ax + b = c$.

1.1 Introduction

Solving linear and quadratic fuzzy equations was carefully examined and investigated by **J. J. Buckley** and **Y. Qu** in *1990* and was *published in the 38th volume of Fuzzy Sets and Systems*. Here, they obtained necessary and sufficient conditions for some linear and quadratic fuzzy equations to have a solution when the parameters are either real or complex fuzzy numbers. Then, the answer, i.e., the solution, will be a real or complex fuzzy number. The authors presented necessary and sufficient conditions for $A + X = C$ and $AX = C$ to have solutions for X when A and C are arbitrary fuzzy subsets of the reals.

This section will consider the fuzzy equivalents of $a + x = c$, $ax = c$, and $ax + b = c$. In each case, we first have a and c (or a, b, c) as real fuzzy numbers and then allow them to be complex.

1.2 Fuzzy Equation

An equation is called a fuzzy equation if its coefficients and variables are fuzzy sets of $\mathbb{R}$ and solutions can be obtained by applying operations of fuzzy arithmetic.

The concept of the equation can be extended to deal with fuzzy quantities in many ways. Let us consider the simple equation $a * x = b$, where $(a, b) \in$, x is a real variable

M. K. Singh, *Applied Fuzzy Mathematics*, Forum for Interdisciplinary Mathematics,
https://doi.org/10.1007/978-981-97-3257-9_1

and $*$ is a group operation on $\mathbb{R}$, such that the unique solution of this equation is $X = a^{-1} * b$, where a^{-1} is the inverse of a.

Theorem 1.1 The equation $A + X = C$ has a solution X if and only if $c_1 - a_1 < c_2 - a_2 < c_3 - a_3$. If a solution exists, X is the triangular fuzzy number $(c_1 - a_1 / c_2 - a_2 / c_3 - a_3)$.

Proof: Taking α –cuts, we obtain $a_i^\alpha + x_i^\alpha = c_i^\alpha$, $i = 1, 2$. Then $x_1 < x_2 < x_3$ and $x_1^\alpha > 0$, $x_2^\alpha < 0$ iff $c_1 - a_1 < c_2 - a_2 < c_3 - a_3$.

If A and C are rectangular complex fuzzy numbers, then let $A = A_1 + iA_2$, $C = C_1 + iC_2$, and $X = X_1 + iX_2$. Then, we must have

$$A + X = C \Rightarrow (A_1 + iA_2) + (X_1 + iX_2) = (C_1 + iC_2) \Rightarrow (A_1 + X_1) + i(A_2 + X_2) = (C_1 + iC_2)$$

It is easy to see that if the above equation holds, then $A_i + X_i = C_i$, $i = 1, 2$

Theorem 1.2 The equation $A + X = C$ has a solution X, a rectangular complex fuzzy number, if and only if the equations $A_i + X_i = C_i$ have a solution, X_i, $i = 1, 2$.

Example 1.1 If A is a real fuzzy number and C is a rectangular complex fuzzy number, then let $X = X_1 + iX_2$, then this is a particular case of the above-considered case, where $A_2 = 0$ (real number 0).

If A is a rectangular complex fuzzy number and C is a real fuzzy number, let $X = X_1 + iX_2$; if the equation has a solution, then $A_2 + X_2 = 0$ (real number zero). Since the last equation has no solution for X_2, $A + X = C$ has no solution for X.

1.3 Fuzzy Equation A + X = C

While solving the fuzzy equation $A + X = C$, it is evident that C—A is not the solution to the equation in the context. To visualize this, let us assume two unique fuzzy numbers represented by closed intervals, $A = [a_1, a_2]$ and $C = [c_1, c_2]$. Then, by using the concept of subtraction of closed intervals, we have

$$\begin{aligned} C - A &= [c_1, c_2] - [a_1, a_2] = [c_1 - a_2, c_2 - a_1] \\ \therefore A + C - A &= [a_1, a_2] + [c_1 - a_2, c_2 - a_1] \\ &= [a_1 + c_1 - a_2, a_2 + c_2 - a_1] \\ &\neq [c_1, c_2] = C \end{aligned}$$

whenever $a_1 \neq a_2$. Thus, we infer that $X = C - A$ is not a solution to the equation $A + X = C$.

1.4 Solution of the Fuzzy Equation A + X = C

Let us assume two fuzzy numbers such that $A = [a_1, a_2], C = [c_1, c_2]$, and $X = [x_1,\ x_2]$ is an unknown fuzzy numbers. Now, from the given equation, we have

$$\begin{aligned} A + X = C &\Rightarrow [a_1, a_2] + [x_1,\ x_2] = [c_1, c_2] \\ &\Rightarrow [a_1 + x_1, a_2 + x_2] = [c_1, c_2] \\ &\Rightarrow a_1 + x_1 = c_1 \text{ and } a_2 + \ x_2 = \ c_2 \\ &\Rightarrow x_1 = \ c_1 - a_1 \text{ and } x_2 = c_2 - a_2 \end{aligned}$$

Since X is an interval, the values of x_1 and x_2 given by the above equations will be the solution of the equation $A + X = C$, provided $x_1 \leq x_2$. In other words, we can say that the equation $A + X = C$ will have a solution if and only if $c_1 - a_1 \leq c_2 - a_2$ viz; if the inequality $c_1 - a_1 \leq c_2 - a_2$ is satisfied, $X = [c_1 - a_1, c_2 - a_2]$ be the solution of the fuzzy equation $A + X = C$.

We, however, know that every fuzzy number can be uniquely expressed by its α-cuts. Hence, let us consider the case when A, C, and X are represented in terms of their α-cuts. Let us consider the case when $\alpha \in]0, 1]$ and $\alpha_A = [\alpha_{a_1}, \alpha_{a_2}]$, $\alpha_C = [\alpha_{c_1}, \alpha_{c_2}]$, and $\alpha_X = [\alpha_{x_1}, \alpha_{x_2}]$ are the respective α-cuts of A, C, and X. After substituting the values of these α-cuts in the concerned interval equation, i.e., $A+X = C \Rightarrow \alpha_{(A+X)} = \alpha_C \Rightarrow \alpha_A + \alpha_X = \alpha_C \Rightarrow \alpha_X = \alpha_C - \alpha_A$

$$\begin{aligned} &\Rightarrow [\alpha_{x_1}, \alpha_{x_2}] = [\alpha_{c_1}, \alpha_{c_2}] - [\alpha_{a_1}, \alpha_{a_2}] \\ &\Rightarrow [\alpha_{x_1}, \alpha_{x_2}] = [\alpha_{c_1} - \alpha_{a_2}, \alpha_{c_2} - \alpha_{a_1}] \end{aligned}$$

we will have a solution if and only if

(i) $\alpha_{c_1} - \alpha_{a_1} \leq \alpha_{c_2} - \alpha_{a_2}$ for all $\alpha \in]0, 1]$, and the solution will be $\alpha_X = [\alpha_{c_1} - \alpha_{a_1}, \alpha_{c_2} - \alpha_{a_2}]$
(ii) If $\alpha \leq \beta$, then $\beta_X \subseteq \alpha_X$ for all $\alpha, \beta \in]0, 1]$ (by property of α-cut), we will have $\alpha_{c_1} - \alpha_{a_1} \leq \beta_{c_1} - \beta_{a_1} \leq \beta_{c_2} - \beta_{a_2} \leq \alpha_{c_2} - \alpha_{a_2}$

Property

(i) It ensures that the interval equation $\alpha_A + \alpha_X = \alpha_C$ has a solution, which is

$$\alpha_X = [\alpha_{c_1} - \alpha_{a_1}, \alpha_{c_2} - \alpha_{a_2}]$$

(b) The second property states that *the solutions of the interval equations for α and β are nested*, i.e., if $\alpha \leq \beta \Rightarrow \beta_X \subseteq \ \alpha_X$. If for all $\alpha \in]0, 1]$, the solution α_X exists and $\alpha_{c_1} - \alpha_{a_1} \leq \beta_{c_1} - \beta_{a_1} \leq \beta_{c_2} - \beta_{a_2} \leq \alpha_{c_2} - \alpha_{a_2}$ holds, then by virtue of the decomposition theorem, we will have

$$X = \bigcup_{\alpha \in]0, 1]} \alpha^X = \bigcup_{\alpha \in]0, 1]} \alpha \cdot \alpha_X$$

Example 1.2 Let A and C be two fuzzy numbers given by

$$\mathrm{A} = \frac{0.2}{[0,1[} + \frac{0.6}{[1,2[} + \frac{0.8}{[2,3[} + \frac{0.9}{[3,4[} + \frac{1}{4} + \frac{0.5}{]4,5]} + \frac{0.1}{]5,6]}$$

and $\mathrm{C} = \frac{0.1}{[0,1[} + \frac{0.2}{[1,2[} + \frac{0.6}{[2,3[} + \frac{0.7}{[3,4[} + \frac{0.8}{[4,5[} + \frac{0.9}{[5,6[} + \frac{1}{6} + \frac{0.5}{]6,7]} + \frac{0.4}{]7,8]} + \frac{0.2}{]8,9]} + \frac{0.1}{]9,10]}$, then solve the fuzzy equation $\mathrm{A} + \mathrm{X} = \mathrm{C}$, for X.

Solution: Now, first of all, we will calculate all relevant α-cuts of A, C, and X in the following table. (Table 1.1)

Remarks 1.1 To calculate the distinct relevant α-cuts of α_A and α_C, we can proceed as follows:-

For α-cuts of α_A:

For α-cut 0.1, we observe that 0.2, 0.6, 0.8, 0.9, 1, 0.5, and 0.1 all are ≥ 0.1. Therefore,

$$\therefore\ 0.1_A = [0,1[\cup[1,2[\cup[2,3[\cup[3,4[\cup[4,4]\cup]4,5]\cup]5,6[=[0,6]$$

When $\alpha = 0.2$, we observe that 0.2, 0.6, 0.8, 0.9, 1, 0.5 all are ≥ 0.2. Therefore,

$$\therefore\ 0.2_A = [0,1[\cup[1,2[\cup[2,3[\cup[3,4[\cup[4,4]\cup]4,5]=[0,5]$$

When $\alpha = 0.3$, we observe that 0.6, 0.8, 0.9, 1, 0.5 all are ≥ 0.3. Therefore,

$$\therefore\ 0.3_A = [1,2[\cup[2,3[\cup[3,4[\cup[4,4]\cup]4,5]=[1,5]$$

When $\alpha = 0.4$, we observe that 0.6, 0.8, 0.9, 1, 0.5 all are ≥ 0.4. Therefore,

$$\therefore\ 0.4_A = [1,2[\cup[2,3[\cup[3,4[\cup[4,4]\cup]4,5]=[1,5]$$

Table 1.1 α-cuts of A, C, and X.

A	$\alpha_C = [\alpha_{c_1}, \alpha_{c_2}]$	$\alpha_A = [\alpha_{a_1}, \alpha_{a_2}]$	$\alpha_X = \alpha_C - \alpha_A = [\alpha_{c_1} - \alpha_{a_1}, \alpha_{c_2} - \alpha_{a_2}]$
1.0	[6, 6]	[4, 4]	[2, 2]
0.9	[5, 6]	[3, 4]	[2, 2]
0.8	[4, 6]	[2, 4]	[2, 2]
0.7	[3, 6]	[2, 4]	[1, 2]
0.6	[2, 6]	[1, 4]	[1, 2]
0.5	[2, 7]	[1, 5]	[1, 2]
0.4	[2, 8]	[1, 5]	[1, 3]
0.3	[2, 8]	[1, 5]	[1, 3]
0.2	[1, 9]	[0, 5]	[1, 4]
0.1	[0, 10]	[0, 6]	[0, 4]

When $\alpha = 0.5$, we observe that 0.6, 0.8, 0.9, 1, 0.5 all are ≥ 0.5. Therefore,

$$\therefore\ 0.5_A = [1, 2[\cup[2, 3[\cup[3, 4[\cup[4, 4]\cup]4, 5]{=}[1, 5]$$

When $\alpha = 0.6$, we observe that 0.6, 0.8, 0.9, 1 all are ≥ 0.6. Therefore,

$$\therefore\ 0.3_A = [1, 2[\cup[2, 3[\cup[3, 4[\cup[4, 4]{=}[1, 4]$$

When $\alpha = 0.7$, we observe that 0.8, 0.9, 1 all are ≥ 0.7. Therefore,

$$\therefore\ 0.7_A = [2, 3[\cup[3, 4[\cup[4, 4]{=}[2, 4]$$

When $\alpha = 0.8$, we observe that 0.8, 0.9, 1 all are ≥ 0.8. Therefore,

$$\therefore\ 0.8_A = [2, 3[\cup[3, 4[\cup[4, 4]{=}[2, 4]$$

When $\alpha = 0.9$, we observe that 0.9, 1 all are ≥ 0.9. Therefore,

$$\therefore\ 0.9_A = [3, 4[\cup[4, 4]{=}[3, 4]$$

When $\alpha = 1$, we observe that $1 \geq 1$. Therefore,

$$\therefore\ 1.0_A = [4, 4]{=}[4, 4]$$

Now to calculate the distinct relevant α-cuts of α_B, we can proceed as follows:-

For α-cut 0.1, we observe that 0.1, 0.2, 0.6, 0.7, 0.8, 0.9, 1, 0.5, 0.4, 0.2 and 0.1 all are ≥ 0.1. Therefore,

$$\therefore 0.1_C = [0, 1[\cup[1, 2[\cup[2, 3[\cup[3, 4[\cup[4, 5[\cup[5, 6[\cup[6, 6]\cup]6, 7]\cup]7, 8]\cup]8, 9]\cup]9, 10]$$
$$= [0, 10]$$

For α-cut 0.2, we observe that 0.2, 0.6, 0.7, 0.8, 0.9, 1, 0.5, 0.4, 0.2 all are ≥ 0.2. Therefore,

$$\therefore\ 0.2_C = [1, 2[\cup[2, 3[\cup[3, 4[\cup[4, 5[\cup[5, 6[\cup[6, 6]\cup]6, 7]\cup]7, 8]\cup]8, 9] = [1, 9]$$

For α-cut 0.3, we observe that 0.6, 0.7, 0.8, 0.9, 1, 0.5, 0.4 all are ≥ 0.3. Therefore,

$$\therefore\ 0.3_C = [2, 3[\cup[3, 4[\cup[4, 5[\cup[5, 6[\cup[6, 6]\cup]6, 7]\cup]7, 8] = [2, 8]$$

For α-cut 0.4, we observe that 0.6, 0.7, 0.8, 0.9, 1, 0.5 all are ≥ 0.4. Therefore,

$$\therefore\ 0.4_C = [2, 3[\cup[3, 4[\cup[4, 5[\cup[5, 6[\cup[6, 6]\cup]6, 7]\cup]7, 8] = [2, 8]$$

For α-cut 0.5, we observe that 0.6, 0.7, 0.8, 0.9, 1, 0.5 all are $\geq$ 0.5. Therefore,

$$\therefore\ 0.5_C = [2, 3[\cup[3, 4[\cup[4, 5]\cup[5, 6[\cup[6, 6]\cup]6, 7] = [2, 7]$$

For α-cut 0.6, we observe that 0.6, 0.7, 0.8, 0.9, 1 all are $\geq$ 0.6. Therefore,

$$\therefore\ 0.6_C = [2, 3[\cup[3, 4[\cup[4, 5[\cup[5, 6]\cup[6, 6] = [2, 6]$$

For α-cut 0.7, we observe that 0.7, 0.8, 0.9, 1 all are $\geq$ 0.7. Therefore,

$$\therefore\ 0.7_C = [3, 4[\cup[4, 5[\cup[5, 6]\cup[6, 6]=[3, 6]$$

For α-cut 0.8, we observe that 0.8, 0.9, 1 all are $\geq$ 0.8. Therefore,

$$\therefore\ 0.8_C = [4, 5[\cup[5, 6[\cup[6, 6]=[4, 6]$$

For α-cut 0.9, we observe that 0.9, 1 all are $\geq$ 0.9. Therefore,

$$\therefore\ 0.9_C = [5, 6[\cup[6, 6]=[5, 6]$$

For α-cut 1, we observe that 1 is $\geq$ 1.0 Therefore,

$$\therefore\ 1.0_C = [6, 6]=[6, 6]$$

Let us observe α_X in the table and A, C in the question. Clearly, universe of discourse X will be [0, 1[, [1, 2[, [2, 2],]2, 3] and]3, 4].

Now [0, 1[is contained in [0, 4], for which corresponding α is 0.1. Therefore, we have $\frac{0.1}{[0,1[}$.

Also [1, 2[is contained in [0, 4], [1, 4], [1, 3], [1, 3], [1, 2], [1, 2],]1, 2] for which corresponding α is 0.1, 0.2, 0.3, 0.4, 0.5, 0.6, 0.7. For union, max α is taken which is obviously 0.7. So $\frac{.7}{[1,2[}$.

Similarly [2, 2] is contained in for α = 0.1, 0.2, 0.3, 0.4, 0.5, 0.6, 0.7, 0.8, 0.9, 1.0; max α for which is 1.0. So, we have $\frac{1}{[2,2]}$.

Again]2, 3] is contained in for α = 0.1, 0.2, 0.3, 0.4; max α for which is α = 0.4. So, we have $\frac{0.4}{]2,3]}$.

Also]3, 4] is contained in for α = 0.1, 0.2; max α for which is α = 0.2. So, we have $\frac{0.2}{]3,4]}$.

Hence, the solution of the given fuzzy equation is the fuzzy number X given by

$$X = \underset{\alpha\,\in\,]0,1]}{\cup} \alpha^X = \underset{\alpha\,\in\,]0,1]}{\cup} \alpha\cdot\alpha_X = \frac{.1}{[0, 1[} + \frac{.2}{]3, 4]} + \frac{.4}{]2, 3]} + \frac{.7}{[1, 2[} + \frac{1}{2}$$

Example 1.3 A and C are two fuzzy numbers whose membership functions are given by

$$A(x) = \begin{cases} \frac{x+3}{3} \text{ if } -3 < x \le 0 \\ \frac{3-x}{3} \text{ if } 0 < x \le 3 \\ 0 \text{ otherwise} \end{cases}$$

$$C(x) = \begin{cases} \frac{x-3}{3} \text{ if } 3 < x \le 6 \\ \frac{9-x}{3} \text{ if } 6 < x \le 9 \\ 0 \text{ otherwise} \end{cases}$$

Then, could you find the solution to the equation A + X = C.
Solution: We know that for any fuzzy set A, its α-cut is defined to be

$$\alpha_A = \{x : A(x) \ge \alpha\}$$

Now, suppose that ∃ functions $g_A(\alpha)$ and $h_A(\alpha)$ such that $A(x) \ge \alpha$ iff, $g_A(\alpha) \le x \le h_A(\alpha) \forall \alpha \in [0, 1]$. Therefore, we have.

$\alpha_A = \{x : g_A(\alpha) \le x \le h_A(\alpha)\} = [g_A(\alpha), h_A(\alpha)]$, where $g_A(\alpha)$ is the lower boundary and $h_A(\alpha)$ is the upper boundary.

To obtain the value of $g_A(\alpha)$ for α_A, let

$$\alpha = \frac{x+3}{3} \Rightarrow x = 3\alpha - 3$$

This means that $g_A(\alpha) = 3\alpha - 3$.
Similarly, the value of $h_A(\alpha)$ for α_A can be obtained from

$$\alpha = \frac{3-x}{3} \Rightarrow x = 3 - 3$$

$$\therefore\ h_A(\alpha) = 3 - 3\alpha$$

$$\therefore\ \alpha_A = [g_A(\alpha), h_A(\alpha)] = [3\alpha - 3, 3 - 3\alpha]$$

Next, we will find the lower boundary and upper boundary for α_C. For this, let

$$\alpha = \frac{x-3}{3} \Rightarrow x = 3\alpha + 3 = g_A(\alpha)$$

Again, let

$$\alpha = \frac{9-x}{3} \Rightarrow x = 9 - 3\alpha = h_A(\alpha)$$

$$\therefore\ \alpha_C = [g_C(\alpha), h_C(\alpha)] = [3\alpha + 3, 9 - 3\alpha]$$

Now from the given fuzzy equation, we have

$$
\begin{aligned}
A + X = C \Rightarrow \alpha_{(A+X)} = \alpha_C \Rightarrow \alpha_A + \alpha_X &= \alpha_C \\
\Rightarrow \alpha_X &= \alpha_C - \alpha_A \\
\Rightarrow \alpha_X &= [3\alpha + 3, 9 - 3\alpha] - [3\alpha - 3, 3 - 3\alpha] \\
\Rightarrow \alpha_X &= [3\alpha + 3 - 3 + 3\alpha,\ 9 - 3\alpha - 3\alpha + 3] \\
\Rightarrow \alpha_X &= [6\alpha,\ 12 - 6\alpha]
\end{aligned}
$$

The lower boundary of α_X is 6α. On equating $6\alpha = x$, we will get the lower boundary as $\alpha = \frac{x}{6}$ for $\alpha \in [0, 1]$. When $\alpha = 0$, we have $x = 0$ and when $\alpha = 1$, we get $x = 6$. Thus, we have

$$X(x) = \frac{x}{6} \text{ for } x \in [0, 6]$$

On equating the upper boundary of α_X, i.e., $12–6\alpha = x$, we get $\alpha = \frac{12-x}{6}$ for $\alpha \in [0, 1]$. When $\alpha = 0$, we get $x = 12$ and when $\alpha = 1$, we have $x = 6$. Thus, we have

$$X(x) = \frac{12 - x}{6} \text{ for } x \in [6, 12]$$

Therefore, the membership function for the fuzzy set X is given by

$$X = \bigcup_{\alpha \in [0, 1]} \alpha^X = \bigcup_{\alpha \in [0, 1]} \alpha \cdot \alpha_X \quad \text{is}$$

$$\therefore X = \begin{cases} \frac{x}{6} \text{ if } 0 \le x \le 6 \\ \frac{12-x}{6} \text{ if } 6 \le x \le 12 \\ 0 \text{ otherwise} \end{cases}$$

As given above, the value of the X is the required solution of the given fuzzy equation $A + X = C$.

Example 1.4 A and C are two fuzzy numbers whose membership functions are given by

$$A(x) = \begin{cases} \frac{x-2}{2} \text{ if } 2 < x \le 4 \\ \frac{6-x}{2} \text{ if } 4 < x \le 6 \\ 0 \text{ otherwise} \end{cases}$$

$$C(x) = \begin{cases} \frac{x-6}{2} \text{ if } 6 < x \le 8 \\ \frac{10-x}{2} \text{ if } 8 < x \le 10 \\ 0 \text{ otherwise} \end{cases}$$

Could you solve the fuzzy equation $A + X = C$?

Solution: We have

$\alpha_A = [g_A(\alpha),\ h_A(\alpha)]$, where $g_A(\alpha)$ is the lower boundary and $h_A(\alpha)$ is the upper boundary of α_A.

To find $g_A(\alpha)$, let

$$\alpha = \frac{x-2}{2} \Rightarrow x = 2\alpha + 2 = g_A(\alpha)$$

Let, $\alpha = \frac{6-x}{2} \Rightarrow x = 6 - 2\alpha = h_A(\alpha)\alpha = \frac{6-x}{2} \Rightarrow x = 6 - 2\alpha = h_A(\alpha)$

Therefore, $\alpha_A = [2\alpha + 2, 6-2\alpha]$

Next, we will find out

$\alpha_C = [g_C(\alpha),\ h_C(\alpha)]$, where $g_C(\alpha)$ is the lower boundary and $h_C(\alpha)$ is the upper boundary of α_C.

Let, $\alpha = \frac{x-6}{2} \Rightarrow\ x = 2\alpha + 6 = g_C(\alpha)$

And let, $\alpha = \frac{10-x}{2} \Rightarrow\ x = 10 - 2\alpha = h_C(\alpha)$

Hence, $\alpha_C = [2\alpha + 6, 10 - 2\alpha]$

We know that

$$\begin{aligned}\alpha_X = \alpha_C - \alpha_A &= [2\alpha + 6, 10 - 2\alpha] - [2\alpha + 2, 6 - 2\alpha] \\ \Rightarrow [\alpha_{x_1}, \alpha_{x_2}] &= [2\alpha + 6 - 6 + 2\alpha, 10 - 2\alpha - 2\alpha - 2] \\ &= [4\alpha,\ 8 - 4\alpha]\end{aligned}$$

Now we will find the membership function for the fuzzy set X. On equating the lower boundary 4α to x, we have

$$\alpha = \frac{x}{4} \quad \text{for} \quad \alpha \in [0, 1].$$

When $\alpha = 0$, we get $x = 0$ and for $\alpha = 1$ gives $x = 4$. Therefore,

$$X(x) = \frac{x}{4} \text{ if } 0 \le x \le 4$$

On equating the upper boundary 8 - 4α equal to x, we get

$$\alpha = \frac{8-x}{4} \quad \text{for} \quad \alpha \in [0, 1].$$

When $\alpha = 0$, we get $x = 8$ and for $\alpha = 1$ gives $x = 4$. Therefore,

$$X(x) = \frac{8-x}{4} \text{ if } 4 \le x \le 8$$

Therefore, the membership function for the fuzzy set X is given by

$$X = \bigcup_{\alpha \in [0,1]} \alpha^X = \bigcup_{\alpha \in [0,1]} \alpha \cdot \alpha_X \qquad \text{is}$$

$$X = \begin{cases} \frac{x}{4} \text{ if } 0 \le x \le 4 \\ \frac{8-x}{4} \text{ if } 4 \le x \le 8 \\ 0 \text{ otherwise} \end{cases}$$

As given above, the value of the X is the required solution of the given fuzzy equation A + X = C.

Example 1.5 In the ordinary arithmetic of real numbers, the equation a = a + b –b holds for any a, b ∈ ℝ. Does this equation hold for fuzzy numbers? Justify your answer and illustrate it with two symmetric triangular fuzzy numbers.

Solution: Let $A(x) = \begin{cases} \frac{x+2}{2} \text{ if } -2 < x \le 0 \\ \frac{2-x}{2} \text{ if } 0 < x \le 2 \\ 0 \text{ otherwise} \end{cases}$ and $B(x) = \begin{cases} \frac{x-2}{2} \text{ if } 2 < x \le 4 \\ \frac{2-x}{2} \text{ if } 4 < x \le 6 \\ 0 \text{ otherwise} \end{cases}$ are two triangular fuzzy numbers.

α-cuts of fuzzy set A is given by

$$\alpha_A = [g_A(\alpha),\ h_A(\alpha)] = [2\alpha - 2, 2 - 2\alpha]$$

α-cuts of fuzzy set B is given by

$$\alpha_B = [g_B(\alpha),\ h_B(\alpha)] = [2\alpha + 2, 6 - 2\alpha]$$

We have

$$\begin{aligned} \alpha_A + \alpha_B &= [2\alpha - 2, 2 - 2\alpha] + [2\alpha + 2, 6 - 2\alpha] = [4\alpha, 8 - 4\alpha] \\ \alpha_A + \alpha_B - \alpha_B &= [4\alpha, 8 - 4\alpha] - [2\alpha + 2, 6 - 2\alpha] = [6\alpha - 6, 6 - 6\alpha] \\ &\ne [2\alpha - 2, 2 - 2\alpha] = \alpha_A \\ &\therefore \alpha_A + \alpha_B - \alpha_B \ne \alpha_A \end{aligned}$$

From the above discussion, we infer that the equation a = a + b –b holds for any a, b ∈ ℝ, but it does not hold good for fuzzy numbers. That is, for fuzzy numbers A + B–B ≠ A, this equation does not hold.

1.5 Solution of the Fuzzy Equation A · X = C

For the sake of simplicity, suppose that A, C, and X are fuzzy numbers over $\mathbb{R}^+$. Further, suppose.that $A = [a_1, a_2]$, $C = [c_1, c_2]$and $X = [x_1, x_2]$ are special fuzzy numbers. On solving equation A. X = C, we observe that $X = \frac{C}{A}$ is not its solution. To visualize this, let us consider.

$\frac{C}{A} = \frac{[c_1, c_2]}{[a_1, a_2]} = [c_1, c_2] \cdot [a_1, a_2]^{-1} = [c_1, c_2] \cdot \left[\frac{1}{a_2}, \frac{1}{a_1}\right] = \left[\frac{c_1}{a_2}, \frac{c_2}{a_1}\right]$, where $0 \notin [a_1, a_2]$

Then,

$$\frac{C}{A} \cdot A = \left[\frac{c_1}{a_2}, \frac{c_2}{a_1}\right] \cdot [a_1, a_2] = \left[\frac{c_1}{a_2} \cdot a_1, \frac{c_2}{a_1} \cdot a_2\right] \neq [c_1, c_2]$$

whenever $c_1 \neq c_2$. Therefore, $X = \frac{C}{A}$ is not a solution of the equation A. X = C.

If we have, then $X = \frac{C}{A} \Rightarrow [x_1, x_2] = \left[\frac{c_1}{a_2}, \frac{c_2}{a_1}\right]$. This gives two ordinary equations of real numbers,

$x_1 = \frac{c_1}{a_2}$ and $x_2 = \frac{c_2}{a_1}$, as the solution of the equation. Since X is an interval, therefore $x_1 \leq x_2$. The equation A. X = C has a solution if and only if $\frac{c_1}{a_2} \leq \frac{c_2}{a_1}$. If this inequality is satisfied, the solution is $X = \left[\frac{c_1}{a_2}, \frac{c_2}{a_1}\right]$.

Since we can express every fuzzy number uniquely in terms of its α-cuts, the given fuzzy numbers are closed intervals; we can apply the above procedure to α-cuts of arbitrary fuzzy numbers.

Let $\alpha \in [0, 1]$ and suppose that $\alpha_A = [\alpha_{a_1}, \alpha_{a_2}]$, $\alpha_C = [\alpha_{c_1}, \alpha_{c_2}]$, and $\alpha_X = [\alpha_{x_1}, \alpha_{x_2}]$ are the respective α-cuts of A, C, and X. After substituting the values of these α-cuts in the concerned interval equation, i.e.,

$$\alpha_A \cdot \alpha_X = \alpha_C \Rightarrow \alpha_{(A \cdot X)} = \alpha_C \Rightarrow \alpha_A \cdot \alpha_X = \alpha_C \Rightarrow \alpha_X = \frac{\alpha_C}{\alpha_A} = \frac{[\alpha_{c_1}, \alpha_{c_2}]}{[\alpha_{a_1}, \alpha_{a_2}]}$$

$$\Rightarrow \alpha_X = [\alpha_{c_1}, \alpha_{c_2}] \cdot \left[\frac{1}{\alpha_{a_2}}, \frac{1}{\alpha_{a_1}}\right]$$

$$= \left[\frac{\alpha_{c_1}}{\alpha_{a_2}}, \frac{\alpha_{c_2}}{\alpha_{a_1}}\right]$$

Then, the fuzzy interval equation A. X = C will have a solution if and only if $\frac{\alpha_{c_1}}{\alpha_{a_2}} \leq \frac{\alpha_{c_2}}{\alpha_{a_1}}$ for each $\alpha \in [0, 1]$.

If $\alpha < \beta$ then it implies that $\frac{\alpha_{c_1}}{\alpha_{a_2}} \leq \frac{\beta_{c_1}}{\beta_{a_2}} \leq \frac{\beta_{c_2}}{\beta_{a_1}} \leq \frac{\alpha_{c_2}}{\alpha_{a_1}}$.

If the solution of the fuzzy interval equation exists, then for $\alpha \in [0, 1]$, the solution will have the form (by the Decomposition Theorem)

$$X = \bigcup_{\alpha \in [0, 1]} {}^{\alpha}X$$

Remarks 1.2 Here A and C are real fuzzy numbers. We need to assume that zero does not belong to the support of A. Therefore, we assume that A > 0 or A < 0.

Theorem 1.3 (a) Suppose that zero does not belong to the support of C. Then, there is a solution X of AX = C if and only if

(i) $a_1c_2 > c_1a_2$ and $a_3c_2 < c_3a_2$ when $A > 0, C \geq 0$
(ii) $a_1c_2 < c_1a_2$ and $a_3c_2 > c_3a_2$ when $A < 0, C \leq 0$
(iii) $a_3c_2 > c_1a_2$ and $a_1c_2 < c_3a_2$ when $A > 0, C \leq 0$
(iv) $a_3c_2 < c_1a_2$ and $a_1c_2 < c_3a_2$ when $A < 0, C \geq 0$
(v) Suppose that zero belongs to the support of $C(c_2 = 0)$. Then, there is a solution X of AX = C if and only if zero belongs to the support of $X(x_2 = 0)$.

Proof: It is beyond the scope of the book.

Example 1.6 Let A and C be two triangular fuzzy numbers having membership functions given as.

$$A(x) = \begin{cases} \frac{x-3}{3} \text{ if } 3 < x \leq 6 \\ \frac{9-x}{3} \text{ if } 6 < x \leq 9 \\ 0 \text{ otherwise} \end{cases}$$

$$C(x) = \begin{cases} \frac{x-9}{3} \text{ if } 9 < x \leq 12 \\ \frac{15-x}{3} \text{ if } 12 < x \leq 15 \\ 0 \text{ otherwise} \end{cases}$$

I've included the solution of the fuzzy equation A. X = C.

Solution: We have.

$\alpha_A = [g_A(\alpha),\ h_A(\alpha)]$, where $g_A(\alpha)$ is the lower boundary and $h_A(\alpha)$ is the upper boundary of α_A.

To find $g_A(\alpha)$, let

$$\alpha = \frac{x-3}{3} \Rightarrow x = 3\alpha + 3 = g_A(\alpha)$$

Let, $\alpha = \frac{9-x}{3} \Rightarrow x = 9 - 3\alpha = h_A(\alpha)$

Therefore, $\alpha_A = [3\alpha + 3, 9 - 3\alpha] = [\alpha_{a_1}, \alpha_{a_2}]$

Next, we will find out.

$\alpha_C = [g_C(\alpha),\ h_C(\alpha)]$, where $g_C(\alpha)$ is the lower boundary and $h_C(\alpha)$ is the upper boundary of α_C.

Let, $\alpha = \frac{x-9}{3} \Rightarrow x = 3\alpha + 9 = g_C(\alpha)$

And let, $\alpha = \frac{15-x}{3} \Rightarrow x = 15 - 3\alpha = h_C(\alpha)$

Hence, $\alpha_C = [3\alpha + 9, 15 - 3\alpha] = [\alpha_{c_1}, \alpha_{c_2}]$

We know that the fuzzy interval equation is $A.X = C \Rightarrow \alpha_{(A \cdot X)} = \alpha_C \Rightarrow \alpha_A \cdot \alpha_X = \alpha_C 1$

$$\Rightarrow \alpha_X = \frac{\alpha_C}{\alpha_A} = \frac{[\alpha_{c_1},\ \alpha_{c_2}]}{[\alpha_{a_1},\ \alpha_{a_2}]}$$

$$\Rightarrow \alpha_X = [\alpha_{c_1},\ \alpha_{c_2}] \cdot \left[\frac{1}{\alpha_{a_2}}, \frac{1}{\alpha_{a_1}}\right]$$

$$= \left[\frac{\alpha_{c_1}}{\alpha_{a_2}}, \frac{\alpha_{c_2}}{\alpha_{a_1}}\right]$$

$$= \left[\frac{3\alpha + 9}{9 - 3\alpha}, \frac{15 - 3\alpha}{3\alpha + 3}\right]$$

Now we will find the membership function for the fuzzy set X. On equating the lower boundary α_X, i.e., $\frac{3\alpha+9}{9-3\alpha}$ to x, we have

$$x = \frac{3\alpha + 9}{9 - 3\alpha}, \text{ for } \alpha \in [0, 1]$$

This implies that

$$\alpha = \frac{9x - 9}{3x + 3}$$

Now, when $\alpha = 0$, we get $x = 1$ and when $\alpha = 1$, we have $x = 2$. Therefore,

$$X(x) = \frac{9x - 9}{3x + 3} \text{ for } 1 \le x \le 2$$

Next, we will equate the upper boundary of α_X, i.e., $\frac{15-3\alpha}{3\alpha+3}$ to x then, we have

$$x = \frac{15 - 3\alpha}{3\alpha + 3} \Rightarrow \alpha = \frac{15 - 3x}{3x + 3}$$

Now, when $\alpha = 0$, we get $x = 5$ and when $\alpha = 1$, we have $x = 2$. Therefore,

$$X(x) = \frac{15 - 3x}{3x + 3} \text{ for } 2 \le x \le 5$$

Hence, the membership function for the fuzzy set X is given by

$$X = \bigcup_{\alpha \in [0, 1]} \alpha^X = \text{ where } \alpha^X = \alpha \cdot \alpha_X$$

Therefore, the solution of the given fuzzy equation is

$$X(x) = \begin{cases} \frac{9x - 9}{3x + 3} \text{ for } 1 \le x \le 2 \\ \frac{15 - 3x}{3x + 3} \text{ for } 2 \le x \le 5 \\ 0 \text{ otherwise} \end{cases}$$

Example 1.7 Let A and C be two fuzzy triangular numbers having membership functions given as.

$$A(x) = \begin{cases} \frac{x - 2}{2} \text{ if } 2 < x \le 4 \\ \frac{6 - x}{2} \text{ if } 4 < x \le 6 \\ 0 \text{ otherwise} \end{cases}$$

$$C(x) = \begin{cases} \frac{x - 6}{2} \text{ if } 6 < x \le 8 \\ \frac{10 - x}{2} \text{ if } 8 < x \le 10 \\ 0 \text{ otherwise} \end{cases}$$

Could you please find the solution to the fuzzy equation A. X = C?

Solution: We have.

$\alpha_A = [g_A(\alpha), \ h_A(\alpha)]$, where $g_A(\alpha)$ is the lower boundary and $h_A(\alpha)$ is the upper boundary of α_A.

To find $g_A(\alpha)$, let

$$\alpha = \frac{x-2}{2} \Rightarrow x = 2\alpha + 2 = g_A(\alpha)$$

Let, $\alpha = \frac{6-x}{2} \Rightarrow x = 6 - 2\alpha = h_A(\alpha)$

Therefore, $\alpha_A = [2\alpha + 2, 6 - 2\alpha] = [\alpha_{a_1}, \alpha_{a_2}]$

Next, we will find out.

$\alpha_C = [g_C(\alpha), \ \ h_C(\alpha)]$, where $g_C(\alpha)$ is the lower boundary and $h_C(\alpha)$ is the upper boundary of α_C.

Let, $\alpha = \frac{x-6}{2} \Rightarrow \ x = 2\alpha + 6 = g_C(\alpha)$

And let, $\alpha = \frac{10-x}{2} \Rightarrow \ x = 10 - 2\alpha = h_C(\alpha)$

Hence, $\alpha_C = [2\alpha + 6, 10 - 2\alpha] = [\alpha_{c_1}, \alpha_{c_2}]$

We know that the fuzzy interval equation is $A \cdot X = C \Rightarrow \alpha_{(A.X)} = \alpha_C \Rightarrow \alpha_A \cdot \alpha_X = \alpha_C$

$$\Rightarrow \alpha_X = \frac{\alpha_C}{\alpha_A} = \frac{[\alpha_{c_1}, \alpha_{c_2}]}{[\alpha_{a_1}, \alpha_{a_2}]}$$

$$\Rightarrow \alpha_X = [\alpha_{c_1}, \alpha_{c_2}] \cdot \left[\frac{1}{\alpha_{a_2}}, \frac{1}{\alpha_{a_1}}\right]$$

$$= \left[\frac{\alpha_{c_1}}{\alpha_{a_2}}, \frac{\alpha_{c_2}}{\alpha_{a_1}}\right]$$

$$= \left[\frac{2\alpha + 6}{6 - 2\alpha}, \frac{10 - 2\alpha}{2\alpha + 2}\right]$$

Now we will find the membership function for the fuzzy set X. On equating the lower boundary of α_X, i.e., $\frac{2\alpha+6}{6-2\alpha}$ to x, we have

$$x = \frac{2\alpha + 6}{6 - 2\alpha}, \ \text{for } \alpha \in [0, 1]$$

This implies that

$$\alpha = \frac{6x - 6}{2x + 2}$$

Now, when $\alpha = 0$, we get $x = 1$ and when $\alpha = 1$, we have $x = 2$. Therefore,

$$X(x) = \frac{6x - 6}{2x + 2} \ \text{for } 1 \le x \le 2$$

Next, we will equate the upper boundary of α_X, i.e., $\frac{10-2\alpha}{2\alpha+2}$ to x then, we have

$$x = \frac{10-2\alpha}{2\alpha+2} \Rightarrow \alpha = \frac{10-2x}{2+2x}$$

Now, when $\alpha = 0$, we get $x = 5$ and when $\alpha = 1$, we have $x = 2$. Therefore,

$$X(x) = \frac{10-2x}{2+2x} \quad \text{for } 2 \le x \le 5$$

Hence, the membership function for the fuzzy set X is given by
$X = \bigcup_{\alpha\in[0,1]} \alpha^X$, where $\alpha^X = \alpha \cdot \alpha_X$
Therefore, the solution of the given fuzzy equation is

$$X(x) = \begin{cases} \frac{6x-6}{2x+2} \text{ for } 1 \le x \le 2 \\ \frac{10-2x}{2+2x} \text{ for } 2 \le x \le 5 \\ 0 \text{ otherwise} \end{cases}$$

Example 1.8 Let A and C be two fuzzy triangular numbers having membership functions given as

$$A(x) = \begin{cases} x-3 \text{ if } 3 < x \le 4 \\ 5-x \text{ if } 4 < x \le 5 \\ 0 \text{ if } x \le 3 \text{ and } x > 5 \end{cases}$$

$$C(x) = \begin{cases} \frac{x-12}{8} \text{ if } 12 < x \le 20 \\ \frac{32-x}{12} \text{ if } 20 < x \le 32 \\ 0 \text{ for } x \le 12 \text{ and } x > 32 \end{cases}$$

I've included the solution to the fuzzy equation A. X = C.

Solution: We have.

$\alpha_A = [g_A(\alpha),\ h_A(\alpha)]$, where $g_A(\alpha)$ is the lower boundary and $h_A(\alpha)$ is the upper boundary of α_A.

To find $g_A(\alpha)$, let

$$\alpha = x - 3 \Rightarrow x = \alpha + 3 = g_A(\alpha)$$

Let, $\alpha = 5 - x \Rightarrow x = 5 - \alpha = h_A(\alpha)$

Therefore, $\alpha_A = [\alpha + 3, 5 - \alpha] = [\alpha_{a_1}, \alpha_{a_2}]$

Next, we will find out

$\alpha_C = [g_C(\alpha),\ h_C(\alpha)]$, where $g_C(\alpha)$ is the lower boundary and $h_C(\alpha)$ is the upper boundary of α_C.

Let, $\alpha = \frac{x-12}{8} \Rightarrow x = 8\alpha + 12 = g_C(\alpha)$

And let, $\alpha = \frac{32-x}{12} \Rightarrow x = 32 - 12\alpha = h_C(\alpha)$

Hence, $\alpha_C = [8\alpha + 12, 32 - 12\alpha] = [\alpha_{C_1}, \alpha_{C_2}]$

We know that the fuzzy interval equation is $A \cdot X = C \Rightarrow \alpha_{(A.X)} = \alpha_C \Rightarrow \alpha_A \cdot \alpha_X = \alpha_C$

$$\Rightarrow \alpha_X = \frac{\alpha_C}{\alpha_A} = \frac{[\alpha_{c_1}, \alpha_{c_2}]}{[\alpha_{a_1}, \alpha_{a_2}]}$$

$$\Rightarrow \alpha_X = [\alpha_{c_1}, \alpha_{c_2}] \cdot \left[\frac{1}{\alpha_{a_2}}, \frac{1}{\alpha_{a_1}}\right]$$

$$= \left[\frac{\alpha_{c_1}}{\alpha_{a_2}}, \frac{\alpha_{c_2}}{\alpha_{a_1}}\right]$$

$$= \left[\frac{8\alpha + 12}{5 - \alpha}, \frac{32 - 12\alpha}{\alpha + 3}\right]$$

Now we will find the membership function for the fuzzy set X. On equating the lower boundary of α_X, i.e., $\frac{8\alpha+12}{5-\alpha}$ to x, we have

$$x = \frac{8\alpha + 12}{5 - \alpha}, \quad \text{for } \alpha \in [0, 1]$$

This implies that

$$\alpha = \frac{5x - 12}{x + 8}$$

Now, when $\alpha = 0$, we get $x = 12/5$ and when $\alpha = 1$, we have $x = 5$. Therefore,

$$X(x) = \frac{5x - 12}{x + 8} \text{ for } 12/5 \le x \le 5$$

Next, we will equate the upper boundary of α_X, i.e., $\frac{32-12\alpha}{\alpha+3}$ to x then, we have

$$x = \frac{32 - 12\alpha}{\alpha + 3} \Rightarrow \alpha = \frac{32 - 3x}{12 + x}$$

Now, when $\alpha = 0$, we get $x = 32/3$ and when $\alpha = 1$, we have $x = 5$. Therefore,

$$X(x) = \frac{32 - 3x}{12 + x} \text{ for } 5 \le x \le 32/3$$

Hence, the membership function for the fuzzy set X is given by

$$X = \bigcup_{\alpha \in [0, 1]} \alpha^X \text{ where } \alpha^X = \alpha \cdot \alpha_X$$

Therefore, the solution of the given fuzzy equation is

$$X(x) = \begin{cases} \frac{5x-12}{x+8} \text{ for } 12/5 \le x \le 5 \\ \frac{32-3x}{12+x} \text{ for } 5 \le x \le 32/3 \\ 0 \text{ otherwise} \end{cases}$$

Example 1.9 In the ordinary arithmetic of real numbers, the equation $a = a \cdot (b/b)$ holds for any $a, b \neq 0 \in \mathbb{R}$. Does this equation hold for fuzzy numbers? Justify your answer and illustrate it by two symmetric triangular fuzzy numbers, i.e., does the equation $A = A \cdot (B/B)$, $0 \notin B$ hold for fuzzy numbers?

Solution: Let $A(x) = \begin{cases} \frac{x+2}{2} \text{ if } -2 < x \le 0 \\ \frac{2-x}{2} \text{ if } \quad 0 < x \le 2 \\ 0 \text{ otherwise} \end{cases}$ and $B(x) = \begin{cases} \frac{x-2}{2} \text{ if } 2 < x \le 4 \\ \frac{6-x}{2} \text{ if } 4 < x \le 6 \\ 0 \text{ otherwise} \end{cases}$ are two triangular fuzzy numbers.

α-cuts of fuzzy set A are given by

$$\alpha_A = \left[g_A(\alpha),\ h_A(\alpha)\right] = [2\alpha - 2,\ 2 - 2\alpha]$$

α-cuts of fuzzy set B are given by

$$\alpha_B = \left[g_B(\alpha),\ h_B(\alpha)\right] = [2\alpha + 2, 6 - 2\alpha]$$

$0 \notin B$.

Hence, we have

$$\alpha_A \cdot \alpha_B \cdot \frac{1}{\alpha_B} = [2\alpha - 2, 2 - 2\alpha] \cdot \left\{[2\alpha + 2, 6 - 2\alpha] \cdot \frac{1}{[2\alpha + 2, 6 - 2\alpha]}\right\}$$

$$\alpha_A \cdot \alpha_B \cdot \frac{1}{\alpha_B} = [2\alpha - 2, 2 - 2\alpha] \cdot \left\{[2\alpha + 2, 6 - 2\alpha] \cdot [2\alpha + 2, 6 - 2\alpha]^{-1}\right.$$

$$\alpha_A \cdot \alpha_B \cdot \frac{1}{\alpha_B} = [2\alpha - 2, 2 - 2\alpha] \cdot [2\alpha + 2, 6 - 2\alpha] \cdot \left[\frac{1}{6 - 2\alpha}, \frac{1}{2\alpha + 2}\right]$$

$$= [(2\alpha - 2)(2\alpha + 2), (2 - 2\alpha)(6 - 2\alpha)] \cdot \left[\frac{1}{6 - 2\alpha}, \frac{1}{2\alpha + 2}\right]$$

We have

$$\begin{aligned}
&[2\alpha - 2, 2 - 2\alpha] \cdot [2\alpha + 2, 6 - 2\alpha] \\
&= [\min\{(2\alpha - 2)(2\alpha + 2), (2\alpha - 2)(6 - 2\alpha), (2 - 2\alpha)(2\alpha + 2), (2 - 2\alpha)(6 - 2\alpha)\} \\
&, \max\{(2\alpha - 2)(2\alpha + 2), (2\alpha - 2)(6 - 2\alpha), (2 - 2\alpha)(2\alpha + 2), (2 - 2\alpha)(6 - 2\alpha)\} \\
&= [(2\alpha - 2)(6 - 2\alpha), (2 - 2\alpha)(6 - 2\alpha)],
\end{aligned}$$

as $(2\alpha - 2)(6 - 2\alpha)$has the lowest value and$(2 - 2\alpha)(6 - 2\alpha)$ has the highest value for $0 \leq x \leq 1$.

Therefore,

$$\alpha_A \cdot \alpha_B \cdot \frac{1}{\alpha_B} = (2\alpha - 2)(6 - 2\alpha), (2 - 2\alpha)(6 - 2\alpha) \cdot \frac{1}{[2\alpha + 2, 6 - 2\alpha]}$$

$$= [(2\alpha - 2)(6 - 2\alpha), (2 - 2\alpha)(6 - 2\alpha)] \cdot \left[\frac{1}{(6 - 2\alpha)}, \frac{1}{(2\alpha + 2)}\right]$$

$$= [\min\left\{\frac{(2\alpha - 2)(6 - 2\alpha)}{6 - 2\alpha}, \frac{(2\alpha - 2)(6 - 2\alpha)}{2\alpha + 2}, \frac{(2 - 2\alpha)(6 - 2\alpha)}{6 - 2\alpha}, \frac{(2 - 2\alpha)(6 - 2\alpha)}{2\alpha + 2}\right\},$$

$$\max\left\{\frac{(2\alpha - 2)(6 - 2\alpha)}{6 - 2\alpha}, \frac{(2\alpha - 2)(6 - 2\alpha)}{2\alpha + 2}, \frac{(2 - 2\alpha)(6 - 2\alpha)}{6 - 2\alpha}, \frac{(2 - 2\alpha)(6 - 2\alpha)}{2\alpha + 2}\right\}]$$

$$= \left[\frac{(2\alpha - 2)(6 - 2\alpha)}{2\alpha + 2}, \frac{(2 - 2\alpha)(6 - 2\alpha)}{2\alpha + 2}\right] \neq [2\alpha - 2, 2 - 2\alpha] = \alpha_A$$

$$\therefore \alpha_A \cdot \alpha_B \cdot \frac{1}{\alpha_B} \neq \alpha_A$$

From the above discussion, we infer that the equation a = a (b/b) holds for any a, b $\in \mathbb{R}$, but it does not hold good for fuzzy numbers. That is, for fuzzy numbers, A. B/B $\neq$ A does not hold.

1.6 Solving Fuzzy Linear Equations A · X + B = C by Using the Method of α-cut:

Let A, B, and C be the fuzzy triangular numbers and let $A = [a_1, a_2, a_3]$, $B = [b_1, b_2, b_3]$, and $C = [c_1, c_2, c_3]$

Then $A \cdot X + B = C$ is a fuzzy linear equation in X. If the solution X exists, it will be a triangular-shaped fuzzy number. Let $X \approx [x_1, x_2, x_3]$. From the crisp equation $ax + b = c$, we immediately obtain that $X = \frac{c-b}{a}$ is not a solution of the equation in concern, i.e., of the equation a. x + b = c, if $a \neq 0$. Here, we have used the crucial facts that $b - b = 0$ and $(1/a)a = 1$ from real numbers to get the solution.

Let us apply the same approach to the fuzzy equation

$$A \cdot X + B = C,$$
$$\Rightarrow A \cdot X + (B - B) = (C - B)$$
$$\Rightarrow (1/A)\{A \cdot X + (B - B)\} = (1/A)(C - B)$$

But the left-hand side of the above equation does not equal X since $B - B \neq 0$ and$(1/A) \cdot A \neq 1$. For example, if B = (1, 2, 3), then B –B = (-2, 0, 2), not zero. Also, if A = (1, 2, 3), then$(1/A) \cdot A \approx ((1/3), 1, 3)$ is a triangular-shaped fuzzy number, not one.

This is a significant obstacle in solving fuzzy equations. Some basic operations we used to solve crisp equations do not hold for fuzzy equations. This is not the only case where such problems occur. This also happens in probability theory if X is a random variable with positive variance, then X – X ≠ 1 and X / X ≠ 1, since both X – X and X / X will have positive variance.

1.7 Classical Method:

This method defines the solution X_c for any $\alpha \in [0, 1]$. Let us suppose that.

$A^\alpha = [a_1(\alpha), a_2(\alpha)]$; $B^\alpha = [b_1(\alpha), b_2(\alpha)]$; $C^\alpha = [c_1(\alpha), c_2(\alpha)]$, and $X_c^\alpha = [x_1(\alpha), x_2(\alpha)]$ denote, respectively, the α-cuts of A, B, C, and X. Substituting these values into the equation A. X + B = C; we find that

$$[a_1(\alpha), a_2(\alpha)][x_1(\alpha), x_2(\alpha)] + [b_1(\alpha), b_2(\alpha)] = [c_1(\alpha), c_2(\alpha)]$$
$$\Rightarrow [a_1(\alpha)x_1(\alpha), a_2(\alpha)\, x_2(\alpha)] + [b_1(\alpha), b_2(\alpha)] = [c_1(\alpha), c_2(\alpha)]$$
$$\Rightarrow [a_1(\alpha)x_1(\alpha) + b_1(\alpha), a_2(\alpha)\, x_2(\alpha) + b_2(\alpha)] = [c_1(\alpha), c_2(\alpha)]$$

We say that this method defines solution X_c when $[x_1(\alpha), x_2(\alpha)]$ defines the α-cuts of a fuzzy number. For the $x_1(\alpha), x_2(\alpha)$ to specify a fuzzy number, we need.

(i) $x_1(\alpha)$ monotonically increasing, $0 \le \alpha \le 1$
(ii) $x_2(\alpha)$ monotonically decreasing, $0 \le \alpha \le 1$ and
(iii) $x_1(1) \le x_2(1)$

If $x_1(1) < 1$, we will get X_c, a trapezoidal-shaped fuzzy number.

Example1.10 Let A = [8, 9, 10]; B = [-3, -2, -1]; C = [3, 5, 7]. Then, could you find the solution of A. X + B = C?

Solution: We have $A^\alpha = [a_1, a_2, a_3] = [a_1 + (a_2 - a_1)\alpha, a_3 - \alpha(a_3 - a_2)] = [8 + \alpha, 10 - \alpha]$; where $a_1 = 8$, $a_2 = 9$ and $a_3 = 1$. Similarly, we have $B^\alpha = [b_1, b_2, b_3] = [-3 + \alpha, -1 - \alpha]$ and $C^\alpha = [c_1, c_2, c_3] = [3 + 2\alpha, 7 - 2\alpha]$

Since A > 0 and C > 0, we must have $X_c > 0$. Now from equation A. X + B = C, we have

$$[8 + \alpha, 10 - \alpha][x_1(\alpha), x_2(\alpha)] + [-3 + \alpha, -1 - \alpha] = [3 + 2\alpha,\ 7 - 2\alpha]$$
$$\Rightarrow [(8 + \alpha)x_1(\alpha), (10 - \alpha)x_2(\alpha)] + [-3 + \alpha, -1 - \alpha] = [3 + 2\alpha, 7 - 2\alpha]$$
$$\Rightarrow [(8 + \alpha)x_1(\alpha) + (-3 + \alpha), (10 - \alpha)x_2(\alpha) + (-1 - \alpha)] = [3 + 2\alpha, 7 - 2\alpha]$$
$$\Rightarrow (8 + \alpha)x_1(\alpha) + (-3 + \alpha) = 3 + 2\alpha \text{ and } (10 - \alpha)x_2(\alpha) + (-1 - \alpha) = 7 - 2\alpha$$
$$\Rightarrow x_1(\alpha) = \frac{6 + \alpha}{8 + \alpha} \quad \textit{and} \quad x_2(\alpha) = \frac{8 - \alpha}{10 - \alpha}$$

We observe that $x_1(\alpha)$ is increasing (its derivative is positive) and $x_2(\alpha)$ is decreasing (its derivative is negative), and $x_1(1) = \frac{7}{9}$ and $x_2(1) = \frac{7}{9}$. Hence, the solution exists with α-cuts of X given by $X_C^\alpha = [x_1(\alpha), x_2(\alpha)] = \left[\frac{6+\alpha}{8+\alpha}, \frac{8-\alpha}{10-\alpha}\right]$.

The fuzzy membership function of X_C is given by

$$X_C(x)\begin{cases} 36x - 27, \frac{3}{4} \le x \le \frac{7}{9} \\ 36 - 45x, \frac{7}{9} \le x \le \frac{4}{5} \\ 0, x \le \frac{3}{4} \; or \; x \ge \frac{4}{5} \end{cases}$$

Example 1.11 Let A = [1, 2, 3]; B = [-3, -2, -1]; C = [3, 4, 5]. Then, could you find the solution of the fuzzy equation A. X + B = C?

Solution: We have $A^\alpha = [a_1, a_2, a_3] = [a_1 + (a_2 - a_1)\alpha, a_3 - \alpha(a_3 - a_2)] = [1 + \alpha, 3 - \alpha]$; where $a_1 = 1$, $a_2 = 2$ and $a_3 = 3$. Similarly, we have $B^\alpha = [b_1, b_2, b_3] = [-3 + \alpha, -1 - \alpha]$ and $C^\alpha = [c_1, c_2, c_3] = [3 + \alpha, 5 - \alpha]$

Since A > 0 and C > 0, we must have $X_c > 0$ and the equation

$$[a_1(\alpha), a_2(\alpha)][x_1(\alpha), x_2(\alpha)] + [b_1(\alpha), b_2(\alpha)] = [c_1(\alpha), c_2(\alpha)] \text{ gives}$$
$$\Rightarrow [a_1(\alpha)x_1(\alpha) + b_1(\alpha), a_2(\alpha)x_2(\alpha) + b_2(\alpha)] = [c_1(\alpha), c_2(\alpha)]$$
$$\Rightarrow [(1 + \alpha)x_1(\alpha) + (-3 + \alpha), (3 - \alpha)x_2(\alpha) + (-1 - \alpha)] = [3 + \alpha, 5 - \alpha]$$
$$\Rightarrow (1 + \alpha)x_1(\alpha) + (-3 + \alpha) = 3 + \alpha \text{ and } (3 - \alpha)x_2(\alpha) + (-1 - \alpha) = 5 - \alpha$$
$$\Rightarrow x_1(\alpha) = \frac{6}{1+\alpha} \quad \text{and } x_2(\alpha) = \frac{6}{3-\alpha}$$

We observe that $x_1(\alpha)$ is decreasing (its derivative is negative) and $x_2(\alpha)$ is increasing (its derivative is positive), and $x_1(1) = 3$ and $x_2(1) = 3$. Hence, the solution does not exist with α-cuts of X given by $X_C^\alpha = [x_1(\alpha), x_2(\alpha)] = \left[\frac{6}{1+\alpha}, \frac{6}{3-\alpha}\right]$.

By using interval arithmetic, we have

$$Xc(\alpha) = [x_1(\alpha), x_2(\alpha)] = \frac{C(\alpha) - B(\alpha)}{A(\alpha)} =$$
$$\frac{[3 + \alpha, 5 - \alpha] - [-3 + \alpha, -1 - \alpha]}{[1 + \alpha, 3 - \alpha]} = \frac{[4 + 2\alpha, 8 - 2\alpha]}{[1 + \alpha, 3 - \alpha]}$$
$$= [4 + 2\alpha, 8 - 3\alpha] \cdot \frac{1}{[1 + \alpha, 3 - \alpha]}$$
$$= [4 + 2\alpha, 8 - 3\alpha]\left[\frac{1}{3 - \alpha}, \frac{1}{1 + \alpha}\right] = \left[\frac{4 + 2\alpha}{3 - \alpha}, \frac{8 - 3\alpha}{1 + \alpha}\right]$$
$$\therefore Xc(\alpha) = [x_1(\alpha), x_2(\alpha)] = \left[\frac{4 + 2\alpha}{3 - \alpha}, \frac{8 - 3\alpha}{1 + \alpha}\right]$$
$$x_1(\alpha) = \frac{4 + 2\alpha}{3 - \alpha} \text{ and } x_2(\alpha) = \frac{8 - 3\alpha}{1 + \alpha} \Rightarrow x_1(1) = 3 \text{ and } x_2(1) = 3$$

$$\therefore X_C(\alpha) = [x_1(\alpha), x_2(\alpha)] = \left[\frac{4 + 2\alpha}{3 - \alpha}, \frac{8 - 3\alpha}{1 + \alpha}\right]$$

$$x_1(\alpha) = \frac{4 + 2\alpha}{3 - \alpha} \quad \text{and } x_2(\alpha) = \frac{8 - 3\alpha}{1 + \alpha} \Rightarrow x_1(1) = 3 \text{ and } x_2(1) = 3$$

So answers by both methods are compatible.

Exercise-1

1. A and C are two fuzzy numbers whose membership functions are given by

$$A(x) = \begin{cases} \frac{x+2}{2} \text{ if } -2 < x \leq 0 \\ \frac{2-x}{2} \text{ if } 0 < x \leq 2 \\ 0 \text{ otherwise} \end{cases}$$

$$C(x) = \begin{cases} \frac{x-2}{2} \text{ if } 2 < x \leq 4 \\ \frac{9-x}{3} \text{ if } 4 < x \leq 6 \\ 0 \text{ otherwise} \end{cases}$$

Then, could you find the solution to the equation A + X = C?

2. Let A and C be two fuzzy triangular numbers having membership functions given by

$$A(x) = \begin{cases} \frac{x-3}{3} \; if \; 3 < x \leq 6 \\ \frac{9-x}{3} \; if \; 6 < x \leq 9 \\ 0 \text{ otherwise} \end{cases}$$

$$C(x) = \begin{cases} \frac{x-9}{3} \; if \; 9 < x \leq 12 \\ \frac{15-x}{3} \; if \; 12 < x \leq 15 \\ 0 \text{ otherwise} \end{cases}$$

I'd like you to please find the solution of the fuzzy equation A. X = C.

3. Choose A and B as any two symmetrical fuzzy numbers, then verify whether $A + B - B = A$ is true or false for it.
4. Choose A and B as any two symmetrical fuzzy numbers, then verify whether $A.(B/B) = A$ is true or false for it.
5. Discuss the theory for solving the fuzzy linear equation $A + X = C$.
6. Discuss the theory for solving the fuzzy linear equation $A.X = C$.
7. Discuss the theory for solving the fuzzy linear equation $A.X + B = C$.
8. With the help of a suitable example, could you show how to solve the equation when the given fuzzy numbers M and N are closed intervals?
9. Solve the interval equation $\alpha_M \cdot \alpha_X = \alpha_N$, where $\alpha_M = [2\alpha + 6, 8 - 2\alpha]$ and $\alpha_N = [2\alpha + 8, 14 - 2\alpha]$.
10. Solve the interval equation $\alpha_P \cdot \alpha_X = \alpha_Q$, where $\alpha_P = [\alpha + 6, 8 - \alpha]$ and $\alpha_Q = [10\alpha + 12, 34 - 12\alpha]$.

Objective-Type Questions and Answers

1 An equation is said to be a fuzzy equation if its coefficients and unknowns are.

(a) real numbers
(b) fuzzy numbers
(c) fuzzy sets
(d) crisp sets.

2 In fuzzy equations of the form $A + X = B$

(a) A and B are unknown fuzzy numbers
(b) A and B are unknown fuzzy sets
(c) A and B are known fuzzy numbers
(d) none of these

3 In the fuzzy equation $A \cdot X = B, X$ is a.

(a) known fuzzy number
(b) an unknown fuzzy number.
(c) known fuzzy set
(d) none of these.

4 One area of fuzzy set theory in which fuzzy numbers and arithmetic operations on fuzzy numbers play a fundamental role is

(a) Fuzzy equations (b) fuzzy relations (c) fuzzy algebra (d) none of these

5 The concept of solving fuzzy linear equations was first proposed by

(a) L. A. Zadeh
(b) J. Buckley
(c) Dubois and Prade
(d) Buckley and Qu.

6 The fuzzy equation $A + X = B$, where $A = [a_1, a_2]$ and $B = [b_1, b_2]$ are two closed Intervals have the solution given by the interval.

(a) $[b_2 - a_2, b_1 - a_1]$
(b) $[b_1 - a_1, b_2 - a_2]$
(c) $[a_2 - b_2, a_1 - b_1]$
(d) none of these.

7 In fuzzy equations of the form $A \cdot X = B$

(a) A and B are known fuzzy numbers
(b) A and B are unknown fuzzy sets
(c) A and B are unknown fuzzy numbers
(d) none of these

8 X in the fuzzy equation $A + X = B$ is

(a) a known fuzzy set
(b) a known fuzzy number.
(c) an unknown fuzzy number
(d) an unknown fuzzy set.

9 If $A = [a_1, a_2]$ and $B = [b_1, b_2]$ are two closed intervals, which may be viewed as special
Fuzzy numbers, then the equation $A + X = B$ has a solution iff.

(a) $b_1 - a_1 < b_2 - a_2$ (b)$b_1 - a_1 \geq b_2 - a_2$
(b) $b_1 - a_1 > b_2 - a_2$(d)$b_1 - a_1 \leq b_2 - a_2$

10 The Fuzzy Equation $A \cdot X = B$, where $A = [a_1, a_2]$ and $B = [b_1, b_2]$ are two closed intervals has the solution given by the Interval

(a) $\left[\frac{b_1}{a_2}, \frac{b_2}{a_1}\right]$
(b) $\left[\frac{b_2}{a_1}, \frac{b_1}{a_2}\right]$
(c) $\left[\frac{a_1}{b_2}, \frac{a_2}{b_1}\right]$
(d) none of these.

11 If $A = [a_1, a_2]$ and $B = [b_1, b_2]$ are two closed Intervals, which may be viewed as special fuzzy numbers, then the equation $A \cdot X = B$ has a solution iff

(a) $\frac{b_1}{a_2} < \frac{b_2}{a_1}$ (b)$\frac{b_1}{a_2} \leq \frac{b_2}{a_1}$ (c)$\frac{b_1}{a_2} > \frac{b_2}{a_1}(d)\frac{b_1}{a_2} \geq \frac{b_2}{a_1}$

12 If $\alpha_A = \left[\alpha_{a_1}, \alpha_{a_2}\right]$, $\alpha_C = \left[\alpha_{c_1}, \alpha_{c_2}\right] \forall\ \alpha \in (0, 1]$, the Solution of the Fuzzy Interval Equation, $\alpha_A \cdot \alpha_X = \alpha_C$ is Possible if and Only if

(a) $\frac{\alpha_{c_1}}{\alpha_{a_2}} > \frac{\alpha_{c_2}}{\alpha_{a_1}}(b)\frac{\alpha_{c_1}}{\alpha_{a_2}} \geq \frac{\alpha_{c_2}}{\alpha_{a_1}}(c)\frac{\alpha_{c_1}}{\alpha_{a_2}} \leq \frac{\alpha_{c_2}}{\alpha_{a_1}}(d)\frac{\alpha_{c_1}}{\alpha_{a_2}} < \frac{\alpha_{c_2}}{\alpha_{a_1}}$

13 The Solution of the Fuzzy Interval Equation $\alpha_A \cdot \alpha_X = \alpha_C$ for α and $\beta \in]0, 1]$ are

(a) Nested
(b) not nested
(c) closed interval
(d) none of these.

14 If $\alpha_A = \left[\alpha_{a_1}, \alpha_{a_2}\right]$, $\alpha_C = \left[\alpha_{c_1}, \alpha_{c_2}\right] \forall\ \alpha \in (0, 1]$, then the solution of the fuzzy interval equation, i.e., $\alpha_A \cdot \alpha_X = \alpha_C$ exists, and it has the form.

(a) $X = \bigcup_{\alpha \in [0, 1]} \alpha^X(b)X = \bigcap_{\alpha \in [0, 1]} \alpha^X(c)X = \bigcup_{\alpha \in (0, 1]} \alpha^X$ (d)$X = \bigcap_{\alpha \in (0, 1]} \alpha^X$

15 If $\alpha_A = [\alpha_{a_1}, \alpha_{a_2}]$, $\alpha_C = [\alpha_{c_1}, \alpha_{c_2}] \forall \alpha \in]0, 1]$, the solution of the fuzzy interval equation $\alpha_A + \alpha_X = \alpha_C$ is possible if and only if

(a) $\alpha_{c_1} - \alpha_{a_1} > \alpha_{c_2} - \alpha_{a_2}$
(b) $\alpha_{c_1} - \alpha_{a_1} < \alpha_{c_2} - \alpha_{a_2}$
(c) $\alpha_{c_1} - \alpha_{a_1} \geq \alpha_{c_2} - \alpha_{a_2}$
(d) $\alpha_{c_1} - \alpha_{a_1} \leq \alpha_{c_2} - \alpha_{a_2}$

16 The fuzzy interval equation $\alpha_A + \alpha_X = \alpha_C$ has a solution

(a) $[\alpha_{c_1} - \alpha_{a_1}, \alpha_{c_2} - \alpha_{a_2}]$
(b) $[\alpha_{c_2} - \alpha_{a_2}, \alpha_{c_1} - \alpha_{a_1}]$
(c) $[\alpha_{a_2} - \alpha_{c_2}, \alpha_{a_1} - \alpha_{c_1}]$
(d) $[\alpha_{a_1} - \alpha_{c_1}, \alpha_{a_2} - \alpha_{c_2}]$

17 The solution of the fuzzy Interval equation $\alpha_A + \alpha_X = \alpha_C$ for α and $\beta \in]0, 1]$ are

(a) Not nested
(b) nested
(c) open interval
(d) none of these

18 In the fuzzy equation $A + X = B$, A and B, the given fuzzy numbers are

(a) Open intervals
(b) left open right closed intervals
(c) Closed intervals
(d) left closed right open intervals

19 If the fuzzy interval equation $\alpha_A + \alpha_X = \alpha_C$ has a solution for all $\alpha, \beta \in (0, 1]$ and are nested, then for all $\alpha \leq \beta$

(a) $\alpha_X \supset \beta_X$ (b) $\alpha_X \supseteq \beta_X$ (c) $\alpha_X \subset \beta_X$ (d) $\alpha_X \subseteq \beta_X$

20 If a solution α_X exists for every $\alpha \in (0, 1]$ and $\alpha_{c_1} - \alpha_{a_1} \leq \alpha_{c_2} - \alpha_{a_2}$ is satisfied for every $\alpha \in (0, 1]$ then the solution X of the fuzzy equation is given by the theorem, known as

(a) First Decomposition Theorem (b) Second Decomposition Theorem
(b) Third Decomposition Theorem (d) None of these

Answers

1. (b) & (c); **2.** (c); **3.** (b); **4.** (a); **5.** (d); **6.** (b); **7.** (a); **8.** (c); **9.** (d); **10.** (a);**11.** (b); **12.** (c); **13.** (a); **14.** (c); **15.** (d); **16.** (a); **17.** (b); **18.** (c); **19.**(b); **20.** (a)

True/False Type Statements and Answers

1 Buckley and Qu investigated fuzzy equations in 1990.
2 In every fuzzy equation, coefficients, as well as unknowns, are all real numbers.

3 Solution of fuzzy equations can be obtained by applying interval fuzzy arithmetic operations and decomposition theorem.
4 Coefficients and unknowns in any fuzzy equations are all fuzzy numbers.
5 For $\alpha \in (0, 1]$ the solution of the fuzzy equation $A + X = B$ by the first decomposition theorem is given by $X = \bigcup_{\alpha \in [0, 1]} \alpha^X$, where$\alpha^X = \alpha \cdot \alpha_X$.
6 The fuzzy equation $A + X = B$ has a solution, if and only if $\alpha_{(A+X)} = \alpha_B \Rightarrow \alpha_A + \alpha_X = \alpha_B$.
7 For $\alpha \in (0, 1)$, the solution of the fuzzy equation $A \cdot X = B$ from the first decomposition theorem is given by $X = \bigcup_{\alpha \in (0, 1]} \alpha^X$, where$\alpha^X = \alpha \cdot \alpha_X$.
8 Let P, Q, and R be any three membership matrices, and if the three relations constraints are connected in such a way that $P^\circ Q = R$, where $\circ$ denotes the max–min composition. This means that $\underset{j \in J}{maxmin}\left(p_{ij}, q_{jk}\right) = r_{ik}$ for all $i \in I$ and $k \in K$. The equation $P \circ Q = R$ encompasses $n \times s$ simultaneous equations. When two of the components in each equation are given, and one is unknown, these equations are referred to as fuzzy relation equations.
9 For α and $\beta \in (0, 1]$, the solution of the fuzzy interval equation $\alpha_A \cdot \alpha_X = \alpha_B$ is not nested.
10 $A \cdot X + B = C$ is a fuzzy linear equation in X.
11 For fuzzy numbers, M and N, $M + N - N \neq M$ does not hold well.
12 $\alpha_X = \left[\alpha_{b_1} - \alpha_{a_1}, \alpha_{b_2} - \alpha_{a_2}\right]$, is not the solution to the interval equation $\alpha_A + \alpha_X = \alpha_B$.
13 If $b_1 - a_1 < b_2 - a_2 < b_3 - a_3$, then the fuzzy equation $A + X = B$ has a solution, and then X is a triangular fuzzy number.
14 The solutions of the interval equations for α and β are nested, i.e., if $\alpha \geq \beta \Rightarrow \beta_X \subseteq \alpha_X$.
15 If $\alpha_A = [2\alpha - 2, 2 - 2\alpha]$ and$\alpha_B = [2\alpha + 2, 6 - 2\alpha]$, then the fuzzy equation $A + X = B$ gives $\alpha_X = [4\alpha, 8 - 4\alpha]$.

Answers

1. (T); 2. (F); 3. (T); 4. (T); 5. (F); 6. (T); 7. (T); 8. (T); 9. (F); 10. (T); 11. (T); 12. (F); 13. (T); 14. (F); 15. (T)

Chapter 2
Graphs, Fuzzy Graphs, and Intuitionistic Fuzzy Graphs

In this chapter, we should be familiar with the meaning of "graph" in very brief and its more general form, the "fuzzy graph" and the "intuitionistic fuzzy graph" in greater detail. Various properties of fuzzy graphs and intuitionistic fuzzy graphs will also be introduced with the help of suitable examples along with their graphs.

2.1 Introduction

In 1736, the *Swiss mathematician* ***Leonhard Euler*** (1707–17,830) introduced the notion of graph theory. A graph is a simple way to view the information needed to relate objects. Nodes denote the objects, and arcs denote relations. The graphic models describe network traffic, rail network, telephone network, etc. The main idea behind this work grew out of a famous problem known as the seven bridges problem of *Konigsberg*. Euler developed some of the fundamental concepts for the theory of graphs. Graphs are discrete structures appearing in many areas of knowledge, such as social science, physical and mathematical sciences, O.R., computer science, and many others. By graphs, we always mean a linear graph because there are no such things as non-linear ones.

When the specifics of the objects or their relationships or both are blurred, it is natural that we need to create a "fuzzy graph model." A fuzzy graph is a uniformly fuzzy binary relationship on the subset.

The term "fuzzy graph" has been used for two different notions. **Zadeh** first introduced it, and the second one by *Rosenfeld* in 1975. However, **Arnold Kauffman** first described the fuzzy graph in 1973. **Rosenfeld's** fuzzy graph is an extension of conventional or classical graph theory. It introduces a fresh perspective that can greatly enhance our understanding of complex systems. In contrast, Zadeh's fuzzy graph is much more relevant to fuzzy control and its applications to industries. In

M. K. Singh, *Applied Fuzzy Mathematics*, Forum for Interdisciplinary Mathematics,
https://doi.org/10.1007/978-981-97-3257-9_2

Zadeh's theory, a fuzzy graph describes a functional mapping between a set of input linguistic variables and an output linguistic variable.

This chapter briefly discusses the concepts and properties of classical graphs as a refresher to demonstrate their generalized application to fuzzy graphs.

Graphs can be used to solve many problems like:

Is it possible to travel all the roads of a town without going to a road twice?

To find the number of colors needed to color the regions of a map of a country.

To find the shortest path between two significant towns, and so on.

2.2 Concepts of Graph Theory and Algebraic Operations on Graphs

2.2.1 Graph

The pair $G = (V, E)$, where $V = \{v_1, v_2, v_3, \ldots, v_n\}$ is a finite nonempty set whose elements are called nodes or points or vertices and $E = \{e_1, e_2, e_3, \ldots, e_n\}$, whose elements are called edges or lines or arcs or arrays or branches, where e_n is identified with the unordered pair $\left(v_i, v_j\right)$ of vertices. v_i is the initial vertex, and v_j is the terminal or final vertex. A graph $G = (V, E)$ in which all edge e is directed is termed a *digraph* or *directed graph*. An arrow at the outer end represents directed edges.

When there is no concern about the direction of an edge, we still write $G = (V, E)$ where E is a set of unordered pairs of elements taken from V. Then $G = (V, E)$ be called an undirected graph. We generally call "V as the vertex set of G" and "E is the edge set of G."

Alternatively, a graph G is usually denoted as $G = (V, E, \emptyset)$, where V is a set of *nodes* or vertices of the graph and $\emptyset$ is a *mapping* from the set of *edge* E to the set of pair of elements of V, *i.e.*, (v_1, v_2) (Figs. 2.1 and 2.2).

Let V be a nonempty set and $R \subseteq V \times V$ then, R is a relation on V. Any relation R on S always defines a graph with node-set S and arc set R.

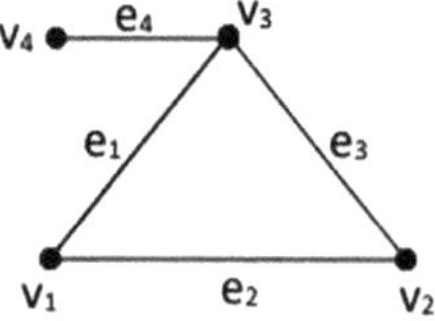

Fig. 2.1 Undirected graphs

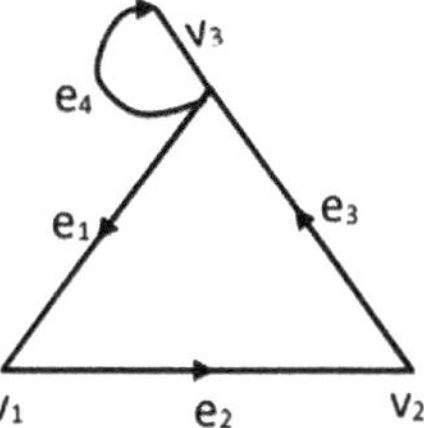

Fig. 2.2 Directed graphs

Fig. 2.3 Parallel edge

2.2.2 Adjacent Nodes

Any pair of nodes connected by an edge in a graph is called an adjacent node, i.e., v_1, v_2, and v_2, v_3 are adjacent nodes. An edge $e = (v_i, v_j)$ is called a *self-loop* if $v_i = v_j$. e_4 is a *loop*. Edges e_i *and* e_j are called *parallel edges* if they have the same pair of vertices as their initial and terminal vertices, or if two nodes of a graph are joined by more than one edge, they are called *parallel edges* (Fig. 2.3).

If vertex v is an end-vertex of some edge e, then we say that e is *incident* on v.

2.2.3 Adjacent Edges

Two edges are considered adjacent if they are incidentally on a common vertex. In Fig. 2.1—e_1, e_4 and e_1, e_2 are adjacent edges.

2.2.4 Weighted Graph

If some weights are assigned to every edge of a graph, the graph is called a weighted graph (Fig. 2.4).

2.2.5 Null Graph

A graph having isolated nodes only is called a null graph. In other words, a graph whose set of edges is empty is referred to as a null graph. In a null graph, each vertex is an isolated vertex (Fig. 2.4).

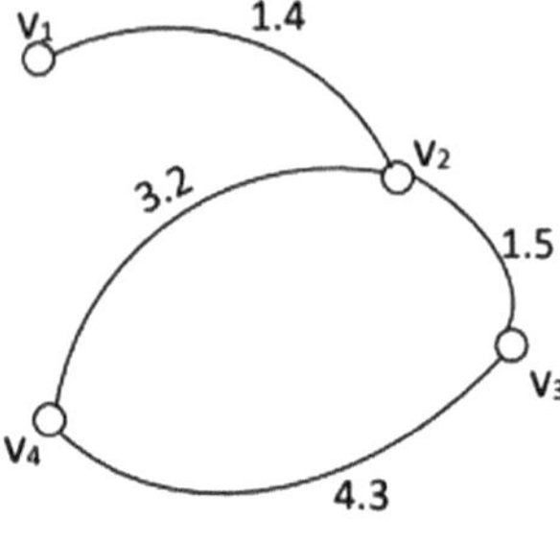

Fig. 2.4 Weighted graphs

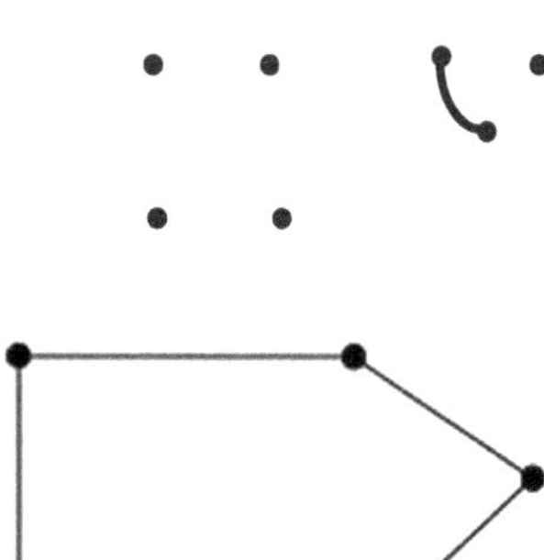

Fig. 2.5 Null graph

Fig. 2.6 Simple graph

Simple graph

2.2.6 *Simple Graph and Multi Graph*

A graph with neither parallel edges nor self-loops is termed a simple graph, and a graph with some parallel edges is called a multi-graph. It is called a **finite graph** if it has a finite number of vertices and edges (Fig. 2.5).

2.2.7 *Pseudo Graph*

A graph in which loops and multiple edges are allowed is called a pseudo graph (Figs. 2.6 and 2.7).

2.2.8 *Degree of a Node*

For an undirected graph G, the degree of a vertex v is denoted by $d(v)$ is the number of edges incident on v, with self-loop counted twice.

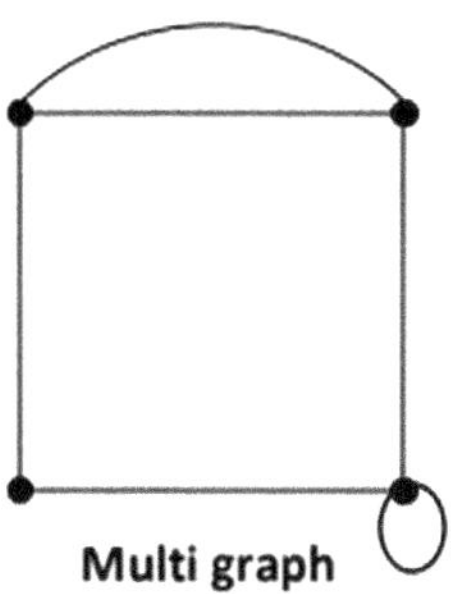

Fig. 2.7 Multi graph

For a digraph G, the **out-degree** of a vertex v is defined as the number of edges of the graph G that begins (initiate) at v, and the ***in-degree*** of v is the number of edges that terminates at v. The *out-degree* is denoted by $\boldsymbol{d}^{+}(\boldsymbol{v})$ and in-degree by the symbol $\boldsymbol{d}^{-}(\boldsymbol{v})$. The sum of in-degree and out-degree is called *the* ***total degree*** *of the vertex v* (Figs. 2.8 and 2.9)

$$d(v_5) = 4, d(v_1) = 4, d(v_4) = 3 \; etc; \; Fig.2.8$$
$$\therefore \; \boldsymbol{d}^{+}(\boldsymbol{v}_1) = 1, \boldsymbol{d}^{-}(\boldsymbol{v}_1) = 2 \text{ and total degree } (v_1) = 3; \; Fig.2.9.$$

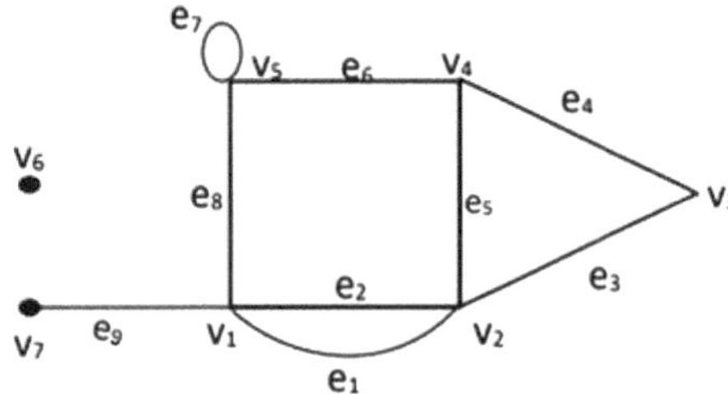

Fig. 2.8 Undirected graph

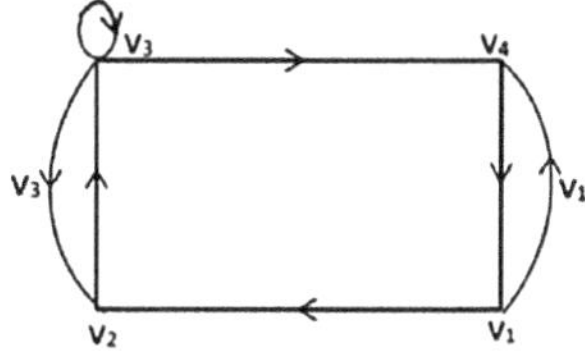

Fig. 2.9 Directed graph

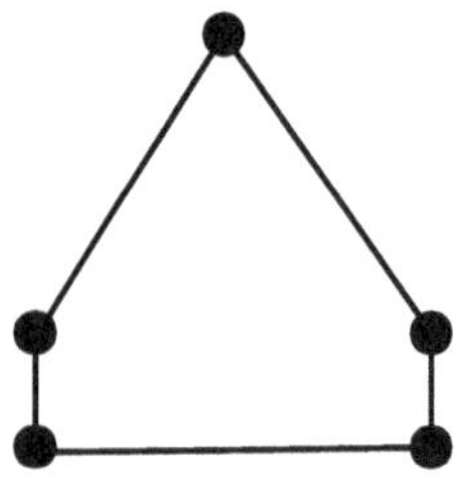

Fig. 2.10 2-regular graph and 1-regular graph

2.2.9 Degree of an Edge

The degree of an edge (u, v) in the underlying graph is denoted by $d_G(u, v)$ and is defined as

$$d_G(u, v) = d_G(u) + d_G(v) - 2$$
$$\therefore d_G(v_3, v_4) = d_G(v_3) + d_G(v_4) - 2 = 2 + 3 - 2 = 3;\ Fig.\ 2.8$$

2.2.10 Regular Graph

A graph G is considered regular if every vertex is adjacent only to vertices having the same degree. In other words, a graph G in which all vertices are of equal degree is called a regular graph If each vertex of graph G has degree k, then the graph is called a k-regular graph.

Both adjoining figures are 2-regular and 1-regular graphs as each vertex has a degree of 2 in the first figure and the second figure is with two vertices and one edge. (Figs. 2.10 and 2.11).

2.2.11 Irregular Graph

A graph G is considered irregular if a vertex is adjacent only to vertices having a distinct degree (Fig. 2.12).

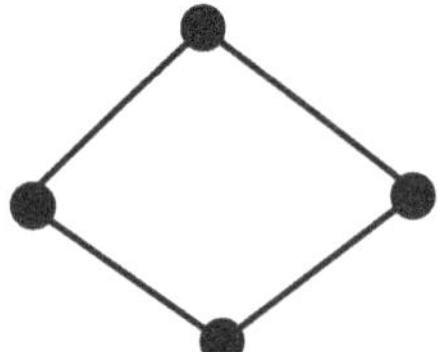

Fig. 2.11 2-regular graph

2.2.12 Complete Graph

A simple graph G with n vertices is said to be a complete graph if the degree of every vertex is $n - 1.K_n$, denotes a complete graph with n vertices. It is also referred to as a **universal graph** (Fig. 2.13).

2.2.13 Order of a Graph

If $G = (V, E)$ is the given graph, then the cardinality of the vertex set V, i.e., $|V|$ of G, is called the order of G (Fig. 2.14).

2.2.14 Subgraph

If $G = (V, E)$ is a graph (directed or undirected), then $G_1 = (V_1, E_1)$ is said to be a subgraph of G, where each edge in E_1 is incident with vertices in V_1. In other words, a graph G_1 is said to be a subgraph of G if all the vertices and edges of G_1 are in G and if the adjacency is preserved in G_1 exactly as in G and is denoted by the symbol $G_1 \subset G$ (Fig. 2.15).

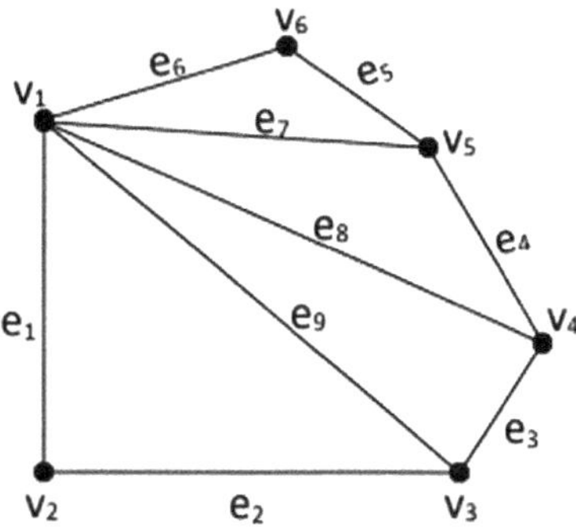

Fig. 2.12 Example of a general Graph G

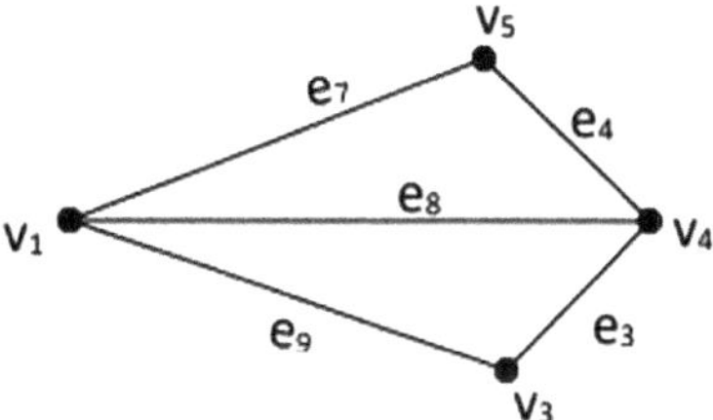

Fig. 2.13 Subgraph of graph G represented by Fig, 2.12 which is irregular and complete

Fig. 2.14 Subgraph of G depicted by Fig. 2.12

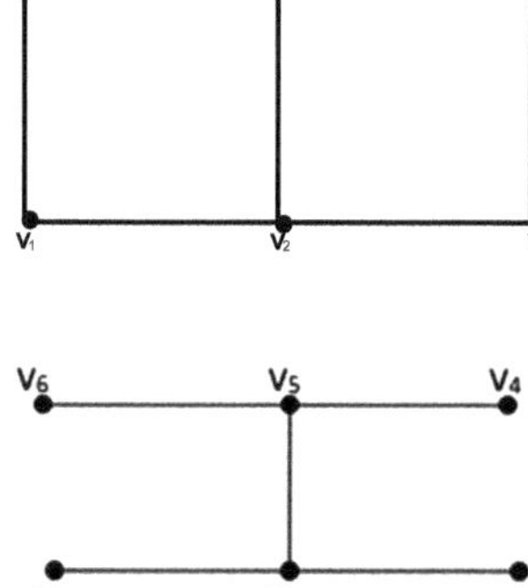

Fig. 2.15 Graph for explaining subgraph

Fig. 2.16 Spanning subgraph of graph G portrayed in Fig. 2.15

From the notion of the subgraph, we can infer the following observations about the subgraph:

(i) Every graph G is a subgraph of itself.
(ii) A null graph is also a subgraph of G.
(iii) A single vertex in a graph G is also a subgraph of G.
(iv) A subgraph of G is also a subgraph of G.
(v) A single edge in G, together with its end vertices, is also a subgraph of G.

2.2.15 *Spanning Subgraph*

Let $G = (V, E)$ be a graph (directed or undirected), and $G_1 = (V_1, E_1)$ be a subgraph of G. Then G_1 is considered a spanning subgraph of G if and only if $V = V_1$. Further, G_1 is said to be a proper subgraph of G if $V_1 \subset V$ and $E_1 \subset E$ (Fig. 2.16).

2.2.16 *Induced Subgraph*

Let $G = (V, E)$ be a graph and $\phi \neq U \subseteq V$. The subgraph of G induced by U is the subgraph whose vertex set is U and which contains all edges from G of the form

$\{u_1, u_2\}$ for $u_1, u_2 \in U$ when G is undirected and is symbolically denoted by $\langle U \rangle$. A subgraph G_1 of a graph $G = (V, E)$ is called an induced subgraph if there exists $\phi \neq U \subseteq V$ where $G_1 = \langle U \rangle$. In other words, G_1 is said to be an induced subgraph of G if its edges set E_1 contains all edges in G whose endpoints belong to vertices in G (Fig. 2.17).

2.2.17 *Walk*

A walk in a graph $G = (V, E)$ is a loop-free finite alternating sequence $v = v_0, e_1, v_1, e_2, v_2, e_3, \ldots, e_{n-1}, v_{n-1}, e_n, v_n = v'$ of vertices and edges from G, starting at vertex v and ending at vertex v' and involving the n edges $e_i = \{v_{i-1}, v_i\}$where $1 \leq i \leq n$. A walk is also called a *chain*. The number of edges in the walk is called the *length of the walk*. The length of this walk is n. When n = 0, there are no edges, i.e., $v = v'$ and the walk is *trivial*. A walk is said to be *closed* if it begins and ends at the same vertex if $v = v'$ (and n > 1) is called a closed walk; otherwise, it is said to be open. In other words, a walk is called an *open walk* if terminal vertices are different in a walk. A walk may repeat both vertices and edges (Fig. 2.18).

2.2.18 *Path*

An open walk in which no vertex appears more than once is called a path; if no vertex of the $v - v'$ walk occurs more than once, then the walk is called a $v - v'$ path. When $v = v'$, the closed path is called a *cycle*. The term cycle always implies the presence

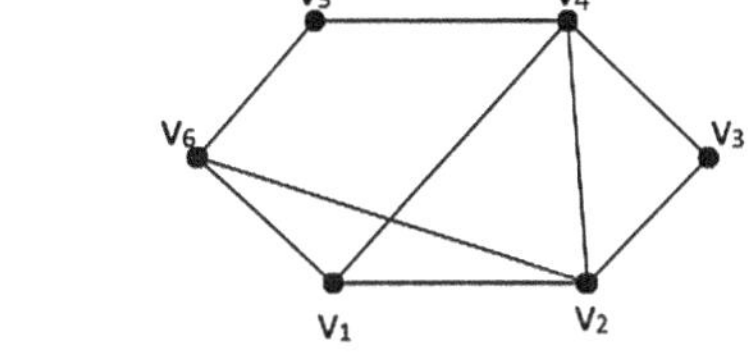

Fig. 2.17 Find the induced subgraph of the Graph G

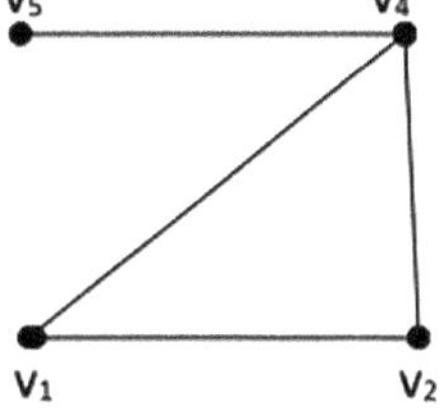

Fig. 2.18 Induced subgraph of the graph G sketched in Fig. 2.17

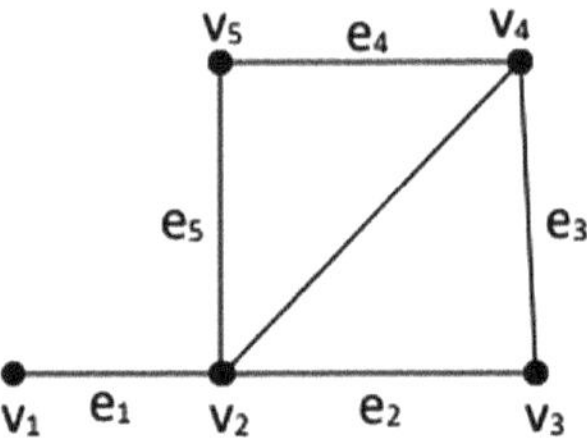

Fig. 2.19 Graph representing Walk/Chain

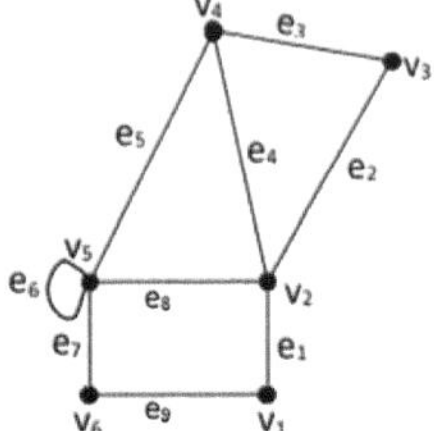

Fig. 2.20 Graph representing path

of at least three edges. The length of a path is defined as the number of edges in the path (Fig. 2.19).

Let n be a non-negative integer and G be an undirected graph. A path of length n from v_0 to v_n in G is a sequence of n edges $e_1, e_2, e_3, \ldots, e_{n-1}, e_n$ of G for which there exists a sequence $v = v_0, v_1, v_2, \ldots v_{n-1}, v_n = v'$ such that e_i has for i = 1, 2, 3, … , n the endpoints v_{i-1} and, when the graph is simple. We denote this path by its vertex sequence $v_0, v_1, v_2, \ldots v_{n-1}, v_n$. The path is a circuit if it begins and ends at the same vertex, i.e., if $u = v$ and has a length greater than zero. (Fig. 2.20).

2.2.19 *Trail*

A walk in graph $G = (V, E)$ in which no edge is repeated is called a trail, or if no edge in the $v - v'$ walk is repeated, then the walk is called a $v - v'$ trail. A closed $v - v'$ trail is called a *circuit* or **tour** in a graph. A closed trail is also called a *tour* (Fig. 2.21).

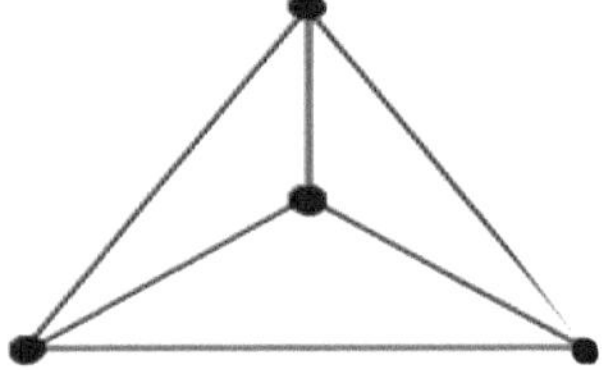

Fig. 2.21 Connected graph

2.2.20 Connected Graph

A graph $G = (V, E)$ is said to be connected if there exists a path between any two distinct vertices of graph G. A graph that is not connected is called a **disconnected graph** (Fig. 2.22).

2.2.21 Distance Between Nodes

Let $G = (V, E)$ is a connected graph, then the distance between the vertices v_1 and v_2 is denoted by $d(v_1, v_{2)}$ and is defined as the length of the shortest path (Fig. 2.23).

(i) $v_1, v_2, v_5, v_6, v_4, v_1, v_3, v_4, v_2$ is a walk as well as a trail. Here vertices (v_1, v_2, v_4) are repeated, while no edge is repeated.
(ii) v_4, v_6, v_4 is a walk. It is also a closed walk.
(iii) v_8 is a walk, trail, and path since no vertex is repeated and there are no edges.
(iv) $v_1, v_2, v_5, v_6, v_4, v_3, v_1$ is a walk, closed walk, trail, closed trail, and cycle. As in the given sequence, no vertex is repeated, no edge is repeated, and the sequence starts at v_1. So it is a closed walk, closed trail, and cycle.
(v) $v_1, v_3, v_4, v_6, v_5, v_2, v_4, v_1$ is not a cycle because vertices v_1 and v_4 are repeated.

2.2.22 Tree

Let $G = (V, E)$ be a loop-free undirected graph. Then, graph G is called a tree if G is connected and has no cycles. In other words, a connected graph G having no

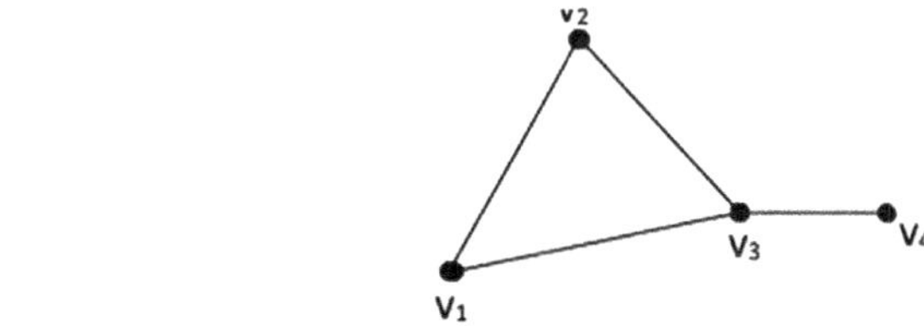

Fig. 2.22 Disconnected graph

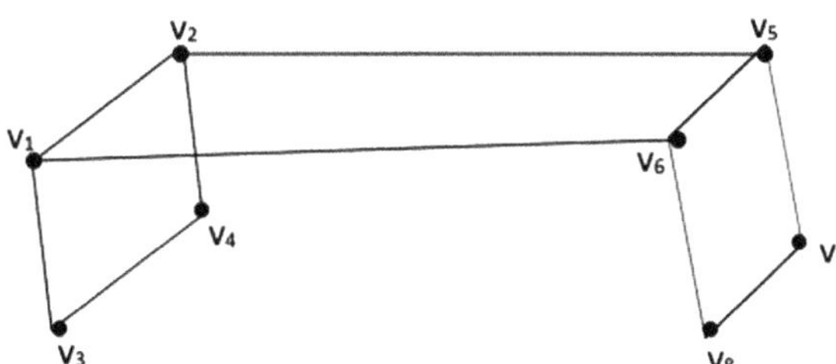

Fig. 2.23 Graph Representing Walk, closed walk, trail, closed trail, path and cycle

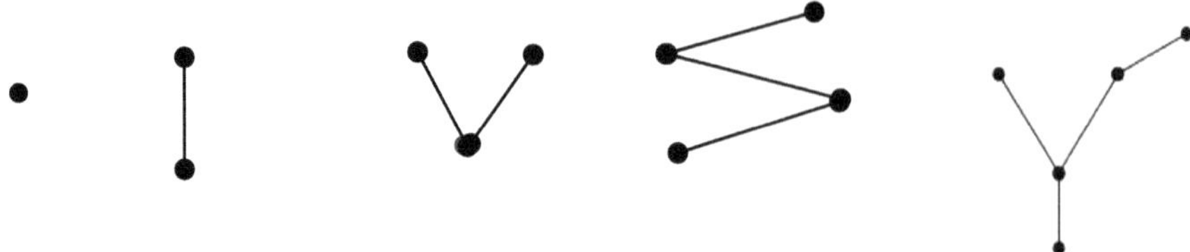

Fig. 2.24 Representing Tree with one, two, three, four and with several vertices

Fig. 2.25 Binary tree

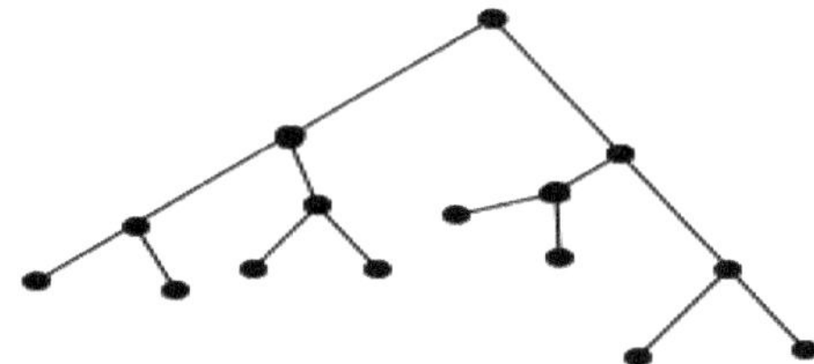

circuit is called a tree. As we know that parallel edges and self-loops form a circuit, it follows that a tree has neither self-loops nor parallel edges (Fig. 2.24).

Are examples of trees with one, two, three, four and with several vertices, respectively?

A tree having no edges but a single vertex is called a degenerate tree. When a graph is a tree, we write T instead of G to highlight the structure. A tree with only one vertex is called a trivial tree. A tree without edges is called a null tree.

A graph G is a tree if and only if there is one and only one path between every pair of vertices of G. A tree with n vertices has precisely $(n-1)$ edges. Every connected graph with n vertices and $(n-1)$ edges is a tree.

2.2.23 Binary Tree

A tree in which there is exactly one vertex of degree two, and each of the remaining vertices is of degree one or three, is called a binary tree (Fig. 2.25).

2.2.24 Forest

It is defined as a collection of trees, or a forest is a graph whose components are all trees. A collection of disjoint trees is said to be a forest. If the root and the corresponding edges connecting the nodes are deleted from a tree, we get a set of disjoint trees. This set of disjoint trees is called a forest. In other words, a forest G is a

Fig. 2.26 Acyclic

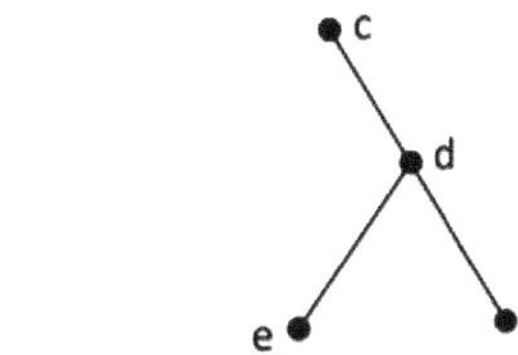

Fig. 2.27 Forest

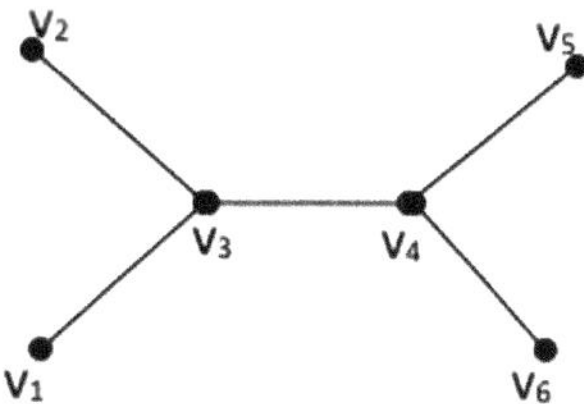

Fig. 2.28 Graph representing vertices, centres, radius and diameter

graph with no cycles; therefore, the connected components of G are trees. (Figs. 2.26 and 2.27).

A crisp graph with no cycles is called an **acyclic** or a **forest**. A connected forest is called a **tree** (Fig. 2.28).

2.2.25 Deletion

Let v_i be a vertex in graph G. Then, $G - v_i$ denotes a subgraph of G obtained by deleting v_i from G. Deletion of a vertex always implies the deletion of all edges incident on that vertex. (Fig. 2.29).

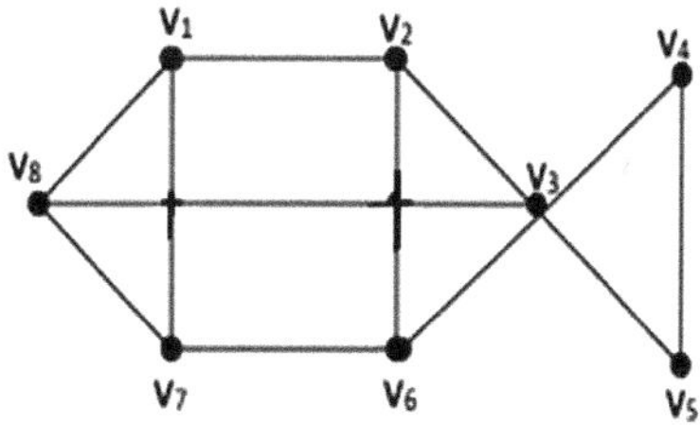

Fig. 2.29 Distance between nodes V_8, V_3 $D(V_8, V_3) = 2$

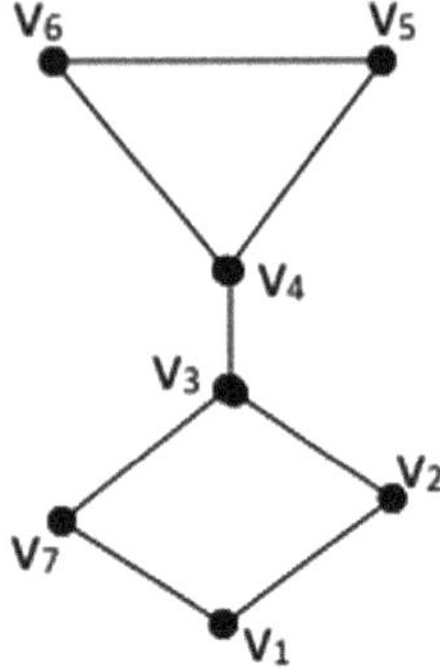

Fig. 2.30 Diameter of the graph G Dim(G) = Δ = 4

2.2.26 *Cut Vertex*

A vertex v of a connected graph $G = (V, E)$ is called a cut vertex or a cut point if its deletion disconnects the remaining graph (Fig. 2.30).

2.2.27 *Bridge*

An edge of a connected graph $G = (V, E)$ is called a bridge if its deletion makes the remaining graph $(G - e)$ disconnected.

2.2.28 *Eccentricity, Center, Radius, and Diameter in a Graph*

The eccentricity of a vertex v in a graph $G = (V, E)$ is defined as the distance between v and the farthest vertex u from v and is denoted by $E(v)$. Thus, we have

$$E(v) = \max_{u \in G} d(v, u)$$

$$E(v_1) = max\{d(v_1, v_2), d(v_1, v_3), d(v_1, v_4), d(v_1, v_5), d(v_1, v_6)\}$$
$$= \{2, 1, 2, 3, 3\} = 3$$

Similarly,

$$E(v_2) = 3, E(v_3) = 2, E(v_4) = 2, E(v_5) = 3, E(v_6) = 3$$

A vertex of minimum eccentricity in graph $G = (V, E)$ is called the *cent*er of the graph G. A graph G may have more than one center. The centers of the graph in Fig. 2.28 are v_3 and v_4.

The eccentricity of the center in a tree is referred to as the *radius* of the tree, and the diameter of a tree is defined as the length of the longest path in a tree. It is optional for the *diameter* to be double of the radius.

Radius in Fig. 2.28 = eccentricity of center = 2 and

Diameter = Δ = length of the longest path in the tree = $\underset{u \in G}{max} E(v) = 3$.

The distance between vertices of a connected graph is a *metric*.

2.2.29 Metric

A metric is a function $m(x, y)$ that satisfies the following three conditions

(i) Non-negativity, i.e., $m(x, y) \geq 0$ *and* $m(x, y) = 0 \Leftrightarrow x = y$
(ii) Symmetry, i.e.,$m(x, y) = m(y, x)$
(iii) Triangular Inequality, i.e., $m(x, z) \leq m(x, y) + m(y, z)$

The distance $d(v_i, v_j)$ between two vertices of a connected graph satisfies conditions (i) and (ii). Since $d(v_i, v_j)$ is the length of the shortest path between vertices, this path cannot be longer than any other path between v_i and v_j, which passes through another vertex v_k, say. Hence

$$d(v_i, v_j) \leq d(v_i, v_k) + d(v_k, v_j)$$

Now, we will discuss the various operations in graphs.

2.2.30 Union of Graphs

Let $G_1 = (V_1, E_1)$, and $G_2 = (V_2, E_2)$ be any two graphs. Then, the union of G_1 and G_2 is denoted by $G_1 \cup G_2$ is a graph with vertex set $V_1 \cup V_2$ and edge set $E_1 \cup E_2$ such that

$$V(G_1 \cup G_2) = V(G_1) \cup V(G_2) = V_1 \cup V_2$$

and

$$E(G_1 \cup G_2) = E(G_1) \cup E(G_2) = E_1 \cup E_2$$

2.2.31 Intersection of Graphs

Let $G_1 = (V_1, E_1)$, and $G_2 = (V_2, E_2)$ be any two graphs such that $V_1 \cap V_2 \neq \emptyset$. Then, the intersection of G_1 and G_2 is denoted by $G_1 \cap G_2$ is a graph with vertex set $V_1 \cap V_2$ and edge set $E_1 \cap E_2$ such that

$$V(G_1 \cap G_2) = V(G_1) \cap V(G_2) = V_1 \cap V_2$$

and

$$E(G_1 \cap G_2) = E(G_1) \cap E(G_2) = E_1 \cap E_2$$

2.2.32 Sum of Two Graphs

Let $G_1 = (V_1, E_1)$ and $G_2 = (V_2, E_2)$ be any two graphs such that $(V_1 \cap V_2) = \emptyset$. Then, the sum of G_1 and G_2 is denoted by $G_1 + G_2$ and is defined as the graph whose vertex set is $V_1 + V_2$, and the edge set of $G_1 + G_2$ consists of those edges in G_1 and G_2 and the edges obtained by joining each vertex of G_1 to each vertex of G_2.

2.2.33 Product of Two Graphs

The Cartesian product of two graphs, G_1, G_2, is denoted by $G_1 \times G_2$ is a graph such that

(i) the vertex set of $G_1 \times G_2$ is the Cartesian product $V(G_1) \times V(G_2)$ and
(ii) any two vertices (u_1, u_2), and (v_1, v_2) are adjacent in $G_1 \times G_2$ if and only if either $u_1 = v_1$ and u_2 is adjacent with v_2 or u_1 is adjacent with v_1 and $u_2 = v_2$.

Fuzzy Graphs

2.3 Fundamental Notions of Fuzzy Graphs and Its Properties

2.3.1 Introduction of Fuzzy Graphs

Fuzzy set theory gives us only an essential and robust representation of quantification of uncertainties, yet a more realistic representation of imprecise concepts expressed in simple language. The mathematical insertion of conventional set theory into fuzzy has become a simple phenomenon. Thus, the knowledge of fuzziness is an enriching one.

In 1736, Euler first suggested the idea of graph theory. A graph is a simple way to view the information needed to relate objects. Nodes denote the objects, and arcs denote relations. The graphic models describe network traffic, rail network, telephone network, etc.

Kaufmann first **introduced** the definition of the fuzzy graph in 1973. Later, ***Azriel Rosenfeld*** proposed the notion of a "*fuzzy graph*" in the realm of mathematics in 1975. Rosenfeld's paper is regarded as the cornerstone of this field of knowledge. When vagueness is involved in describing objects and their relationships, we require a fuzzy graph model.

Let $A = \{(x, A(x)) : x \in X\}$ and $B = \{(y, B(y)) : y \in Y\}$ be two fuzzy sets on X and Y, where $X, Y \subseteq R$. Then, $R = \{[(x, y), R(x, y) : (x, y) \in X \times Y]\}$ is a fuzzy relation on A and B if

$$R(x, y) \leq A(x), \forall (x, y) \in X \times Y$$

$$R(x, y) \leq B(x), \forall (x, y) \in X \times Y$$

or, equivalently

$$R(x, y) \leq \min\{A(x), B(y)\}, \forall (x, y) \in X \times Y \tag{2.1}$$

This fuzzy relation can always represent a *weighted graph* or fuzzy graph, where the *arc* $(x, y) \in X \times Y$ has *weight* $R(x, y) \in [0, 1]$. Elements of the fuzzy relation R are termed the *"nodes"* of the fuzzy graph depicted by this fuzzy relation. The degrees of membership of the elements of the related fuzzy sets A and B convey or express the *"strength"* of or the flow in the respective edges of the graph, whereas the degrees of membership of the respective pairs in the relation R express the *"flows"* or *"capacities"* of the graph. Equation (2.1) describes an exciting graph feature: "In no case, the flows in the edges of the graph can exceed the flows in the respective nodes, i.e., pair of edges."

2.3.2 Weighted Graph

A fuzzy relation $\lambda : V \times V \rightarrow [0, 1]$ can always define a weighted or fuzzy graph where the $arc(x, y) \in V \times V$ has weight $\lambda(x, y) \in [0, 1]$. We would like to discuss only the undirected graph. We assume that V is a finite set unless otherwise stated. We assume that our fuzzy relation is reflexive and symmetric so that all arcs can be regarded as unordered pairs of nodes with no loops.

2.3.3 Fuzzy Graph

A fuzzy graph $G = (V, \gamma, \lambda)$ is a triple consisting of a nonempty set V besides a pair of functions $\gamma : V \rightarrow [0, 1]$ *and* $\lambda : V \times V \rightarrow [0, 1], i.e., \lambda : E \rightarrow [0, 1]$, where$E = V \times V$

$$\text{such that } \forall\, x, y \in V, \lambda(x, y) \leq \gamma(x) \wedge \gamma(y)$$

or equivalently

$$\forall\, x, y \in V, \lambda(x, y) \leq min\{\gamma(x), \gamma(y)\}$$

The fuzzy set γ is called the **fuzzy vertex set** of G, and λ is called the **fuzzy edge set** of G. λ is a fuzzy relation on γ. In our discussion, elements of V (a finite set) are called the nodes or vertices of G, and the pair of vertices is edges in G, and the fuzzy graph has no loops. Further, λ is reflexive, i.e., $\lambda(x, x) \leq min\{\gamma(x), \gamma(x)\} = \gamma(x), i.e., \lambda(x, x) = \gamma(x) \forall\, x$

$$\text{and symmetric, i.e., } \lambda(x, y) = \lambda(y, x) \forall (x, y)$$

Let us suppose that the underlying classical graph is denoted by the symbol $G^* = (\lambda^*, \gamma^*)$, where

$\lambda^* = \{u \in V : \lambda(u) > 0\}$ and $\gamma^* = \{(u, v) \in V \times V : \gamma(u, v) > 0\}$, where $\lambda^* = V$.

We will denote the fuzzy graph simply by the notation G or (γ, λ) to represent the fuzzy graph

$$G = (V, \gamma, \lambda).$$

The classical graph $G^* = (\lambda^*, \gamma^*)$ is a particular case of the fuzzy graph $G = (\gamma, \lambda)$, with each vertex and edge of (λ^*, γ^*) having a degree of membership 1.

2.3.4 Arcs or Edges

An unordered pair of nodes is considered arcs or edges.

2.3.5 Adjacent Edges

Two distinct edges, say λ_1 and λ_2, are said to be adjacent if they have a common end-vertex. That is $\lambda_1 \cap \lambda_2 \neq \emptyset$.

Example 2.1 Suppose that $V = \{x, y, z\}$ and define a fuzzy set $\gamma : V \to [0, 1]$ as; $\gamma(x) = 0.4, \gamma(y) = 0.9$ and $\gamma(z) = 0.7$ and we define $\lambda : V \times V \to [0, 1]$ by $\lambda(xy) = 0.4, \lambda(yz) = 0.6, (xz) = 0.1$. Then, we have

$$\lambda(xy) = 0.4 \leq 0.4 \wedge 0.9 = \gamma(x) \wedge \gamma(y) = min\{\gamma(x), \gamma(y)\}$$

$$\lambda(yz) = 0.6 \leq 0.9 \wedge 0.7 = \gamma(y) \wedge \gamma(z) = min\{\gamma(y), \gamma(z)\}$$

$$\lambda(xz) = 0.1 \leq 0.4 \wedge 0.7 = \gamma(x) \wedge \gamma(z) = min\{\gamma(x), \gamma(z)\}$$

Thus, we get that

$$\forall\, x, y \in V, \lambda(x, y) \leq \gamma(x) \wedge \gamma(y)$$

or

$$\forall\, x, y \in V, \lambda(x, y) \leq min\{\gamma(x), \gamma(y)\}$$

This shows that G or $(\gamma, \lambda)or\ G = (V, \gamma, \lambda)$ is a fuzzy graph.
However, if we choose $\lambda(xy) = 0.6$ then, we have (Fig. 2.31)

$$\lambda(xy) = 0.6 \nleq 0.4 \wedge 0.9 = \gamma(\text{x}) \wedge \gamma(\text{y}) = \min\{\gamma(\text{x}), \gamma(\text{y})\}$$

This shows that G or (γ, λ) is no longer a fuzzy graph.

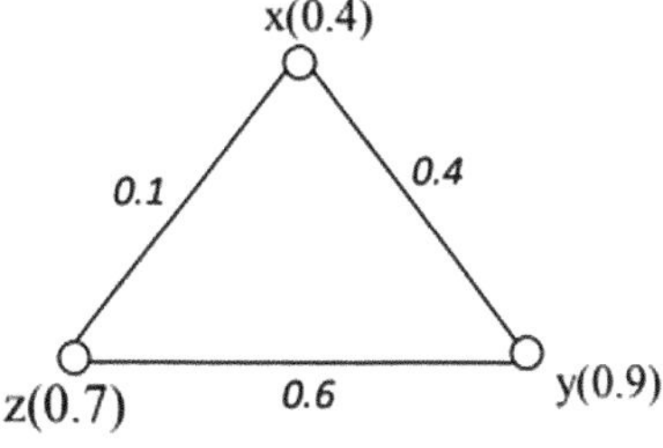

Fig. 2.31 Fuzzy graph representing Example 2.1

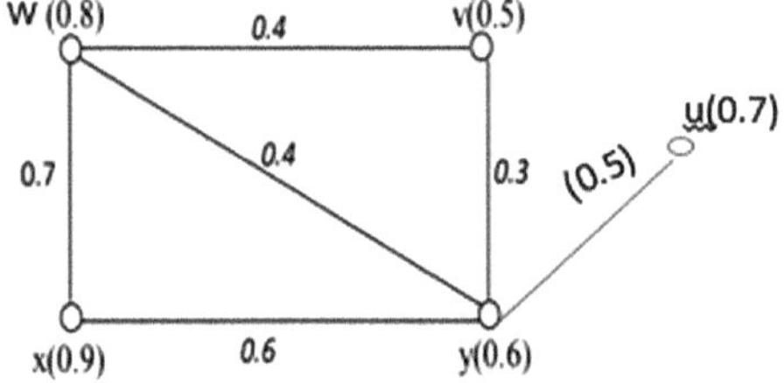

Fig. 2.32 Representing fuzzy graph of Example 2.2

Example 2.2 Suppose that $V = \{u, v, w, x, y\}$ and a fuzzy set $\gamma : V \to [0, 1]$ is defined as $\gamma(u) = 0.7, \gamma(v) = 0.5$, $\gamma(w) = 0.8, \gamma(x) = 0.9$, and $\gamma(y) = 0.6$, and we define $\lambda : V \times V \to [0, 1]$ by $\lambda(uy) = 0.5, \lambda(vy) = 0.3, \lambda(wy) = 0.4, \lambda(vw) = 0.4, \lambda(xy) = 0.6$ and $\lambda(wx) = 0.7$. Then, we have

$$\lambda(wx) = 0.7 \leq 0.8 \wedge 0.9 = \gamma(w) \wedge \gamma(x)$$

$$\lambda(xy) = 0.6 \leq 0.8 \wedge 0.6 = \gamma(x) \wedge \gamma(y)$$

$$\lambda(vw) = 0.4 \leq 0.5 \wedge 0.8 = \gamma(v) \wedge \gamma(w)$$

$$\lambda(wy) = 0.4 \leq 0.8 \wedge 0.6 = \gamma(w) \wedge \gamma(y)$$

$$\lambda(vy) = 0.3 \leq 0.5 \wedge 0.6 = \gamma(v) \wedge \gamma(y)$$

$$\lambda(uy) = 0.5 \leq 0.7 \wedge 0.6 = \gamma(u) \wedge \gamma(y)$$

Thus, we get that $\forall\, x, y \in V, \lambda(x, y) \leq \gamma(x) \wedge \gamma(y) \Rightarrow G\ or\ (\gamma, \lambda)$ is a fuzzy graph (Fig. 2.32).

Example 2.3 Suppose that $V = \{u, v, w, x\}$ and a fuzzy set $\gamma : V \to [0, 1]$ is defined as $\gamma(u) = 0.5, \gamma(v) = 0.7$, $\gamma(w) = 0.9, \gamma(x) = 0.8$ and we define $\lambda : V \times V \to [0, 1]$ by $\lambda(uv) = 0.4$, $\lambda(wx) = 0.7, \lambda(ux) = 0.5, \lambda(uw) = 0.3$. Then, we get that $\forall\, x, y \in V, \lambda(x, y) \leq \gamma(x) \wedge \gamma(y) \Rightarrow G\ or(\gamma, \lambda)$ is a fuzzy graph. However, if we redefine $\lambda(uv) = 0.6$ then it is no longer a fuzzy graph?

From the notion of the fuzzy graph, we infer that an unweighted graph (V, E) is trivially a fuzzy graph with $\gamma(x) = 1, \forall\, x \in V$ and $\lambda(x, y) = 0, \forall\, x, y \in V$. Also we write (V, λ) to represent a (Fig. 2.33).

fuzzy graph with $\gamma(x) = 1, \forall\, x \in V$.

Example 2.4 Let us assume that $V = \{u, v, w, x, y\}$ and a fuzzy set $\gamma : V \to [0, 1]$ is defined as

	u	v	w	x	y
γ	0.3	0.6	0.8	0.9	0.8

and $\lambda : V \times V \to [0, 1]$ is defined as (Fig. 2.34).

	uv	vw	vx	wx	wy	xy	uy
λ	0.3	0.5	0.4	0.8	0.8	0.6	0.2

Then, we observe that $G = (V, \gamma, \lambda)$ is a fuzzy graph.

Example 2.5 Let us assume that $V = \{u, v, w, x, y, z\}$ and a fuzzy set $\gamma : V \to [0, 1]$ is defined as

	u	v	w	x	y	z
γ	0.4	0.6	0.8	0.1	0.2	0.3

and $\lambda : V \times V \to [0, 1]$ is defined as

	uv	vw	vz	wx	wy	xy	yz	wz	uz
λ	0.4	0.6	0.3	0.1	0.2	0.1	0.2	0.3	0.3

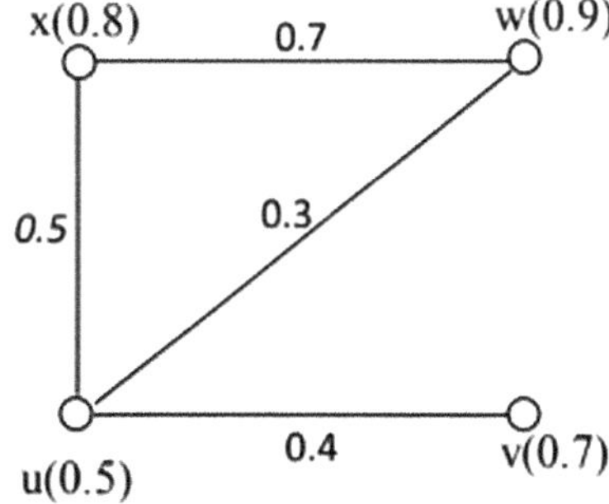

Fig. 2.33 Representing fuzzy graph of Example 2.3

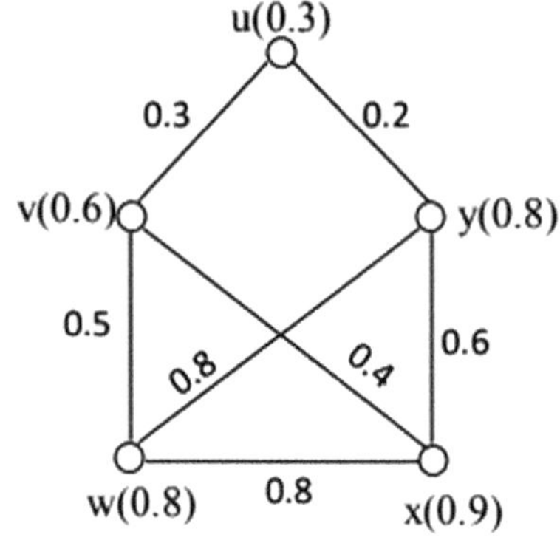

Fig. 2.34 Representing fuzzy graph given in Example 2.4

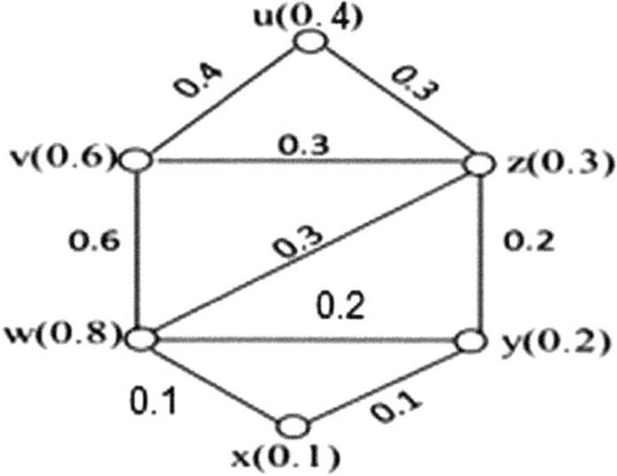

Fig. 2.35 Representing fuzzy graph of Example 2.5

Then, we can easily verify that $G = (V, \gamma, \lambda)$ is a fuzzy graph (Fig. 2.35).

Example 2.6 Let us assume that $V = \{u, v, w, x\}$ and a fuzzy set $\gamma : V \to [0, 1]$ is defined as

	u	v	w	x
γ	0.9	0.7	0.5	1

and $\lambda : V \times V \to [0, 1]$ is defined as

	uv	vw	uw	ux	vx	wx
λ	0.7	0.5	0.5	0.9	0.7	0.5

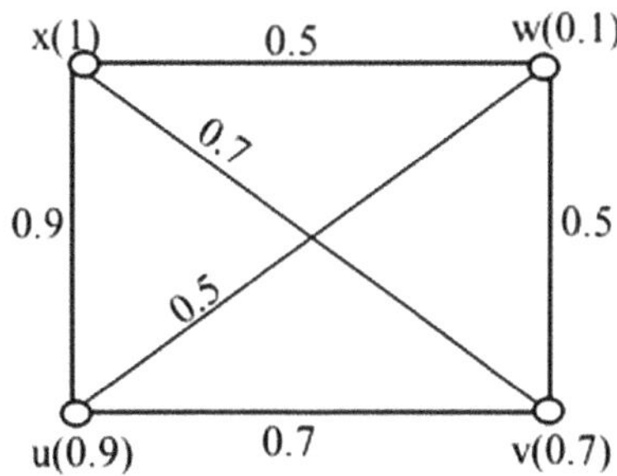

Fig. 2.36 Representing fuzzy graph as in Example 2.6

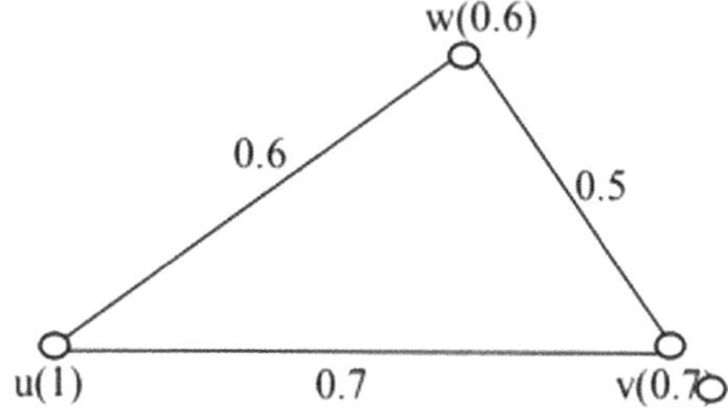

Fig. 2.37 Representing fuzzy graph as in Example 2.7

Then, it can be easily verified that $G = (V, \gamma, \lambda)$ is a fuzzy graph (Fig. 2.36).

Example 2.7 Let us assume that $V = \{u, v, w\}$ and a fuzzy set $\gamma : V \to [0, 1]$ is defined as

	u	v	w
γ	1	0.7	0.6

and $\lambda : V \times V \to [0, 1]$ is defined as

	uv	vw	uw
λ	0.7	0.5	0.6

Then, we can easily verify that $G = (V, \gamma, \lambda)$ is a fuzzy graph (Fig. 2.37).

2.3.6 *Alternative Definition of Fuzzy Graph*

Let us assume that V represents the classical set of nodes. Then a fuzzy graph is a pair of functions $G = (\gamma, \lambda)$ defined by the relation as

$$G(x_m, x_n) : \{[(x_m, x_n), \lambda(x_m, x_n)] : (x_m, x_n) \in V \times V\}$$

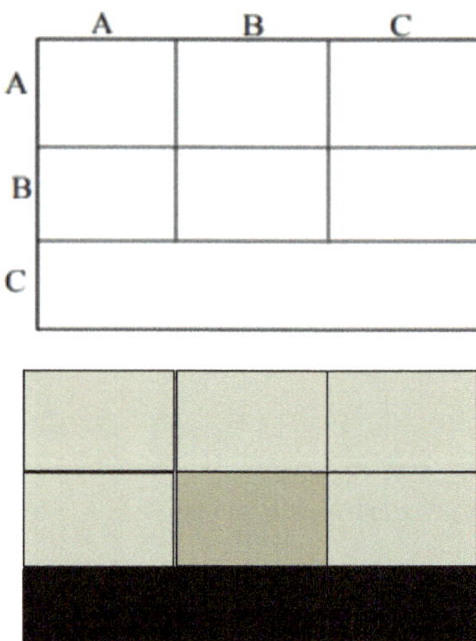

Fig. 2.38 Representing fuzzy graph is shaded black/white

Or A fuzzy graph is a pair $G = (\gamma, \lambda), \gamma$ is a fuzzy subset of V and, λ is a symmetric fuzzy relation on γ. The elements of V are called the nodes or vertices of G, and the pair of vertices is the edges in G. The underlying crisp graph of the fuzzy graph $G = (\gamma, \lambda)$ is denoted as $G^* = (V, E)$ where $E \subseteq V \times V$. The crisp graph (V, E) is a particular case of the fuzzy graph G with each vertex and edge of (V, E) having a degree of membership one.

Example 2.8 Let $V = \{v_1, v_2, v_3, v_4, v_5\}$, then a fuzzy graph G can be described as

$$G(v_i, v_j) = \left\{ \begin{array}{l} [(v_1, v_2), 0.3], [(v_1, v_3), 0.8], [(v_1, v_1), 0.7], [(v_2, v_1), 0.5], [(v_3, v_1), 0.2], \\ [(v_3, v_2), 0.6], [(v_4, v_3), 0.8], \left[(v_1, v_5), 0.9\right], \left[(v_3, v_5), 0.2\right], \left[(v_5, v_2), 0.3\right] \end{array} \right\}$$

Example 2.9 Suppose there is a set $V = \{s_1, s_2, s_3\}$, and R is a fuzzy relation defined in $V \times V$. The fuzzy relation R is depicted in the adjoining Fig. 2.38. In the figure, the brightness of the shaded portion represents the strength of the relation. Relation (s_1, s_2) is brighter/stronger than that of the relation (s_1, s_3). The corresponding fuzzy graph is displayed in Fig. 2.38. In this case, the strength of the relation is shown by the thickness of the line.

2.3.7 *Fuzzy Subgraph*

We call a fuzzy graph $H = (V, \mu, \beta)$ a fuzzy subgraph of $G = (V, \gamma, \lambda)$ if $\mu(x) = \gamma(x) \forall\, x \in \mu^*$ *and* $\beta(x, y) = \lambda(x, y) \forall\, (x, y) \in \beta^*$.

Or, A fuzzy graph $H = (V, \mu, \beta)$ is called a fuzzy subgraph of $G = (V, \gamma, \lambda)$ if $\mu \subseteq \gamma$ and $\beta \subseteq \lambda$; that is $\mu(x) \leq \gamma(x)$ for every $x \in V$ and $\beta(e) \leq \lambda(e)$ for every $e \in E$.

or A fuzzy graph $H(v_i, v_j)$ is called a fuzzy subgraph of $G(v_i, v_j)$ if

$$H(v_i, v_j) \leq G(v_i, v_j)\, \forall (v_i, v_j) \in V \times V$$

Example 2.10 Let $G(v_1, v_2)$ be a fuzzy graph as defined in Example 2.8. Then the fuzzy subgraph $H(v_i, v_j)$ be given by

$$H(v_i, v_j) = \left\{ \begin{array}{c} [(v_1, v_2), 0.2], [(v_1, v_3), 0.5], [(v_1, v_1), 0.4], [(v_2, v_1), 0.3], [(v_3, v_1), 0.2], \\ [(v_3, v_2), 0.3], [(v_1, v_5), 0.6], [(v_3, v_5), 0.1], [(v_5, v_2), 0.3] \end{array} \right\}$$

2.3.8 *Partial Fuzzy Subgraph*

Let us suppose that we have a fuzzy graph $G = (V, \gamma, \lambda)$. Then a fuzzy graph $H = (V, \mu, \beta)$ is said to be a partial fuzzy subgraph of G if $\mu \subseteq \gamma$ *and* $\beta \subseteq \lambda$, i.e.,

$$\mu(x) \leq \gamma(x) \forall\, x \in V$$

and

$$\beta(x, y) \leq \lambda(x, y) \forall\, x, y \in V$$

Example 2.11 Let us define $\gamma(u) = 0.3, \gamma(v) = 0.1, \gamma(w) = 0.5, \gamma(x) = 0.8, \gamma(y) = 0.4; \lambda(u, y) = 0.3, \lambda(y, v) = 0.2, \lambda(v, w) = 0.1, \lambda(w, x) = 0.4$. Then, this is an example of a partial fuzzy subgraph of the fuzzy graph $G = (V, \gamma, \lambda)$ defined in Example 2.4 (Fig. 2.39).

As $\mu(x) \leq \gamma(x), \forall\, x \in V$.
and $\beta(x, y) \leq \lambda(x, y) \forall\, x, y \in V$.

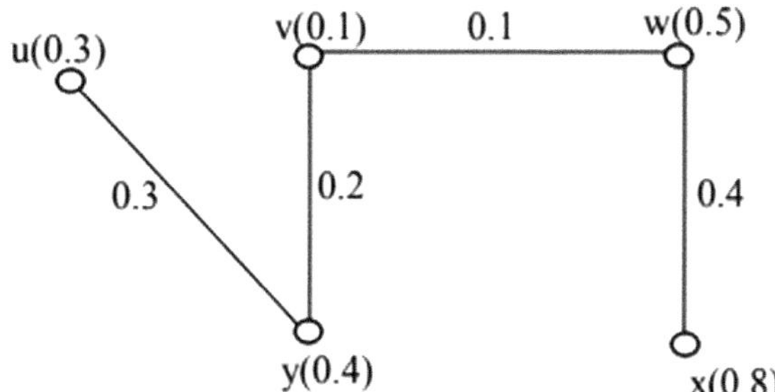

Fig. 2.39 Partial fuzzy graph

2.3.9 *Maximal Partial Fuzzy Subgraph*

Let $G = (V, \gamma, \lambda)$ is a fuzzy graph and μ is a fuzzy subset of V such that $\mu \subseteq \gamma$. Then the partial fuzzy subgraph of (γ, λ) induced by μ is the maximal partial fuzzy subgraph of (γ, λ) that has **fuzzy node-set** μ. This is the partial fuzzy subgraph (μ, β), where

$$\beta(u, v) = \mu(u) \wedge \mu(v) \wedge \lambda(u, v), \forall\, u, v \in V$$

Example 2.12 Let $G = (V, \gamma, \lambda)$ is a fuzzy graph in which vertices and arcs are defined as follows (Figs. 2.40 and 2.41):

	u	v	w	x	y
γ	0.6	0.4	0.7	0.9	0.5

	vw	wx	xy	yu	wy	vy
λ	0.3	0.6	0.5	0.4	0.3	0.2

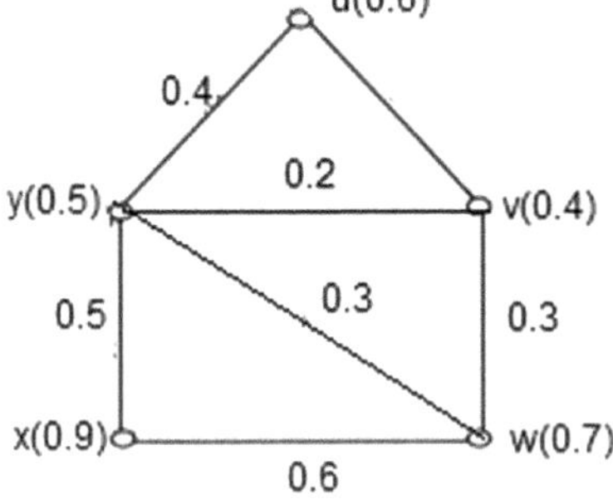

Fig. 2.40 Fuzzy graph for constructing maximal partial fuzzy graph

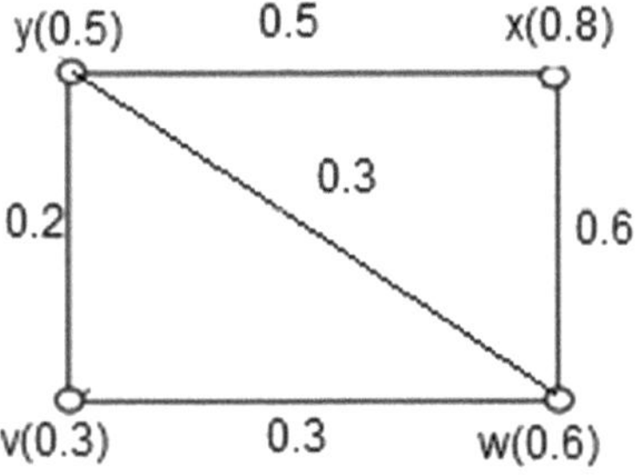

Fig. 2.41 Maximal partial fuzzy graph

Let μ and β are defined as

	v	w	x	y
μ	0.3	0.6	0.8	0.5

	vw	*wx*	*xy*	*vy*	*wy*
β	0.3	0.6	0.5	0.2	0.3

Then, we have

$$\beta(v, w) = \mu(v) \wedge \mu(w) \wedge \lambda(v, w) \Rightarrow 0.3 = 0.3 \wedge 0.6 \wedge 0.3 \text{ holds good.}$$

$$\beta(v, y) = \mu(v) \wedge \mu(y) \wedge \lambda(v, y) \Rightarrow 0.2 = 0.3 \wedge 0.5 \wedge 0.2 \text{ holds good.}$$

Similarly, we can verify for the other pair of elements of V.

2.3.10 Induced Fuzzy Subgraph

The fuzzy graph $H = (V_1, \mu, \beta)$ is called a fuzzy subgraph of $G = (V, \gamma, \lambda)$ induced by V_1 if, $\mu(x) = \gamma(x), \forall\, x \in V_1$ and $\beta(xy) = \lambda(xy), \forall\, x, y \in V_1$. The symbol $\langle V_1 \rangle$ usually denotes the fuzzy subgraph induced by V_1.

Example 2.13 Let $V = \{u, v, w, x, y\}$. Define the fuzzy set γ on V as $\gamma(u) = 0.4, \gamma(v) = 1, \gamma(w) = 0.7\ \gamma(x) = 1, \gamma(y) = 0.8$ and $\lambda : V \times V \rightarrow [0, 1]$ such that $\lambda(uv) = 0.2, \lambda(vw) = 0.7\ , \lambda(vx) = 0.8, \lambda(xy) = 0.9, \lambda(wy) = 0.5, \lambda(wx) = 0.5$; Then, $G = (V, \gamma, \lambda)$ is a fuzzy graph.

Also, if, we have $\mu(v) = 1, \mu(w) = 0.7, \mu(x) = 1, \mu(y) = 0.8$ and $\lambda : V_1 \times V_1 \rightarrow [0, 1]$.

Be such that $\beta(vw) = 0.7, \beta(vx) = 0.8, \beta(xy) = 0.9, \beta(wy) = 0.5, \beta(wx) = 0.5$. Then H is the induced fuzzy subgraph of G induced by V_1 (Fig. 2.42).

2.3.11 Spanning Fuzzy Subgraph

Let $G = (V, \gamma, \lambda)$ be a fuzzy graph. Then a partial fuzzy subgraph $H = (V, \mu, \beta)$ is said to span G if $\gamma = \mu$; and in this case, (μ, β) is referred to as the spanning fuzzy subgraph of the fuzzy graph (γ, λ) (Figs. 2.43 and 2.44).

In other words, fuzzy subgraph (γ, λ) spans fuzzy graph $G = (\gamma, \lambda)$ if the node sets of $H = (\mu, \beta)$ and $G = (\gamma, \lambda)$ are equal; they differ only in their arc weights (Figs. 2.44 and 2.45).

Fig. 2.42 Fuzzy graph for constructing induced fuzzy subgraph

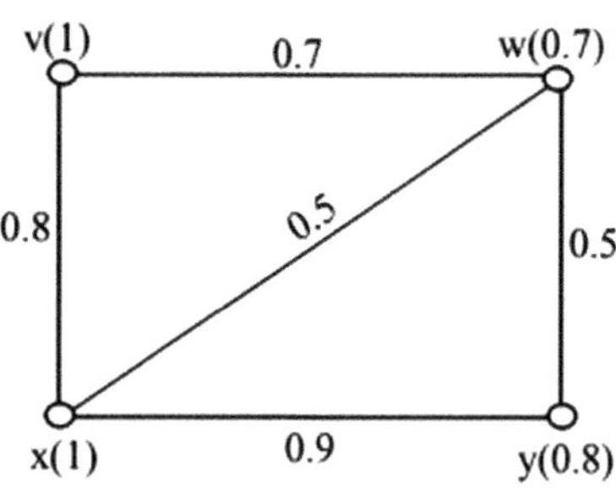

Fig. 2.43 Induced fuzzy subgraph of the graph represented in Fig. 2.42

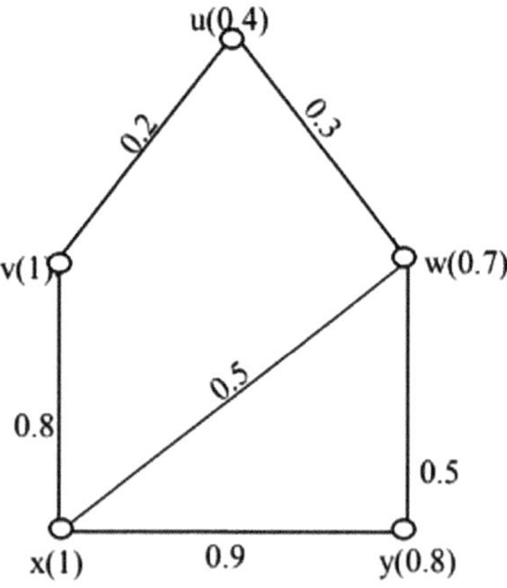

Fig. 2.44 Spanning fuzzy subgraph of the graph represented in Fig. 2.42

Example 2.14 A spanning fuzzy subgraph of $G(v_i, v_j)$ discussed in Example 2.8 is given by $H(v_i, v_j) = \left\{ \begin{array}{l} [(v_1, v_2), 0.1], [(v_1, v_3), 0.3], [(v_2, v_1), 0.2], \\ [(v_3, v_2), 0.3], [(v_1, v_5), 0.4], [(v_5, v_2), 0.2] \end{array} \right\}$.

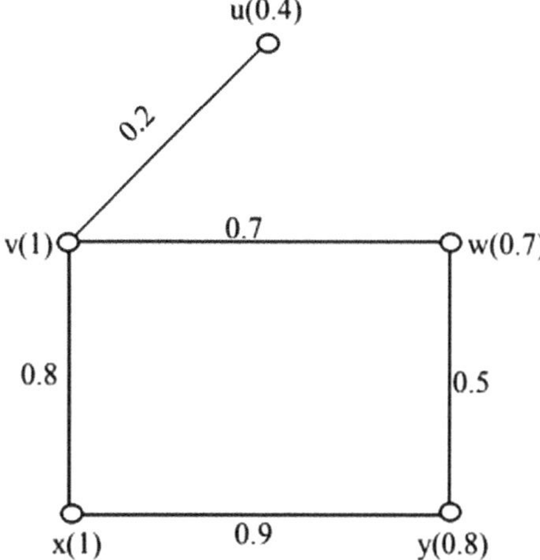

Fig. 2.45 Spanning fuzzy subgraph

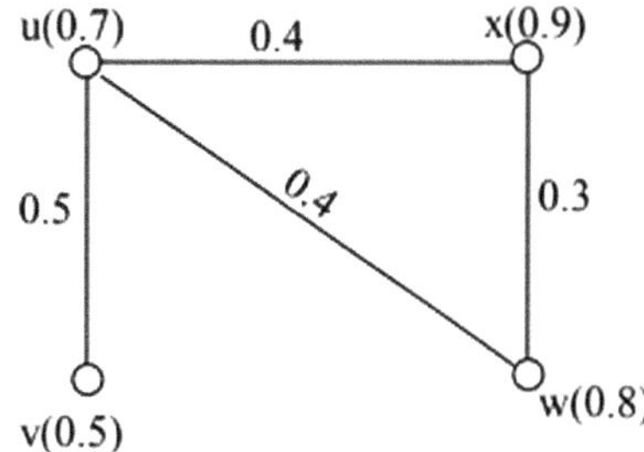

Fig. 2.46 Graph representing order, size, degree and total degree of a node

2.3.12 *Threshold Graph*

Let us assume that $G = (V, \gamma, \lambda)$ is a fuzzy graph. For any threshold, assume that $\gamma^t = \{x \in \gamma^* : \gamma(x) \geq t\}$ and $\lambda^t = \{uv \in \lambda^* : \lambda(uv) \geq t\}$. Since $\lambda(u, v) \leq \gamma(u) \wedge \gamma(v)$.

for all $u, v \in V$. We have $\lambda^t \subseteq \gamma^t \times \gamma^t$, i.e., $\lambda^t \subseteq \{xy : \gamma(x) \geq t, \gamma(y) \geq t\}$ and hence $H = (\gamma^t, \lambda^t)$ is a fuzzy graph with vertex set γ^t and edge set λ^t.

Then H is referred to as the threshold graph of the fuzzy graph G corresponding to t.

Remarks 2.1

(i) Let $G : (\gamma, \lambda)$ be a fuzzy graph. If $0 \leq t_1 \leq t_2 \leq 1$, then $(\gamma^{t_2}, \lambda^{t_2})$ is a subgraph of $(\gamma^{t_1}, \lambda^{t_1})$.

(ii) Let $H = (\mu, \beta)$ be a partial fuzzy subgraph of $G = (\gamma, \lambda)$. For any threshold, (μ^t, β^t) is a subgraph of (γ^t, λ^t).

2.3.13 Isolated Vertex

A vertex x in a fuzzy graph $G = (\gamma, \lambda)$ is called an isolated vertex if $\lambda(x, y) = 0$ for all $x \neq y$.

2.3.14 Adjacent Vertex

In a fuzzy graph $G = (\gamma, \lambda)$, two vertices, x and y are said to be adjacent or neighbors if $\lambda(x, y) > 0$, and the set of all neighbors of x is denoted by $N(x)$.

2.3.15 Order of a Fuzzy Graph

Let us assume that we have a fuzzy graph $G = (\gamma, \lambda)$ with the underlying set V. Then the order of the fuzzy graph G is denoted symbolically by $\circ(G)$or by the symbol p and is defined as

$$p = \circ(G) = \sum_{x \in V} \gamma(x)$$

2.3.16 Size of the Fuzzy Graph

Let us have a fuzzy graph $G = (\gamma, \lambda)$ with the underlying set V. Then the size of the fuzzy graph G is denoted by the symbol $S(G)$ or by the letter q and is defined as

$$q = S(G) = \sum_{x, y \in V} \lambda(x, y)$$

The notion of the degree of a node has been extended by **R. T. Yeh & S. Y. Bang** in 1975 as follows:

2.4 Degree of a Node and Edge in a Fuzzy Graph

2.4.1 Degree of a Vertex (Node) in a Fuzzy Graph

Let us suppose that we have a fuzzy graph $G = (\gamma, \lambda)$ with the underlying set V. Then the degree of a node, say "u" in a fuzzy graph G is denoted by the symbol $d_G(u)$or by the symbol $d(u)$ and is defined as

$$d_G(u) = \sum_{(u, v) \in V} \lambda(u, v)$$

or equivalently,

$$d_G(u) = \sum_{u \neq v} \lambda(u, v), \forall(u, v) \in V \text{ and } \lambda(u, v) = 0, \forall(u, v) \notin V$$

2.4.2 Minimum Degree of a Node

The minimum degree of a node u in a fuzzy graph $G = (\gamma, \lambda)$ is denoted by the symbol $\delta_u(G)$ and is defined as

$$\delta_u(G) = \min_{u \in V} \{d(u)\}$$

2.4.3 Maximum Degree of a Node

The maximum degree of a node u in a fuzzy graph $G = (\gamma, \lambda)$ is denoted by the symbol $\Delta_u(G)$ and is defined as

$$\Delta_u(G) = \max_{u \in V} \{d(u)\}$$

2.4.4 Total Degree of a Node in a Fuzzy Graph

Let $G = (\gamma, \lambda)$ be a fuzzy graph. Then the total degree of the vertex u is symbolically denoted by $td_G(u)$ and is defined as

$$td_G(u) = \sum_{(u,v) \in V} \{\lambda(u, v) + \gamma(u)\} = d_G(u) + \gamma(u), \forall (u, v) \in V$$

Example 2.15 Let us suppose that $V = \{u, v, w, x\}$ and a fuzzy set $\gamma : V \to [0, 1]$ is defined as $\gamma(u) = 0.7, \gamma(v) = 0.5, \gamma(w) = 0.8, \gamma(x) = 0.9$ and we define λ : $V \times V \to [0, 1]$ by $\lambda(uv) = 0.5, \lambda(wx) = 0.3, \lambda(ux) = 0.4, \lambda(uw) = 0.4$.

Then the **order of *G*, size of G, degree** and **the total degree of a node, and minimum** and **maximum degree of a node** are given by

$$p = \mathrm{o}(G) = \sum_{u, v \in V} \lambda(u, v) = \lambda(uv) + \lambda(wx) + \lambda(ux) + \lambda(uw)$$
$$= 0.5 + 0.3 + 0.4 + 0.4 = 1.6$$

$$q = S(G) = \sum_{x \in V} \gamma(x) = \gamma(u) + \gamma(v) + \gamma(w) + \gamma(x)$$
$$= 0.7 + 0.5 + 0.8 + 0.9 = 2.9$$

$$d_G(u) = \sum_{(u,v) \in V} \lambda(u, v) = \lambda(uv) + \lambda(ux) + \lambda(uw)$$
$$= 0.5 + 0.4 + 0.4 = 1.3$$

$$d_G(v) = 0.5$$

$$d_G(w) = \lambda(uw) + \lambda(wx) = 0.4 + 0.3 = 0.7$$

$$d_G(x) = \lambda(ux) + \lambda(wx) = 0.4 + 0.3 = 0.7$$

$$td_G(u) = \sum_{(u,v) \in V} \lambda(u, v) + \gamma(u) = d_G(u) + \gamma(u) = 1.3 + 0.7 = 2.0$$

Similarly, we can get

$$td_G(v) = d_G(v) + \gamma(v) = 0.5 + 0.5 = 1.0$$

$$td_G(w) = d_G(w) + \gamma(w) = 0.7 + 0.8 = 1.5$$

$$td_G(x) = d_G(x) + \gamma(x) = 0.7 + 0.9 = 1.6$$

$$\begin{aligned}\delta_u(G) =& min\{d_G(u), d_G(v), d_G(w), d_G(x)\}\\ &= min\{1.3, 0.5, 0.7, 0.7\} = 0.5\end{aligned}$$

$$\begin{aligned}\Delta_u(G) =& max\{d_G(u), d_G(v), d_G(w), d_G(x)\}\\ &= max\{1.3, 0.5, 0.7, 0.7\} = 1.3\end{aligned}$$

2.4.5 Degree of an Edge

Let us have a fuzzy graph $G = (\gamma, \lambda)$ with the underlying set V. Then the degree of an edge (u, v) in the fuzzy graph G is denoted by the symbol $d_G(u, v)$ and is defined as

$$d_G(u, v) = d_G(u) + d_G(v) - 2\lambda(u, v)$$

2.4.6 Minimum Degree of an Edge

The ***minimum degree of an edge*** (u, v) in the fuzzy graph G is denoted by the symbol $\delta_E(G)$ and is defined by the formula as

$$\delta_E(G) = \wedge\{d_G(u, v) : (u, v) \in E\}$$

2.4.7 Maximum Degree of an Edge

The ***maximum degree of an edge*** (u, v) in the fuzzy graph G is denoted by the symbol $\triangle_E\ (G)$ and is defined by the formula as

$$\Delta_E(G) = \vee\{d_G(u, v) : (u, v) \in E\}$$

Example 2.16 Let the fuzzy graph $G = (\gamma, \lambda)$ be defined as $\gamma(x_1) = 0.7, \gamma(x_2) = 0.5, \gamma(x_3) = 0.6, \gamma(x_4) = 0.8, \gamma(x_5) = 0.6$, and $\lambda(x_1, x_2) = 0.7, \lambda(x_2, x_3) = 0.4, \lambda(x_3, x_4) = 0.6, \lambda(x_4, x_5) = 0.4$,

$$\lambda(x_1, x_5) = 0.5, \lambda(x_3, x_5) = 0.3 \text{ where } V = \{x_1, x_2, x_3, x_4, x_5\}$$

$$d_G(x_1) = \lambda(x_1, x_2) + \lambda(x_1, x_5) = 0.7 + 0.5 = 1.2$$

$$d_G(x_2) = \lambda(x_1, x_2) + \lambda(x_2, x_3) = 0.7 + 0.4 = 1.1$$

$$\begin{aligned} d_G(x_3) &= \lambda(x_2, x_3) + \lambda(x_3, x_4) + \lambda(x_3, x_5) \\ &= 0.4 + 0.6 + 0.3 = 1.3 \end{aligned}$$

$$d_G(x_4) = \lambda(x_3, x_4) + \lambda(x_4, x_5) = 0.6 + 0.4 = 1.0$$

$$d_G(x_5) = \lambda(x_4, x_5) + \lambda(x_1, x_5) + \lambda(x_3, x_5) = 0.4 + 0.5 + 0.3 = 1.2$$

$$d_G(x_1, x_2) = d_G(x_1) + d_G(x_2) - 2\lambda(x_1, x_2) = 1.2 + 1.1 - 2 \times (0.7) = 0.9$$

$$d_G(x_2, x_3) = d_G(x_2) + d_G(x_3) - 2\lambda(x_2, x_3) = 1.1 + 1.3 - 2 \times (0.4) = 1.6$$

$$d_G(x_3, x_4) = d_G(x_3) + d_G(x_4) - 2\lambda(x_3, x_4) = 1.3 + 1.0 - 2 \times (0.6) = 1.1$$

$$d_G(x_4, x_5) = d_G(x_4) + d_G(x_5) - 2\lambda(x_4, x_5) = 1.0 + 1.2 - 2 \times (0.4) = 1.4$$

$$d_G(x_1, x_5) = d_G(x_1) + d_G(x_5) - 2\lambda(x_1, x_5) = 1.2 + 1.2 - 2 \times (0.5) = 1.4$$

$$d_G(x_3, x_5) = d_G(x_3) + d_G(x_5) - 2\lambda(x_3, x_5) = 1.3 + 1.2 - 2 \times (0.3) = 1.9$$

$$\delta_E(G) = \wedge\{d_G(x_1, x_2) : (x_1, x_2) \in E\} = \min\{0.9, 1.6, 1.1, 1.4, 1.4, 1.9\} = 0.9$$

$$\Delta_E(G) = \vee\{d_G(x_1, x_2) : (x_1, x_2) \in E\} = \max\{0.9, 1.6, 1.1, 1.4, 1.4, 1.9\} = 1.9$$

2.4.8 Total Degree of an Edge

Let us assume that we have a fuzzy graph $G = (\gamma, \lambda)$ with the underlying set V. Then the total degree of an edge (u, v) in the fuzzy graph G is symbolically denoted by $td_G(u, v)$ and is defined as (Figs. 2.49 and 2.50)

$$td_G(u, v) = d_G(u) + d_G(v) - \lambda(u, v)$$

2.4.9 Minimum Total Degree of an Edge

The minimum total degree of an edge (u, v) in the fuzzy graph G is denoted by the symbol $\delta_{tE}(G)$ and is defined by the formula as

$$\delta_{tE}(G) = \wedge\{td_G(u, v) : (u, v) \in E\}$$

2.4.10 Maximum Total Degree of an Edge

The maximum total degree of an edge (u, v) in the fuzzy graph G is denoted by the symbol Δ_{tE} (G) and is defined by the formula as

$$\Delta_{tE}(G) = \vee\{td_G(u, v) : (u, v) \in E\}$$

Example 2.17 Let the fuzzy graph $G = (\gamma, \lambda)$ be defined as $\gamma(x_1) = 0.7$, $\gamma(x_2) = 0.5$, $\gamma(x_3) = 0.6$, $\gamma(x_4) = 0.8$, $\gamma(x_5) = 0.6$, and $\lambda(x_1, x_2) = 0.7$, $\lambda(x_2, x_3) = 0.4$, $\lambda(x_3, x_4) = 0.6$, $\lambda(x_4, x_5) = 0.4$, $\lambda(x_1, x_5) = 0.5$, $\lambda(x_3, x_5) = 0.3$; where $V = \{x_1, x_2, x_3, x_4, x_5\}$. See Fig. 2.47.

$$d_G(x_1) = 1.2, d_G(x_2) = 1.1, d_G(x_3) = 1.3, d_G(x_4) = 1.0, d_G(x_5) = 1.2$$

$$td_G(x_1, x_2) = d_G(x_1) + d_G(x_2) - \lambda(x_1, x_2) = 1.2 + 1.1 - (0.7) = 1.6$$

$$td_G(x_2, x_3) = d_G(x_2) + d_G(x_3) - \lambda(x_2, x_3) = 1.1 + 1.3 - (0.4) = 2.0$$

$$td_G(x_3, x_4) = d_G(x_3) + d_G(x_4) - \lambda(x_3, x_4) = 1.3 + 1.0 - (0.6) = 1.7$$

$$td_G(x_4, x_5) = d_G(x_4) + d_G(x_5) - \lambda(x_4, x_5) = 1.0 + 1.2 - (0.4) = 1.8$$

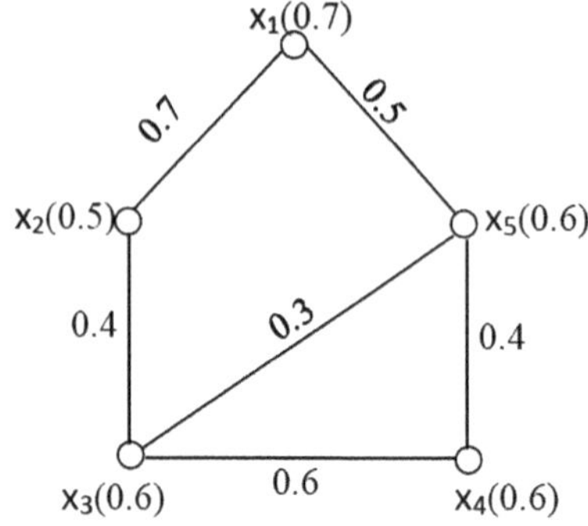

Fig. 2.47 Graph representing degree, min & max degree of an edge

$$td_G(x_1, x_5) = d_G(x_1) + d_G(x_5) - \lambda(x_1, x_5) = 1.2 + 1.2 - (0.5) = 1.9$$

$$td_G(x_3, x_5) = d_G(x_3) + d_G(x_5) - \lambda(x_3, x_5) = 1.3 + 1.2 - (0.3) = 2.2$$

$$\delta_{tE}(G) = \wedge\{d_G(x_1, x_2) : (x_1, x_2) \in E\} = \min\{1.6, 2.0, 1.7, 1.8, 1.9, 2.2\} = 1.6$$

$$\Delta_{tE}(G) = \vee\{d_G(x_1, x_2) : (x_1, x_2) \in E\} = \max\{1.6, 2.0, 1.7, 1.8, 1.9, 2.2\} = 2.2$$

2.4.11 *Scalar Cardinality*

Let us have a fuzzy graph $G = (\gamma, \lambda)$ *and* $H \subseteq G$. Then the scalar cardinality of H is symbolically denoted by $|H|$ and is defined by the rule as

$$|H| = \sum_{u \in H} \gamma(u)$$

The scalar cardinality of V is denoted by p and is called the **order of** G.

Example 2.18 Let us consider the fuzzy graph $G = (\gamma, \lambda)$ defined by λ; where $V = \{x_1, x_2, x_3, x_4\}$. Let $H \subseteq G$, then the scalar cardinality of H is given by (Fig. 2.48).

$$|H| = \sum_{x \in H} \gamma(x) = \gamma(x_1) + \gamma(x_2) + \gamma(x_3) + \gamma(x_4) = 0.5 + 0.7 + 0.9 + 0.8 = 2.9$$

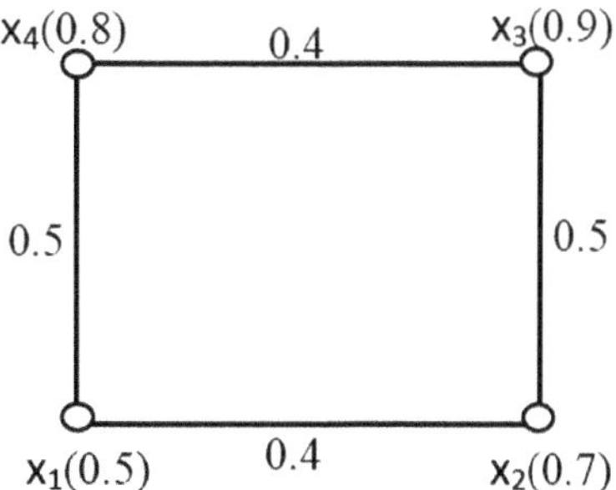

Fig. 2.48 Regular fuzzy graph but not totally regular fuzzy graph

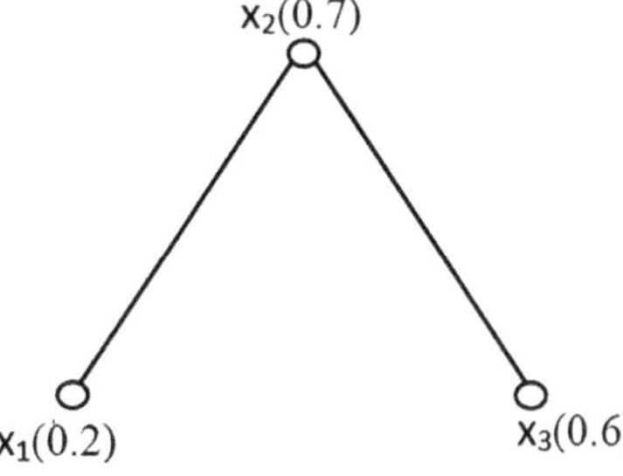

Fig. 2.49 Not regular fuzzy graph but totally regular fuzzy graph

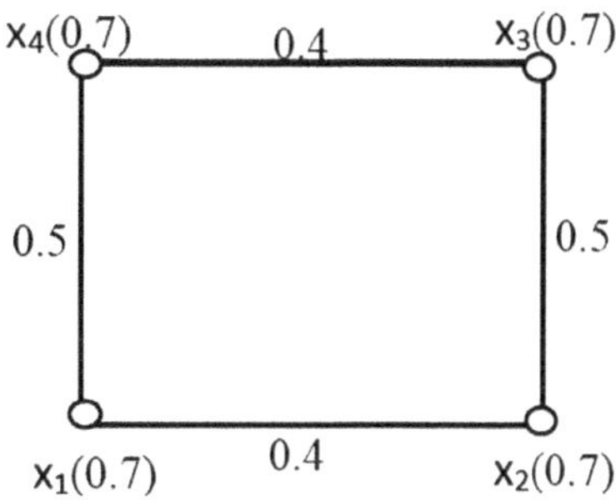

Fig. 2.50 Totally regular as well as regular fuzzy graph

2.5 Regular and Irregular Fuzzy Graphs and Their Type

2.5.1 Regular Fuzzy Graph

A fuzzy graph $G = (\gamma, \lambda)$ is said to be a regular fuzzy graph if each node has the same degree. If each node has the same degree, then the fuzzy graph is said to be of degree k or is termed $k-$ regular fuzzy graph.

If each vertex of the fuzzy graph $G = (\gamma, \lambda)$ has the same total degree k, then the fuzzy graph G is referred to as $k-$ *regular fuzzy graph* or a **totally regular fuzzy graph** of total degree k.

Example 2.19 Let us consider the fuzzy graph $G = (\gamma, \lambda)$, whose vertices and edges are defined as $\gamma(\mathrm{x}_1) = 0.5, \gamma(\mathrm{x}_2) = 0.7, \gamma(\mathrm{x}_3) = 0.9, \gamma(\mathrm{x}_4) = 0.8$; and edges as $\lambda(\mathrm{x}_1, \mathrm{x}_2) = 0.4, \lambda(\mathrm{x}_2, \mathrm{x}_3) = 0.5, \lambda(\mathrm{x}_3, \mathrm{x}_4) = 0.4, \lambda(\mathrm{x}_4, \mathrm{x}_1) = 0.5$, (Fig.2.48)

Here, we observe that

$$d_G(x_1) = \lambda(x_1, x_2) + \lambda(x_4, x_1) = 0.4 + 0.5 = 0.9 = d_G(x_2) = d_G(x_3) = d_G(x_4)$$

That is, $d_G(v) = 0.9, \forall v \in G \Rightarrow G$ is a regular fuzzy graph.

$$td_G(x_1) = d_G(x_1) + \gamma(x_1) = 0.9 + 0.5 = 1.4$$

$$td_G(x_2) = d_G(x_2) + \gamma(x_2) = 0.9 + 0.7 = 1.6$$

Thus, $td_G(x_1) \neq td_G(x_2) \Rightarrow G$ is not totally regular fuzzy graph.

Example 2.20 Consider the fuzzy graph $G = (\gamma, \lambda)$, whose vertices and edges are defined as $\gamma(x_1) = 0.2, \gamma(x_2) = 0.7, \gamma(x_3) = 0.6$; and edges as $\lambda(x_1, x_2) = 0.4, \lambda(x_2, x_3) = 0.5$. (Fig. 2.49)

$$d_G(x_1) = \lambda(x_1, x_2) + \lambda(x_1, x_3) = 0.4 + 0.5 = 0.9$$

$$d_G(x_2) = \lambda(x_1, x_2) = 0.4 \; and \; d_G(x_3) = \lambda(x_3, x_2) = 0.5$$

Thus, $d_G(x_1) \neq d_G(x_2) \neq d_G(x_3) \Rightarrow G$ is not a regular fuzzy graph.
However, we have

$$td_G(x_1) = d_G(x_1) + \gamma(x_1) = 0.9 + 0.2 = 1.1$$

$$td_G(x_2) = d_G(x_2) + \gamma(x_2) = 0.4 + 0.7 = 1.1$$

$$td_G(x_3) = d_G(x_3) + \gamma(x_3) = 0.5 + 0.6 = 1.1$$

This shows that $td_G(x_1) = td_G(x_2) = td_G(x_3) = 1.1 \Rightarrow G$ is a totally regular fuzzy graph.

Thus, the above fuzzy graph is an example of a graph that is totally regular but not regular.

Example 2.21 Let us consider the fuzzy graph, whose vertices and edges are defined as $\gamma(x_1) = \gamma(x_2) = \gamma(x_3) = \gamma(x_4) = 0.7; and \; \lambda(x_1, x_2) = 0.4, \; \lambda(x_2, x_3) = 0.5, \lambda(x_3, x_4) = 0.4, \lambda(x_4, x_1) = 0.5, (Fig. \; 2.50)$

Now, we observe that

$d_G(x_1) = d_G(x_2) = d_G(x_3) = d_G(x_4) = 0.9 \Rightarrow G$ is a regular tuzzy path

$$td_G(x_1) = d_G(x_1) + \gamma(x_1) = 0.9 + 0.7 = 1.6$$

$$td_G(x_2) = d_G(x_2) + \gamma(x_2) = 0.9 + 0.7 = 1.6$$

$$td_G(x_3) = d_G(x_3) + \gamma(x_3) = 0.9 + 0.7 = 1.6$$

$$td_G(x_4) = d_G(x_4) + \gamma(x_4) = 0.9 + 0.7 = 1.6$$

Thus, for all $u \in td_G(u) = 1.6 \Rightarrow G$ is totally regular as well as regular.

2.5.2 *Irregular Fuzzy Graph*

A fuzzy graph $G = (\gamma, \lambda)$ is said to be an irregular fuzzy graph if a node is adjacent to a node with distinct degrees.

Example 2.22 Consider the fuzzy graph $G = (\gamma, \lambda)$ whose vertices and edges are defined as $\gamma(x_1) = 0.5, \gamma(x_2) = 0.7, \gamma(x_3) = 0.5, \gamma(x_4) = 0.3, \gamma(x_5) = 0.6$, and $\lambda(x_1, x_2) = 0.3, \lambda(x_2, x_3) = 0.5$,

$\lambda(x_3, x_4) = 0.4, \lambda(x_4, x_5) = 0.3, \lambda(x_1, x_5) = 0.4;$
where $V = \{x_1, x_2, x_3, x_4, x_5\}$.

$$d_G(x_1) = \lambda(x_1, x_2) + \lambda(x_1, x_5) = 0.3 + 0.4 = 0.7$$

$$d_G(x_2) = \lambda(x_1, x_2) + \lambda(x_2, x_3) = 0.3 + 0.5 = 0.8$$

In this case x_1, x_2 are adjacent nodes and are such that $d_G(x_1) \neq d_G(x_2) \Rightarrow G$ is an irregular
fuzzy graph (Fig. 2.51).

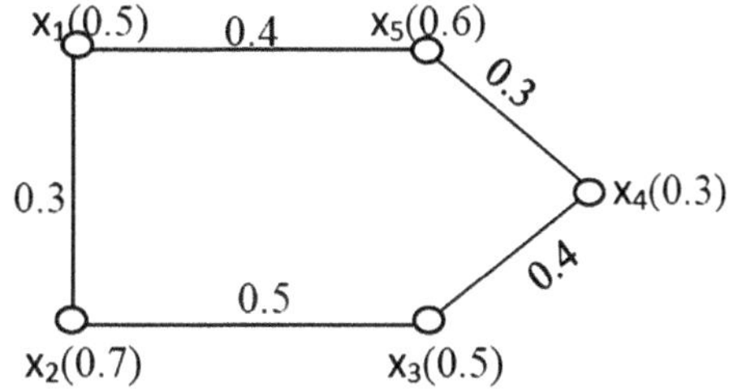

Fig. 2.51 Irregular fuzzy graph

2.5.3 Totally Irregular Fuzzy Graph

A fuzzy graph $G = (\gamma, \lambda)$ is considered a totally irregular fuzzy graph if a node is adjacent to a node with distinct total degrees.

Example 2.23 Let us consider the fuzzy graph $G = (\gamma, \lambda)$, whose vertices and edges are defined as $\gamma(x_1) = 0.8$, $\gamma(x_2) = 0.7$, $\gamma(x_3) = 0.6$, $\gamma(x_4) = 0.5$ and $\lambda(x_1, x_2) = 0.1$, $\lambda(x_1, x_3) = 0.3$,

$$\lambda(x_3, x_4) = 0.3, \lambda(x_2, x_4) = 0.2$$

where $V = \{x_1, x_2, x_3, x_4\}$.

$$d_G(x_1) = \lambda(x_1, x_2) + \lambda(x_1, x_3) = 0.1 + 0.3 = 0.4$$

$$d_G(x_2) = \lambda(x_1, x_2) + \lambda(x_2, x_4) = 0.1 + 0.2 = 0.3$$

$$d_G(x_3) = \lambda(x_1, x_3) + \lambda(x_3, x_4) = 0.3 + 0.3 = 0.6$$

$$d_G(x_4) = \lambda(x_3, x_4) + \lambda(x_2, x_4) = 0.3 + 0.2 = 0.5$$

$$td_G(x_1) = d_G(x_1) + \gamma(x_1) = 0.4 + 0.8 = 1.2$$

$$td_G(x_2) = d_G(x_2) + \gamma(x_2) = 0.3 + 0.7 = 1.0$$

$$td_G(x_3) = d_G(x_3) + \gamma(x_3) = 0.6 + 0.6 = 1.2$$

$$td_G(x_4) = d_G(x_4) + \gamma(x_4) = 0.5 + 0.5 = 1.0$$

In this fuzzy graph, we observe that $\exists$ nodes $x_1, x_2 \in G$ and are adjacent nodes such that

$td_G(x_1) \neq td_G(x_2) \Rightarrow G$ is a totally irregular fuzzy graph.

Further, we also observe that $\exists$ nodes $x_3, x_4 \in G$ and are adjacent nodes such that

$d_G(x_3) \neq d_G(x_4) \Rightarrow G$ is an irregular fuzzy graph. Thus, this is an example of a fuzzy graph that is totally irregular as well as irregular (Fig. 2.52).

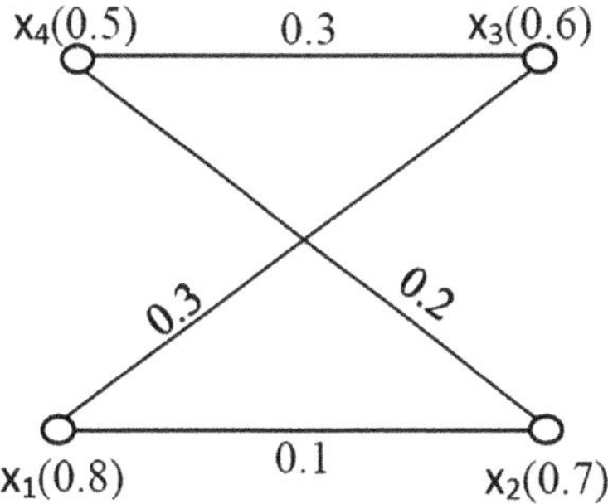

Fig. 2.52 Totally irregular fuzzy graph

2.5.4 *Strongly Irregular Fuzzy Graph*

A fuzzy graph $G = (\gamma, \lambda)$ is said to be a strongly irregular fuzzy graph if every pair of nodes has distinct degrees.

Example 2.24 Consider the fuzzy graph $G = (\gamma, \lambda)$, whose vertices and edges are defined as $\gamma(x_1) = 0.7, \gamma(x_2) = 0.4, \gamma(x_3) = 0.6, \gamma(x_4) = 0.8, \gamma(x_5) = 1.0$, and $\lambda(x_1, x_2) = 0.4, \lambda(x_2, x_3) = 0.6,$

$\lambda(x_3, x_4) = 0.5, \lambda(x_4, x_5) = 0.8, \lambda(x_1, x_5) = 0.4$; where $V = \{x_1, x_2, x_3, x_4, x_5\}$. Obviously, it

is a cycle of length five.

$$d_G(x_1) = \lambda(x_1, x_2) + \lambda(x_1, x_5) = 0.4 + 0.4 = 0.9$$

$$d_G(x_2) = \lambda(x_1, x_2) + \lambda(x_2, x_3) = 0.4 + 0.6 = 1.0$$

$$d_G(x_3) = \lambda(x_2, x_3) + \lambda(x_3, x_4) = 0.6 + 0.5 = 1.1$$

$$d_G(x_4) = \lambda(x_4, x_5) + \lambda(x_3, x_4) = 0.8 + 0.5 = 1.3$$

$$d_G(x_5) = \lambda(x_4, x_5) + \lambda(x_1, x_5) = 0.8 + 0.4 = 1.2$$

Every pair of nodes of graph $G = (\gamma, \lambda)$ has distinct degrees (Fig. 2.53).

2.5.5 *Not Strongly Irregular Fuzzy Graph*

A fuzzy graph $G = (\gamma, \lambda)$ is said to be a not strongly irregular fuzzy graph if a pair of nodes has the same degree.

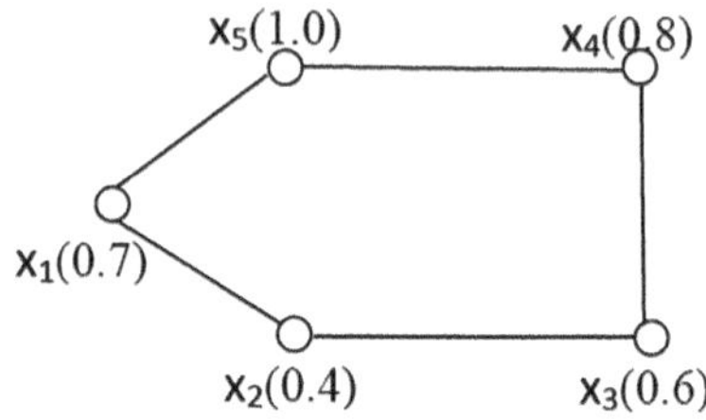

Fig. 2.53 Strongly irregular fuzzy graph

Example 2.25 Let us consider the fuzzy graph $G = (\gamma, \lambda)$, whose vertices and edges are defined as $\gamma(x_1) = 0.3$, $\gamma(x_2) = 0.2$, $\gamma(x_3) = 0.4$, $\gamma(x_4) = 0.6$ and $\lambda(x_1, x_2) = 0.2$, $\lambda(x_2, x_3) = 0.1$,

$$\lambda(x_3, x_4) = 0.2, \lambda(x_1, x_4) = 0.3 \text{ where } V = \{x_1, x_2, x_3, x_4\}$$

$$d_G(x_1) = \lambda(x_1, x_2) + \lambda(x_1, x_4) = 0.2 + 0.3 = 0.5$$

$$d_G(x_2) = \lambda(x_1, x_2) + \lambda(x_2, x_3) = 0.2 + 0.1 = 0.3$$

$$d_G(x_3) = \lambda(x_2, x_3) + \lambda(x_3, x_4) = 0.1 + 0.2 = 0.3$$

$$d_G(x_4) = \lambda(x_3, x_4) + \lambda(x_1, x_4) = 0.2 + 0.3 = 0.5$$

In this way, we observe that in the fuzzy graph $G = (\gamma, \lambda)$, there are two pairs of nodes, $\{x_1, x_4\}$ and (Fig. 2.54).

$\{x_2, x_3\}$, that have the same degree, and hence, $G = (\gamma, \lambda)$ is not a strongly irregular fuzzy graph.

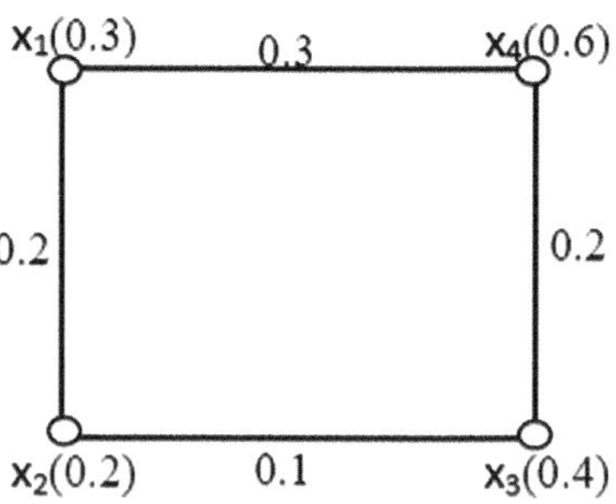

Fig. 2.54 Not-strongly irregular fuzzy graph

2.5.6 *Highly Irregular Fuzzy Graph*

A fuzzy graph $G = (\gamma, \lambda)$ is considered a highly irregular fuzzy graph if every node in G is adjacent to the nodes having distinct degrees.

Example 2.26 Let us consider the fuzzy graph $G = (\gamma, \lambda)$, whose vertices and edges are defined as $\gamma(x_1) = 0.8, \gamma(x_2) = 0.5, \gamma(x_3) = 0.4, \gamma(x_4) = 0.6$ and $\lambda(x_1, x_2) = 0.6, \lambda(x_2, x_3) = 0.3,$

$$\lambda(x_3, x_4) = 0.4, \lambda(x_1, x_4) = 0.8$$

where $V = \{x_1, x_2, x_3, x_4\}$

$$d_G(x_1) = \lambda(x_1, x_2) + \lambda(x_1, x_4) = 0.6 + 0.8 = 1.4$$

$$d_G(x_2) = \lambda(x_1, x_2) + \lambda(x_2, x_3) = 0.6 + 0.3 = 0.9$$

$$d_G(x_3) = \lambda(x_2, x_3) + \lambda(x_3, x_4) = 0.3 + 0.4 = 0.7$$

$$d_G(x_4) = \lambda(x_3, x_4) + \lambda(x_1, x_4) = 0.4 + 0.8 = 1.2$$

From the above calculation, we observe that in the fuzzy graph $G = (\gamma, \lambda)$, there are nodes adjacent to the nodes that have distinct degrees, and hence $G = (\gamma, \lambda)$ is a highly irregular fuzzy graph (Fig. 2.55).

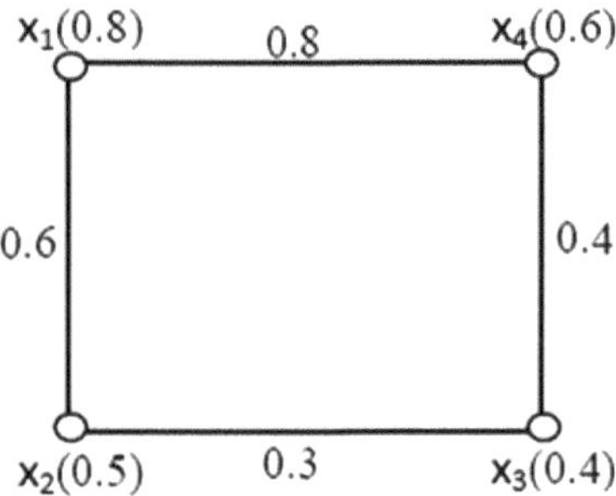

Fig. 2.55 Highly irregular fuzzy graph

2.5.7 *Neighborly Irregular Fuzzy Graph*

A fuzzy graph $G = (\gamma, \lambda)$ is said to be a neighborly irregular fuzzy graph if every two adjacent nodes of a fuzzy graph G have distinct degrees.

Example 2.27 Let us consider the fuzzy graph $G = (\gamma, \lambda)$, whose vertices and edges are defined as $\gamma(x_1) = 0.9, \gamma(x_2) = 1.0, \gamma(x_3) = 0.8, \ \gamma(x_4) = 0.8$ and $\lambda(x_1, x_2) = 0.9, \lambda(x_2, x_3) = 0.8,$

$\lambda(x_3, x_4) = 0.4, \lambda(x_1, x_4) = 0.7$; where $V = \{x_1, x_2, x_3, x_4\}$.

$$d_G(x_1) = \lambda(x_1, x_2) + \lambda(x_1, x_4) = 0.9 + 0.7 = 1.6$$

$$d_G(x_2) = \lambda(x_1, x_2) + \lambda(x_2, x_3) = 0.9 + 0.5 = 1.4$$

$$d_G(x_3) = \lambda(x_2, x_3) + \lambda(x_3, x_4) = 0.8 + 0.7 = 1.5$$

$$d_G(x_4) = \lambda(x_3, x_4) + \lambda(x_1, x_4) = 0.4 + 0.7 = 1.1$$

Going through the above facts, we observe that in the fuzzy graph $G = (\gamma, \lambda)$, every pair of adjacent nodes has a distinct degree, and hence $G = (\gamma, \lambda)$ is a neighborly irregular fuzzy graph (Fig. 2.56).

Property 2.1 A highly irregular fuzzy graph need not be a neighborly irregular fuzzy graph; Could you cite an illustrative example supporting the assertion? (Fig. 2.57)

Solution: Let us define a fuzzy graph $G = (\gamma, \lambda)$, whose vertices and edges are defined as $\gamma(x_1) = 0.6, \gamma(x_2) = 0.7, \gamma(x_3) = 0.4, \gamma(x_4) = 0.8, \gamma(x_5) = 0.5$ and $\lambda(x_1, x_2) = 0.3, \lambda(x_2, x_3) = 0.3,$

$$\lambda(x_3, x_4) = 0.4, \lambda(x_1, x_4) = 0.3, \lambda(x_4, x_5) = 0.5$$

where $V = \{x_1, x_2, x_3, x_4, x_5\}$

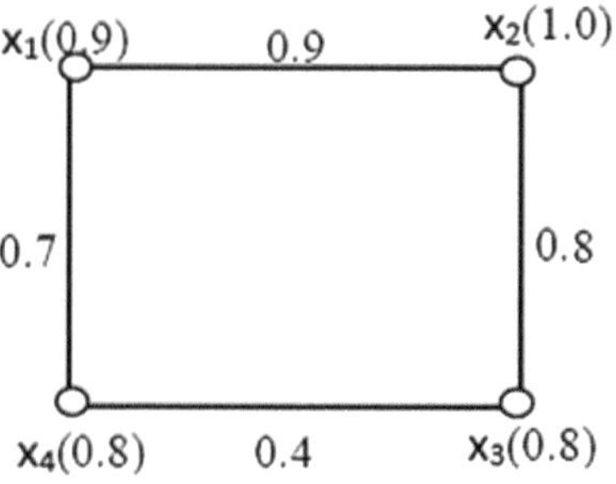

Fig. 2.56 Neighbourly irregular fuzzy graph

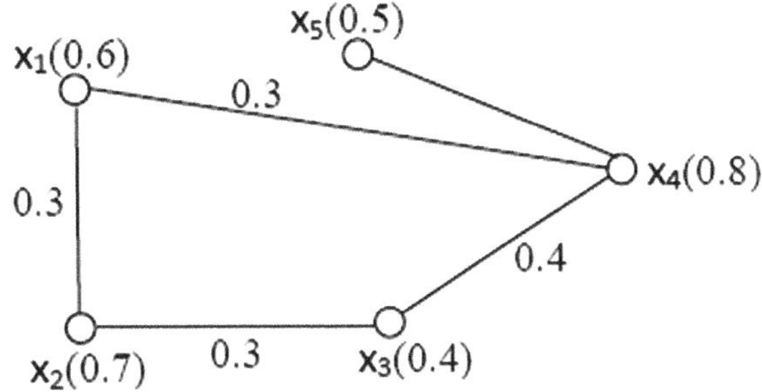

Fig. 2.57 Highly irregular fuzzy graph G that is not neighbourly irregular

$$d_G(x_1) = \lambda(x_1, x_2) + \lambda(x_1, x_4) = 0.3 + 0.3 = 0.6$$

$$d_G(x_2) = \lambda(x_1, x_2) + \lambda(x_2, x_3) = 0.3 + 0.3 = 0.6$$

$$d_G(x_3) = \lambda(x_2, x_3) + \lambda(x_3, x_4) = 0.3 + 0.4 = 0.7$$

$$d_G(x_4) = \lambda(x_3, x_4) + \lambda(x_1, x_4) + \lambda(x_4, x_5) = 0.4 + 0.3 + 0.5 = 1.2$$

$$d_G(x_5) = \lambda(x_4, x_5) = 0.5$$

From the above description of the fuzzy graph, we infer that all the adjacent nodes have distinct degrees. This implies that the fuzzy graph $G = (\gamma, \lambda)$ is highly irregular. But nodes x_1 and x_2 exist such that $d_G(x_1) = d_G(x_2) = 0.6 \Rightarrow$ the fuzzy graph G is not neighborly irregular.

Property 2.2 A neighborly irregular fuzzy graph doesn't need to be a highly irregular fuzzy graph; could you give an example supporting the assertion?

Solution: Let us define a fuzzy graph $G = (\gamma, \lambda)$, whose vertices and edges are defined as $\gamma(x_1) = 0.5, \gamma(x_2) = 0.3, \gamma(x_3) = 0.4, \gamma(x_4) = 0.4, \gamma(x_5) = 0.6$ and $\lambda(x_1, x_2) = 0.5, \lambda(x_2, x_3) = 0.2$,

$\lambda(x_3, x_4) = 0.4, \lambda(x_1, x_4) = 0.3, \lambda(x_3, x_5) = 0.3, \lambda(x_4, x_5) = 0.3$; where $V = \{x_1, x_2, x_3, x_4, x_5\}$

$$d_G(x_1) = \lambda(x_1, x_2) + \lambda(x_1, x_4) = 0.5 + 0.3 = 0.8$$

$$d_G(x_2) = \lambda(x_1, x_2) + \lambda(x_2, x_3) = 0.5 + 0.2 = 0.7$$

$$\begin{aligned} d_G(x_3) &= \lambda(x_2, x_3) + \lambda(x_3, x_4) + \lambda(x_3, x_5) \\ &= 0.2 + 0.4 + 0.3 = 0.9 \end{aligned}$$

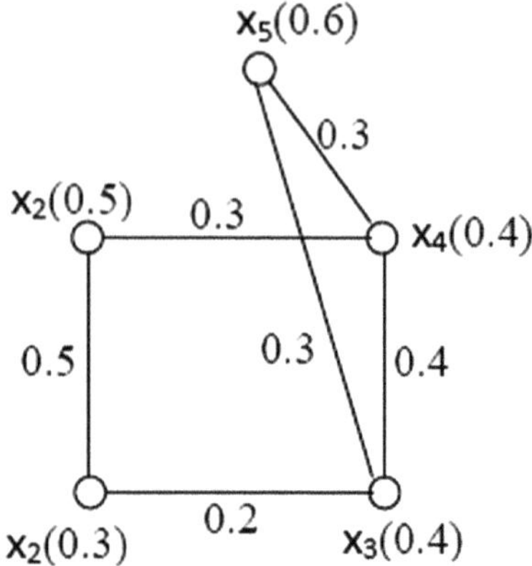

Fig. 2.58 Neighbourly Irregular Fuzzy Graph that is not highly irregular fuzzy graph

$$d_G(x_4) = \lambda(x_3, x_4) + \lambda(x_4, x_5) = 0.4 + 0.3 = 0.7$$

$$d_G(x_5) = \lambda(x_3, x_5) + \lambda(x_4, x_5) = 0.3 + 0.3 = 0.6$$

In this example, we observe that no two adjacent nodes of the fuzzy graph $G = (\gamma, \lambda)$ have the same degree. Therefore, G is a neighborly irregular fuzzy graph. However, to the node x_3, adjacent nodes x_2 and x_4 exist; such that $d_G(x_2) = d_G(x_4) = 0.8$, and therefore, G is not a highly irregular fuzzy graph (Fig. 2.58).

2.6 Some Typical Exemplars of Fuzzy Graphs, Paths, and Connectedness in It

2.6.1 *Strong Fuzzy Graph*

A fuzzy graph $G = (\gamma, \lambda)$ is said to be a strong fuzzy graph if

$$\lambda(x, y) = \gamma(x) \wedge \gamma(y), \forall (x, y) \in \lambda^* = E \subseteq V \times V$$

Example 2.28 Let us define a fuzzy graph $G = (\gamma, \lambda)$ by defining the vertices and edges as: $\gamma(x_1) = 0.5, \gamma(x_2) = 0.7, \gamma(x_3) = 0.4, \gamma(x_4) = 0.6$ and $\lambda(x_1, x_2) = 0.5, \lambda(x_2, x_3) = 0.4, \lambda(x_3, x_4) = 0.4,$

$\lambda(x_4, x_1) = 0.5$; where $V = \{x_1, x_2, x_3, x_4\}$.

Then, we get that

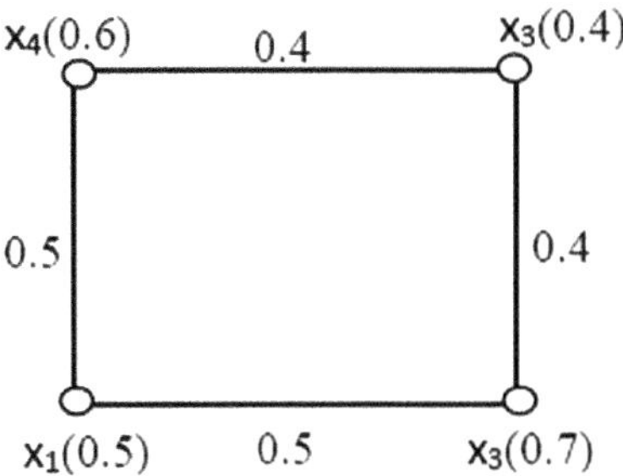

Fig. 2.59 Strong fuzzy graph

$$\begin{aligned}\lambda(x_1, x_2) &= \gamma(x_1) \wedge \gamma(x_2)\\ &= 0.5 \wedge 0.7 = 0.5\\ \lambda(x_3, x_4) &= \gamma(x_3) \wedge \gamma(x_4)\\ &= 0.4 \wedge 0.6 = 0.4\end{aligned}$$

and so on for other nodes and edges.

Thus, we observe that $\lambda(x, y) = \gamma(x) \wedge \gamma(y), \forall(x, y) \in \lambda^* \Rightarrow G$ is a strong fuzzy graph (Fig. 2.59).

2.6.2 Complete Fuzzy Graph

A fuzzy graph $G = (\gamma, \lambda)$ with underlying set V is considered a complete fuzzy graph if it satisfies

$\lambda(x, y) = \gamma(x) \wedge \gamma(y), \forall x, y \in \gamma^*$ where γ^* is referred to as the nonempty set V of nodes and is denoted by the symbol K_γ.

Example 2.29 Let us define a fuzzy graph $G = (\gamma, \lambda)$ by defining the vertices and edges as: $\gamma(x_1) = 0.7, \gamma(x_2) = 0.8, \gamma(x_3) = 0.6, \gamma(x_4) = 0.5$ and $\lambda(x_1, x_2) = 0.7, \lambda(x_2, x_3) = 0.6, \lambda(x_3, x_4) = 0.5,$

$\lambda(x_2, x_4) = 0.5, \lambda(x_1, x_4) = 0.5, \lambda(x_1, x_3) = 0.6$; where $V = \{x_1, x_2, x_3, x_4\}$.
Then, we get from $\lambda(x, y) = \gamma(x) \wedge \gamma(y), \forall x, y \in \gamma^*$ that

$$\begin{aligned}\lambda(x_1, x_2) &= \gamma(x_1) \wedge \gamma(x_2), \forall x_1, x_2 \in \gamma^*\\ &= 0.7 \wedge 0.8 = 0.7\end{aligned}$$

$$\lambda(x_2, x_3) = \gamma(x_2) \wedge \gamma(x_3) = 0.8 \wedge 0.6 = 0.6$$

$$\lambda(x_3, x_4) = \gamma(x_3) \wedge \gamma(x_4) = 0.6 \wedge 0.5 = 0.5$$

$$\lambda(x_4, x_1) = \gamma(x_4) \wedge \gamma(x_1) = 0.5 \wedge 0.7 = 0.5$$

$$\lambda(x_2, x_4) = \gamma(x_2) \wedge \gamma(x_4) = 0.8 \wedge 0.5 = 0.5$$

$$\lambda(x_1, x_3) = \gamma(x_1) \wedge \gamma(x_3) = 0.7 \wedge 0.6 = 0.6$$

Thus, $\lambda(x, y) = \gamma(x) \wedge \gamma(y), \forall x, y \in V \Rightarrow G$ is a complete fuzzy graph (Fig. 2.60).

Example 2.30 Let us define a fuzzy graph $G = (\gamma, \lambda)$ by defining the vertices and edges as: $\gamma(x) = 0.5$, $\gamma(y) = 0.4$ *and* $\gamma(z) = 0.7$; $\lambda(x, y) = 0.4$, $\lambda(y, z) = 0.4$, $\lambda(z, x) = 0.5$. Then, we get that.

$$\begin{aligned} \lambda(x, y) &= \gamma(x) \wedge \gamma(y) \\ &= 0.5 \wedge 0.0.4 = 0.4 \end{aligned}$$

Similarly, one can very for others nodes.

Thus, $\lambda(x, y) = \gamma(x) \wedge \gamma(y), \forall x, y \in V \Rightarrow G$ is a complete fuzzy graph.

Property 2.3 With the help of a suitable example, could you show that every complete fuzzy graph is a strong fuzzy graph but not conversely?

Example 2.31 Let us assume that $V = \{u, v, w, x, y\}$ and a fuzzy set $\gamma : V \to [0, 1]$ is defined as.

	u	v	w	x	y
γ	0.6	0.9	0.6	0.7	0.8

and $\lambda : V \times V \to [0, 1]$ is defined as

	uv	vw	vx	wx	wy	xy	uy	vy	ux	uw
λ	0.6	0.6	0.7	0.6	0.6	0.7	0.6	0.8	0.6	0.6

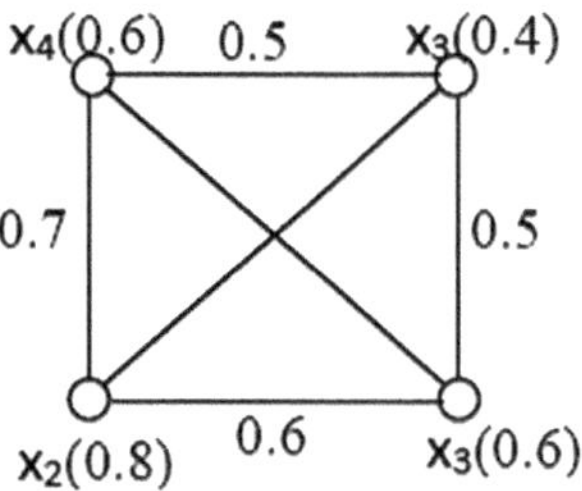

Fig. 2.60 Complete fuzzy graph

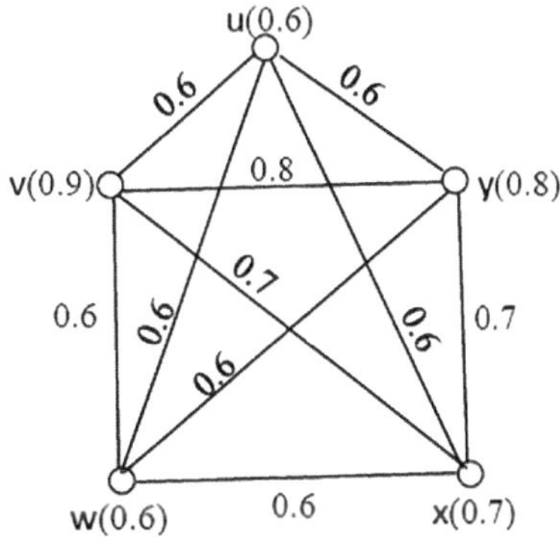

Fig. 2.61 Complete as well as strong fuzzy graph

Then, we observe that $G = (V, \gamma, \lambda)$ is a fuzzy graph that is both complete and strong.

Now, consider the fuzzy graph defined as follows (Fig. 2.61)

	u	v	w	x	y
γ	0.6	0.9	0.6	0.7	0.8

	uv	vw	wx	vx	xy	uy	ux	uw
λ	0.6	0.6	0.6	0.6	0.7	0.6	0.6	0.6

Here, we observe that $G = (V, \gamma, \lambda)$ is a fuzzy graph which is a strong fuzzy sgraph but is not a complete fuzzy graph (Fig. 2.62).

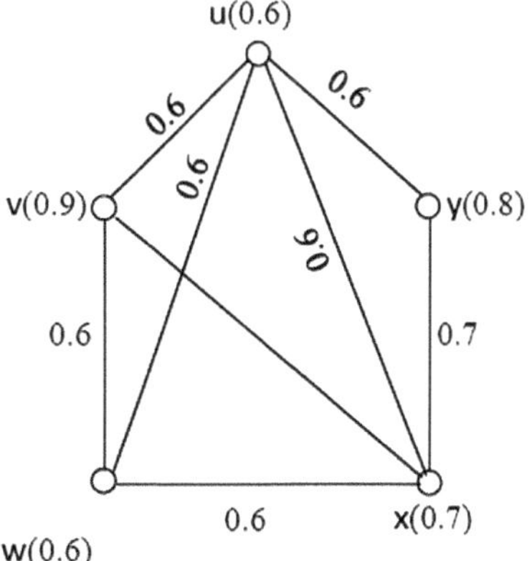

Fig. 2.62 Strong fuzzy graph that is not a complete fuzzy graph

2.6.3 Path of a Fuzzy Graph

In a fuzzy graph $G = (\gamma, \lambda)$, a path p of length n is a sequence of distinct nodes/vertices $v_0, v_1, v_2, \ldots, v_n$ such that $\lambda(v_{i-1}, v_i) > 0$; $i = 1, 2, 3, \ldots, n (1 \leq i \leq n)$.

Here $n \geq 0$ is called the length of the path p. The consecutive pairs (v_{i-1}, v_i) are called the **arcs** or **edges** of the path. A single node v may also be considered as a path. We call the path p a cycle if $v_0 = v_n$ and $n \geq 3$.

2.6.4 Strength of the Path

In a fuzzy graph, the strength of the path p is defined as the weight of the weakest arc of the path. The degree of membership of the weakest arc is called its strength.

In other words, the strength of the path p is defined as $\bigwedge_{i=1}^{n} \lambda(v_{i-1}, v_i)$. It is the **strength of the weakest edge** and is symbolically denoted by $d(p)$ *or* $S(p)$.

2.6.5 Cycle and Fuzzy Cycle

If in a path p, $v_0 = v_n$ and $n \geq 3$ then the path p is called a cycle and if p contains more than one weakest arc it is referred to as a fuzzy cycle. In other words, a fuzzy graph G is called a fuzzy cycle if and only if (γ^*, λ^*) is a cycle and there does not exist a unique $(x, y) \in \lambda^*$ such that $\lambda(x, y) = \wedge\{\lambda(u, v) : (u, v) \in \lambda^*\}$. That is, it contains more than one weakest arc.

Example 2.32 Let us consider a fuzzy graph $G = (\gamma, \lambda)$, whose nodes and arcs are defined as $\gamma(u) = 0.6$, $\gamma(v) = 1, \gamma(w) = 0.9$; $\lambda(uv) = 0.5, \lambda(vw) = 0.5$ and $\lambda(wu) = 0.6$, respectively. Then, there is a fuzzy cycle as it contains more than one weakest arc (Fig. 2.63).

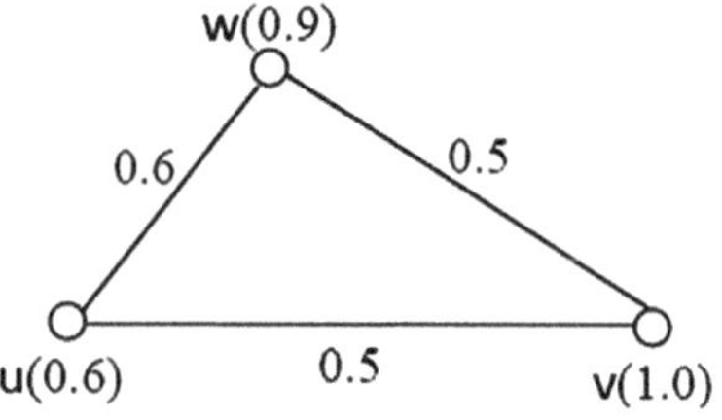

Fig. 2.63 Fuzzy cycle

2.6.6 Strength of Connectedness

In a fuzzy graph $G = (\gamma, \lambda)$, the strongest path joining any two vertices u and v is that path which has strength $\lambda^{\infty}(u, v)$, where $\lambda^{\infty}(u, v)$ is usually referred to as the strength of connectedness between the nodes u and v.

Example 2.33 Let G $= (\gamma, \lambda)$ be a fuzzy graph in which vertices and arches are defined as follows:

	u	v	w	x	y
γ	0.4	1	0.8	1	0.9

	uv	uw	vw	wy	xy	xv	xw
λ	0.2	0.3	0.7	0.7	0.9	0.8	0.7

In this fuzzy graph the strongest path joining v (Fig. 2.64).
and y is the path: v, x, y with $\lambda^{\infty}(v, y) = 0.8$
Also,

$$\lambda^{\infty}(u, v) = \lambda^{\infty}(u, w) = \lambda^{\infty}(u, x) = \lambda^{\infty}(u, y) = 0.3$$

$$\lambda^{\infty}(v, w) = 0.7, \lambda^{\infty}(v, x) = 0.8, \lambda^{\infty}(w, x) = 0.7 = \lambda^{\infty}(w, y) \textit{ and } \lambda^{\infty}(x, y) = 0.9$$

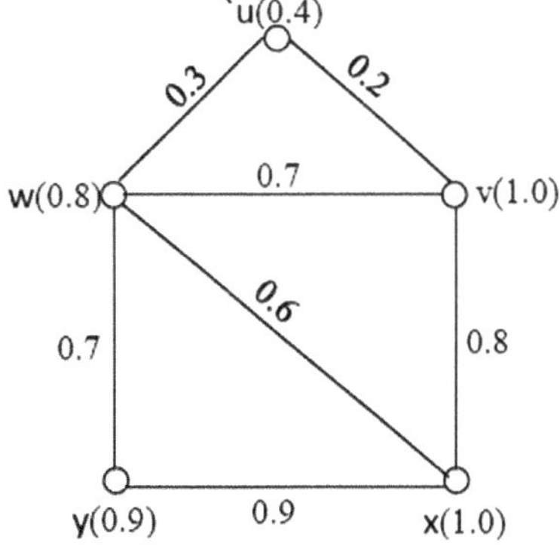

Fig. 2.64 Strongest path joining v and y

2.6.7 Connected Path

Two nodes u(first) and v(last) in a fuzzy graph G = (γ, λ) that are joined by a path are said to be connected. Connected is a reflexive and symmetric relation, and it is also transitive. We will say that u and v are connected if and only if $\lambda^{\infty}(v_0, v_n) > 0$.

The equivalence classes of vertices under this relation are called *connected components* of the fuzzy graph under consideration.

2.6.8 Strength of Connectedness Between Nodes

The strength of connectedness between two nodes u and v in a fuzzy graph G = (γ, λ) is defined as the maximum of the strengths of all paths between u and v and is symbolically denoted by $CONN_G(u, v)$ or $\lambda^{\infty}(u, v)$ and is defined as

$$CONN_G(u, v) = max\{\lambda(u, v_0) \wedge \lambda(v_0, v_1) \wedge \lambda(v_1, v_2) \wedge \ldots \wedge \lambda(v_n, v) : v_0, v_1, v_2, \ldots, v_n, v \in V\}$$

and we say that a path of length n connects u and v. That is,

$$\lambda^n(u, v) = Sup\{\lambda(u, v_0) \wedge \lambda(v_0, v_1) \wedge \lambda(v_1, v_2) \wedge \ldots \wedge \lambda(v_n, v) : v_0, v_1, v_2, \ldots, v_n, v \in V\}$$

Example 2.34 Let G = (γ, λ) be a fuzzy graph, in which nodes and arcs are defined as (Fig. 2.65).

	u	v	w	x
γ	0.9	0.8	0.7	0.6

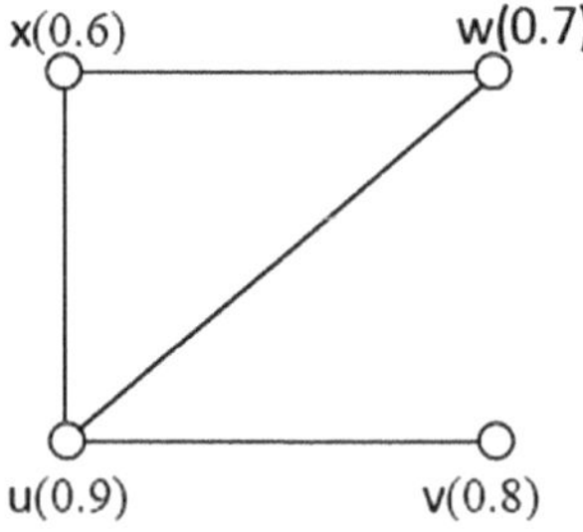

Fig. 2.65 Connected fuzzy graph

	uv	uw	wx	xv
λ	0.2	0.4	0.4	0.3

Then, the fuzzy graph G is a connected fuzzy graph as $CONN_G(u, v) = 0.2 > 0$, and the path from u to w is $\{u, v, w, x\}$ and is of length 3.

2.6.9 *Connected Fuzzy Graph*

A fuzzy graph G $= (\gamma, \lambda)$ is said to be connected if

$$\lambda^n(u, v) > 0 \text{ for all } u, v \in \gamma^*$$

or,

$$CONN_G(u, v) > 0, \text{ for all } u, v \in \gamma^*$$

2.7 Strong Path and Neighborly Edge Fuzzy Graph

2.7.1 *Strong Arc (Strong Edge)*

An arc (u, v) in a fuzzy graph G $= (\gamma, \lambda)$ is said to be strong if $\gamma(u, v) \geq \lambda^\infty(u, v)$, and the node v is said to be **the strong neighbor** of u.

2.7.2 *Strong Arc Fuzzy Graph*

In a fuzzy graph, if every arc is a strong arc, the fuzzy graph is referred to as a strong arc fuzzy graph.

2.7.3 *Strong Path*

A path in a fuzzy graph, in which every arc is a strong arc, is called a strong path; if the path contains n strong edges, it is denoted by the symbol P_n.

2.7.4 Neighborly Edge Irregular Fuzzy Graph

Let $G = (V, \gamma, \lambda)$ be a connected fuzzy graph on the underlying graph $G^* = (V, \gamma^*, \lambda^*)$. The fuzzy graph G is said to be a neighborly edge irregular fuzzy graph if every pair of adjacent arcs/edges has distinct degrees.

Example 2.35 Let us consider the fuzzy graph $G = (\gamma, \lambda)$ whose vertices and edges are defined as

	a	b	c	d
γ	0.9	1.0	0.8	0.7

	ab	bc	cd	da
λ	0.9	0.5	0.7	0.7

where $V = \{a, b, c, d\}$. Then, we have

$$d_G(a) = \lambda(a, b) + \lambda(a, d) = 0.9 + 0.7 = 1.6$$

$$d_G(b) = \lambda(a, b) + \lambda(b, c) = 0.9 + 0.5 = 1.4$$

$$d_G(c) = \lambda(b, c) + \lambda(c, d) = 0.5 + 0.7 = 1.2$$

$$d_G(d) = \lambda(c, d) + \lambda(a, d) = 0.7 + 0.7 = 1.4$$

$$d_G(a, b) = d_G(a) + d_G(b) - 2\lambda(a, b) = 1.6 + 1.4 - 2 \times (0.9) = 1.2$$

$$d_G(b, c) = d_G(b) + d_G(c) - 2\lambda(b, c) = 1.4 + 1.2 - 2 \times (0.5) = 1.6$$

$$d_G(c, d) = d_G(c) + d_G(d) - 2\lambda(c, d) = 1.2 + 1.4 - 2 \times (0.7) = 1.2$$

$$d_G(d, a) = d_G(d) + d_G(a) - 2\lambda(d, a) = 1.4 + 1.6 - 2 \times (0.7) = 1.6$$

From above, we see that every pair of adjacent edges has distinct degrees. Hence, G is a neighborly edge irregular fuzzy graph (Fig. 2.66)

Fig. 2.66 Neighborly edge irregular fuzzy graph

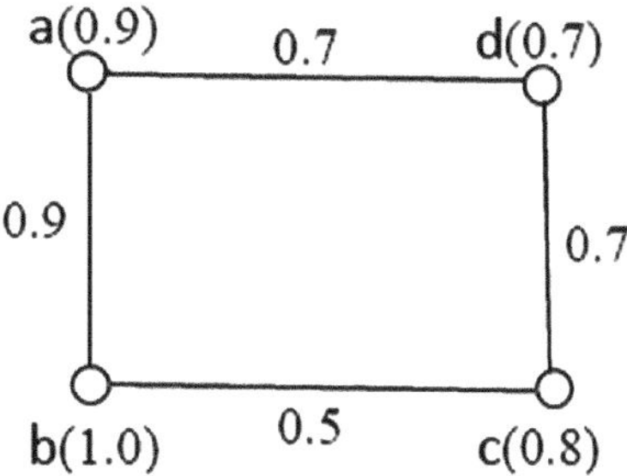

2.7.5 *Neighborly Edge Totally Irregular Fuzzy Graph*

Let $G = (V, \gamma, \lambda)$ be a connected fuzzy graph on the underlying graph $G^* = (V, \gamma^*, \lambda^*)$. The fuzzy graph G is considered a neighborly edge totally irregular fuzzy graph if every pair of adjacent arcs has a distinct total degree.

Example 2.36 Consider the fuzzy graph $G = (\gamma, \lambda)$ whose vertices and edges are defined as:

	a	b	C	d	e
γ	0.5	0.2	0.4	0.6	0.8

	ab	bc	cd	de	ae
λ	0.2	0.4	0.3	0.6	0.2

where $V = \{a, b, c, d, e\}$. This is a cycle of length five. Then, we have (Fig. 2.67)

$$d_G(a) = \lambda(a, b) + \lambda(a, e) = 0.2 + 0.2 = 0.4$$

$$d_G(b) = \lambda(a, b) + \lambda(b, c) = 0.2 + 0.4 = 0.6$$

$$d_G(c) = \lambda(b, c) + \lambda(c, d) = 0.4 + 0.3 = 0.7$$

$$d_G(d) = \lambda(d, e) + \lambda(c, d) = 0.6 + 0.3 = 0.9$$

$$d_G(e) = \lambda(d, e) + \lambda(a, e) = 0.6 + 0.2 = 0.8$$

$$d_G(a, b) = d_G(a) + d_G(b) - 2\lambda(a, b) = 0.4 + 0.6 - 2 \times (0.2) = 0.6$$

$$d_G(b, c) = d_G(b) + d_G(c) - 2\lambda(b, c) = 0.6 + 0.7 - 2 \times (0.4) = 0.5$$

$$d_G(c, d) = d_G(c) + d_G(d) - 2\lambda(c, d) = 0.7 + 0.9 - 2 \times (0.3) = 1.0$$

$$d_G(d, e) = d_G(d) + d_G(e) - 2\lambda(d, e) = 0.9 + 0.8 - 2 \times (0.6) = 0.5$$

$$d_G(e, a) = d_G(e) + d_G(a) - 2\lambda(e, a) = 0.8 + 0.4 - 2 \times (0.2) = 0.8$$

$$td_G(a, b) = d_G(a, b) + \lambda(a, b) = 0.6 + 0.2 = 0.8$$

$$td_G(b, c) = d_G(b, c) + \lambda(b, c) = 0.5 + 0.4 = 0.9$$

$$td_G(c, d) = d_G(c, d) + \lambda(c, d) = 1.0 + 0.3 = 1.3$$

$$td_G(d, e) = d_G(d, e) + \lambda(d, e) = 0.5 + 0.6 = 1.1$$

$$td_G(e, a) = d_G(e, a) + \lambda(e, a) = 0.8 + 0.2 = 1.0$$

Thus, from above, we observe that every pair of adjacent edges have a distinct total degree, and hence the fuzzy graph G is a neighborly edge totally irregular fuzzy graph.

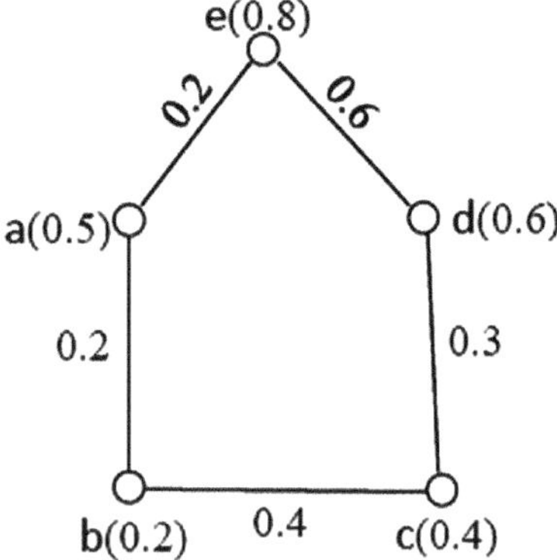

Fig. 2.67 Neighbourly edge totally irregular fuzzy graph

2.7.6 *Strongly Edge Irregular Fuzzy Graph*

Let $G = (V, \gamma, \lambda)$ be a connected fuzzy graph on the underlying graph $G^* = (V, \gamma^*, \lambda^*)$. The fuzzy graph G is referred to as a strongly edge irregular fuzzy graph if every pair of arcs has a distinct degree or no two arcs have the same degree.

Example 2.37 Consider the fuzzy graph $G = (\gamma, \lambda)$ on the underlying graph $G(V, E)$, a cycle of length five (Fig. 2.68).

Let the vertices and edges of the fuzzy graph be defined as:

	a	b	c	d	e
γ	0.7	0.4	0.6	0.8	1

	ab	bc	cd	de	ae
λ	0.4	0.6	0.5	0.9	0.4

$$d_G(a, b) = d_G(a) + d_G(b) - 2\lambda(a, b) = 0.8 + 1 - 2 \times (0.4) = 1$$

$$d_G(b, c) = d_G(b) + d_G(c) - 2\lambda(b, c) = 1.0 + 1.1 - 2 \times (0.6) = 0.9$$

$$d_G(c, d) = d_G(c) + d_G(d) - 2\lambda(c, d) = 1.1 + 1.4 - 2 \times (0.5) = 1.5$$

$$d_G(d, e) = d_G(d) + d_G(e) - 2\lambda(d, e) = 1.4 + 1.2 - 2 \times (0.9) = 0.8$$

$$d_G(e, a) = d_G(e) + d_G(a) - 2\lambda(e, a) = 1.2 + 0.8 - 2 \times (0.4) = 1.2$$

Thus, we see that no two arcs have the same degrees. It follows that G is a strong edge irregular fuzzy graph.

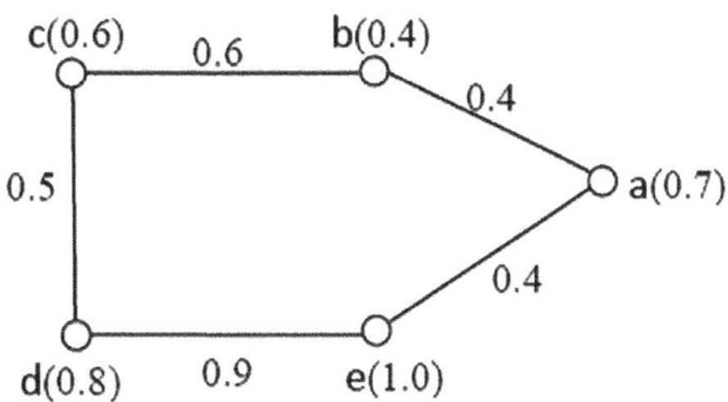

Fig. 2.68 Strongly edge irregular fuzzy graph

2.7.7 Strongly Edge Totally Irregular Fuzzy Graph

Let $G = (\gamma, \lambda)$ be a connected fuzzy graph on graph $G = (V, E)$. Then G is called a strongly edge totally irregular fuzzy graph if no two arcs have the same total degree or we can say that every pair of arcs have distinct total degrees.

Example 2.38 Consider the fuzzy graph $G = (\gamma, \lambda)$ on graph $G(V, E)$, a cycle of length five. Let the vertices and edges of the fuzzy graph be defined as (Fig. 2.69).

	a	b	c	d	e
γ	0.7	0.4	0.6	0.8	1

	ab	bc	cd	de	ea
λ	0.4	0.6	0.5	0.9	0.4

$$d_G(a) = 0.8, d_G(b) = 1, d_G(c) = 1.1, d_G(d) = 1.4, d_G(e) = 1.2$$

$$d_G(a, b) = 1, d_G(b, c) = 0.9, d_G(c, d) = 1.5, d_G(d, e) = 0.8, d_G(e, a) = 1.2$$

$$td_G(a, b) = d_G(a, b) + \lambda(a, b) = 1 + 0.4 = 1.4$$

$$td_G(b, c) = d_G(b, c) + \lambda(b, c) = 0.9 + 0.6 = 1.5$$

$$td_G(c, d) = d_G(c, d) + \lambda(c, d) = 1.5 + 0.5 = 2$$

$$td_G(d, e) = d_G(d, e) + \lambda(d, e) = 0.8 + 0.9 = 1.7$$

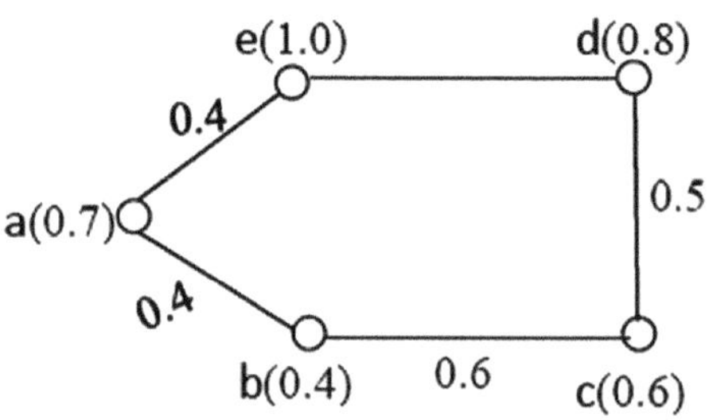

Fig. 2.69 Strongly edge totally irregular fuzzy graph

$$td_G(e, a) = d_G(e, a) + \lambda(e, a) = 1.2 + 0.4 = 1.6$$

Thus, we observe that every pair of arcs in the fuzzy graph G has distinct total degrees. Therefore, G is a strong edge totally irregular fuzzy graph.

Remarks 2.2

(i) Strongly edge totally irregular fuzzy graph need not be strongly edge irregular fuzzy graph.
(ii) Strongly edge irregular fuzzy graph need not be strongly edge totally irregular.

Theorem 2.1 Show that if G *is* a strong edge irregular fuzzy graph and λ is a constant function, G is a strong edge totally irregular fuzzy graph.

Proof Suppose $G = (\gamma, \lambda)$ is a connected fuzzy graph on the underlying graph $G = (V, E)$. We further assume that λ is a constant function, such that, $\lambda(a, b) = k, \forall(a, b) \in E$, where k is a constant. Let (γ, λ) be a strongly edge irregular fuzzy graph, and (a, b) and (c, d) be any pair of arcs in E. Since (γ, λ) is a strongly edge irregular fuzzy graph, pair of arcs (a, b) and (c, d) in E exist such that

$$\begin{aligned} d_G(a, b) \neq d_G(c, d) &\Rightarrow d_G(a, b) + k \neq d_G(c, d) + k \\ &\Rightarrow d_G(a, b) + \lambda(a, b) \neq d_G(c, d) + \lambda(c, d) \end{aligned}$$

where (a, b) and (c, d) are any pair of arcs in E. Hence G is a strong edge, totally irregular fuzzy graph.

Theorem 2.2 Show that if (γ, λ) is a strongly edge totally irregular fuzzy graph and λ is a constant function, then (γ, λ) is a strongly edge irregular fuzzy graph.

Proof Let (γ, λ) be a connected fuzzy graph and λ is a constant function, i.e., $\lambda(a, b) = k, \forall(a, b) \in E.$, where k is a constant. Let (γ, λ) is a strong edge totally irregular fuzzy graph. This implies that

$$td_G(a, b) \neq td_G(c, d)$$

where (a, b) and (c, d) are any pair of arcs in E.

Thus,

$$\begin{aligned} td_G(a, b) \neq td_G(c, d) &\Rightarrow d_G(a, b) + \lambda(a, b) \neq d_G(c, d) + \lambda(c, d) \\ &\Rightarrow d_G(a, b) + k \neq d_G(c, d) + k \\ &\Rightarrow d_G(a, b) \neq d_G(c, d) \end{aligned}$$

where (a, b) and (c, d) are any pair of arcs in E, it follows that (γ, λ) is a strongly edge irregular fuzzy graph.

2.8 μ—Length, Eccentricity of a Node and Central Node

2.8.1 *μ—Length*

In a fuzzy graph G = (γ, λ), for any path $p = v_0, v_1, v_2, \ldots, v_n$, the $\mu-$ length is denoted by the symbol $\ell(p)$ and is defined as the sum of the reciprocals of p' s arc weights, i.e.,

$$\ell(p) = \sum_{i=1}^{n} \frac{1}{\mu(x_{i-1}, x_i)}$$

If $n = 0$, we define $\ell(p) = 0$. Evidently, for $n \geq 1$ we have $\ell(p) \geq 1$.

2.8.2 *μ–Distance*

The μ-distance between any two vertices v_i, v_j of a fuzzy graph $G = (\gamma, \lambda)$ is denoted by the symbol $\delta(v_i, v_j)$ and is defined as the smallest $\mu-$ length of any path p from v_i to v_j for all $v_i, v_j \in G$.

Theorem 2.3 Could you show that $\delta(u, v)$ is a metric?

Proof

(i) $\delta(u, v) = 0$ if and only if, since $\ell(p) = 0$ if and only if p has length 0.
(ii) $\delta(u, v) = \delta(v, u)$, since we know that the reversal of a path is again a path and μ is symmetric.
(iii) Since the interconnection of a path from u to v and from v to w is a path from u to w, and ℓ is an additive for the interconnection of paths.

In the classical case, or the non-fuzzy case, $\ell(p)$ is just the length n of the path p. Since all the μ's are 1. And, therefore $\delta(u, v)$ reduces to the usual definition of distance, i.e., the length of the shortest path between u and v.

Based on the concept of this metric, *P. Bhattacharya* proposed the notions of eccentricity and center in fuzzy graphs in the year 1987.

2.8.3 *The Eccentricity of a Node*

Let $G = (\gamma, \lambda)$ be a connected fuzzy graph, and v be a node in G. Then, the eccentricity of v is denoted by $e(v)$ and is defined as

$$e(v) = max\ \delta(u, v), \forall\ u \in G$$

2.8.4 *Radius and Diameter of a Fuzzy Graph*

Let $G = (\gamma, \lambda)$ be a connected fuzzy graph. Then, the radius of the fuzzy graph G is denoted by $r(G)$ and is defined as the minimum eccentricity of the nodes, i.e.,

$$r(G) = min\{e(v) : v \in G\}$$

And the diameter of G is denoted by $d(G)$ and is defined as the maximum eccentricity of the nodes, i.e.,

$$r(G) = max\{e(v) : v \in G\}$$

2.8.5 *Central Node*

A node v in a connected fuzzy graph $G = (\gamma, \lambda)$ is called a central node if

$$e(v) = r(G)$$

We call the fuzzy subgraph of $G = (V, \gamma, \lambda)$ induced by the central nodes of G, the center of G.

Example 2.39 Let $G_1 = (\gamma_1, \lambda_1)$, and $G_2 = (\gamma_2, \lambda_2)$ are two fuzzy graphs whose Figures are drawn as shown in **2.70** and **2.71** and whose vertices and edges are defined as

	v_1	v_2	v_3	v_4
γ_1	0.4	0.6	0.3	0.7

	v_1v_2	v_2v_3	v_3v_4	v_4v_1
λ_1	1	0.5	0.5	1

	v_1	v_2	v_3	v_4
γ_2	0.3	0.2	0. 4	0.8

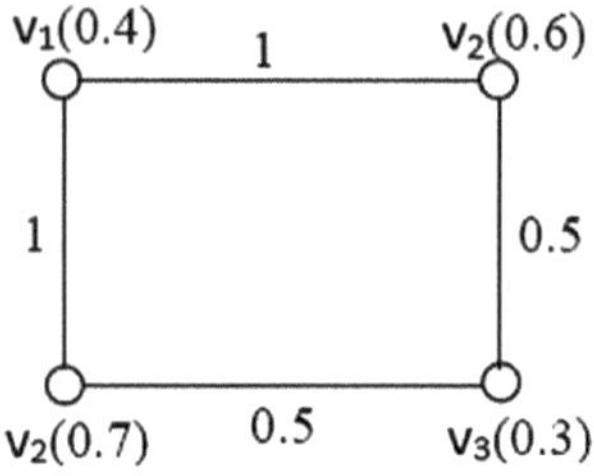

Fig. 2.70 Graph representing eccentricity, radius & central node

	v_1v_2	v_1v_3	v_1v_4	v_2v_3	v_2v_4	v_3v_4
λ_2	0.5	0.25	1	0.5	1	1

From Fig. 2.70, we have from the graph, $\delta(v_1v_3) = 3, \delta(v_2v_4) = 2$

$e(v_1) = e(v_3) = 3, e(v_4) = e(v_2) = 2, r(G_1) = 2$ and $d(G_1) = 3$.
Hence v_2 and $v_4$are central nodes.

From figure (2.71), we have from the graph $G_2, \delta(v_1v_2) = 2, \delta(v_2v_4) = 1, e(v_1) = e(v_2) = e(v_3) = 2,$

$e(v_4) = 1, r(G_2) = 1$ and $d(G_2) = 2$. Here v_4 is a central node.

2.8.6 Complement of a Fuzzy Graph

The complement of a fuzzy graph $G = (\gamma, \lambda)$ is denoted by $\overline{G : (\overline{\gamma}, \overline{\lambda})}$ and is defined as $\overline{\gamma} = \gamma$ and $\overline{\lambda}(u, v) = \gamma(u) \wedge \gamma(v) - \lambda(u, v), \forall\, u, v \in \gamma$ (Fig. 2.71).

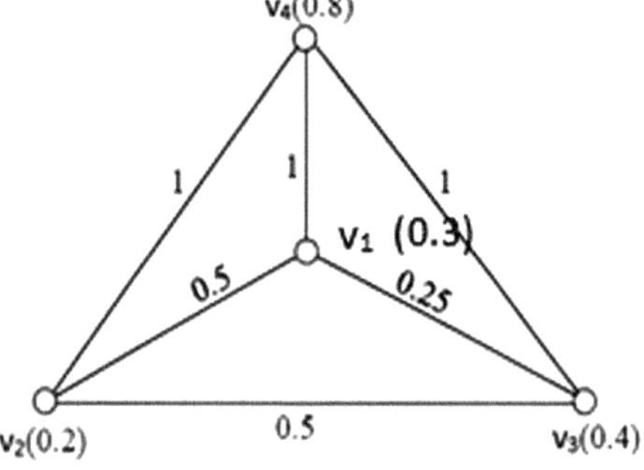

Fig. 2.71 Graph representing central node

2.8.7 μ-Complement of a Fuzzy Graph

The μ-complement of a fuzzy graph $G = (\gamma, \lambda)$ is denoted by $G^{\mu} : (\gamma^{\mu}, \lambda^{\mu})$ where $\gamma^{\mu} = \gamma$, and

$$\lambda^{\mu}(u, v) = \begin{cases} \lambda(u, v) if \lambda(u, v) = 0 \\ \gamma(u) \wedge \gamma(v) - \lambda(u, v) if \lambda(u, v) > 0 \end{cases}$$

Example 2.40 Let us consider a fuzzy graph $G = (\gamma, \lambda)$ in which nodes and edges are defined as follows (Figs. 2.72 and 2.73).

	u	v	w	x	y
γ	0.4	0.7	0.8	0.5	0.6

	uv	vw	wx	xy	xv
λ	0.3	0.6	0.3	0.4	0.2

$$\overline{\lambda}(u, v) = \gamma(u) \wedge \gamma(v) - \lambda(u, v) = 0.4 \wedge 0.7 - 0.3 = 0.4 - 0.3 = 0.1$$

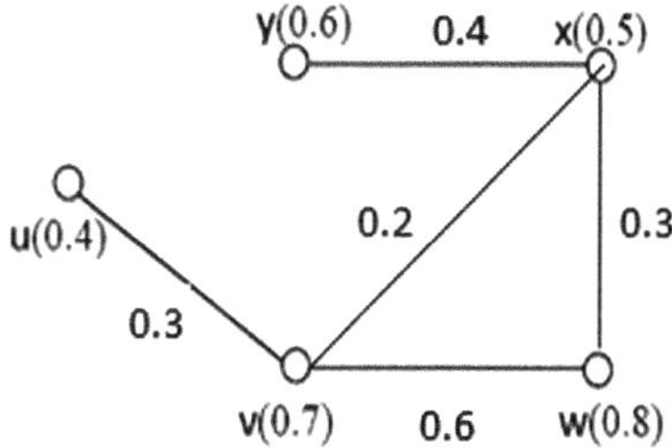

Fig. 2.72 Fuzzy graph G for finding the complement

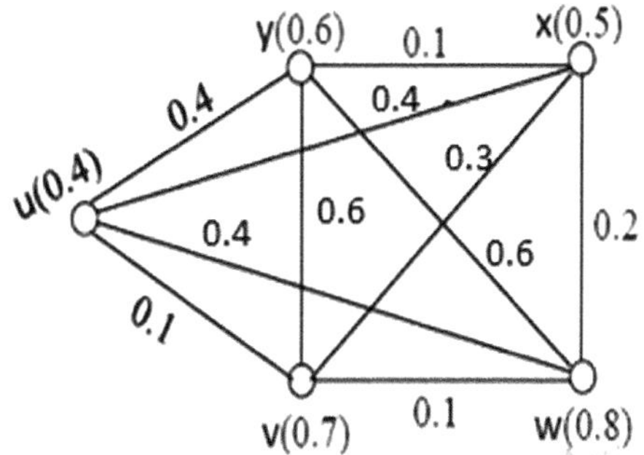

Fig. 2.73 Complement of fuzzy graph depicted in Fig. 2.72 $G = \overline{G}$

$$\overline{\lambda}(v, w) = \gamma(v) \wedge \gamma(w) - \lambda(v, w) = 0.7 \wedge 0.8 - 0.6 = 0.7 - 0.6 = 0.1$$

$$\overline{\lambda}(w, x) = \gamma(w) \wedge \gamma(x) - \lambda(w, x) = 0.8 \wedge 0.5 - 0.3 = 0.5 - 0.3 = 0.2$$

$$\overline{\lambda}(x, y) = \gamma(x) \wedge \gamma(y) - \lambda(x, y) = 0.5 \wedge 0.6 - 0.4 = 0.5 - 0.4 = 0.1$$

$$\overline{\lambda}(x, v) = \gamma(x) \wedge \gamma(v) - \lambda(x, v) = 0.5 \wedge 0.7 - 0.2 = 0.5 - 0.2 = 0.3$$

$$\overline{\lambda}(v, y) = \gamma(v) \wedge \gamma(y) - \lambda(v, y) = 0.7 \wedge 0.6 - 0 = 0.6$$

$$\overline{\lambda}(u, y) = \gamma(u) \wedge \gamma(y) - \lambda(u, y) = 0.4 \wedge 0.6 - 0 = 0.4$$

$$\overline{\lambda}(w, y) = \gamma(w) \wedge \gamma(y) - \lambda(w, y) = 0.8 \wedge 0.6 - 0 = 0.6$$

$$\overline{\lambda}(u, x) = \gamma(u) \wedge \gamma(x) - \lambda(u, x) = 0.4 \wedge 0.5 - 0 = 0.4$$

$$\overline{\lambda}(u, w) = \gamma(u) \wedge \gamma(w) - \lambda(u, w) = 0.4 \wedge 0.8 - 0 = 0.4$$

Thus, we have the complement of the fuzzy graph G that is, $\overline{G}$ as

	u	v	w	x	y
$\overline{\gamma} = \gamma^{\mu}$	0.4	0.7	0.8	0.5	0.6

	uv	vw	wx	xy	xv	wy	ux	uw	uy	vy
$\overline{\lambda}$	0.1	0.1	0.2	0.1	0.3	0.6	0.4	0.4	0.4	0.6

	uv	vw	wx	xy	xv
λ^{μ}	0.1	0.1	0.2	0.1	0.3

The fundamental notions of operations on fuzzy graphs, such as the union of fuzzy graphs, join of fuzzy graphs, Cartesian product of fuzzy graphs, and composition of fuzzy graphs, have been introduced in the realm of fuzzy graph theory by **J. N. Mordeson** & **C. S. Peng** in 1994 (Information Science, V. 79) (Figs. 2.73 and 2.74).

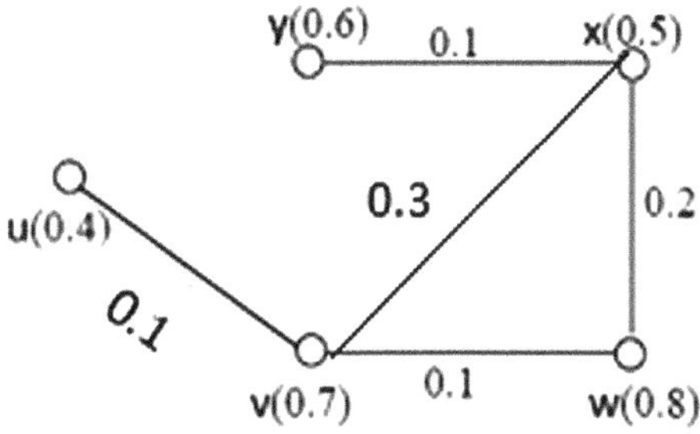

Fig. 2.74 Representing G^{μ}, *the* μ−Complement of a fuzzy graph

2.9 Operations on Fuzzy Graphs

2.9.1 Union of Two Fuzzy Graphs

Let us assume that $G_1 : (\gamma_1, \lambda_1)$ and $G_2 : (\gamma_2, \lambda_2)$ be two fuzzy graphs with the underlying classical graphs $G_1^* : (V_1, E_1)$, and $G_2^* : (V_2, E_2)$ with $V_1 \cap V_2 = \emptyset$ and let $G^* = G_1^* \cup G_2^* = (V_1 \cup V_2, E_1 \cup E_2)$ be the union of G_1^* and G_2^*. Then the union of two fuzzy graphs G_1 and G_2 is a fuzzy graph denoted by the symbol $G = G_1 \cup G_2 : (\gamma_1 \cup \gamma_2, \lambda_1 \cup \lambda_2)$ on $G^* : (V, E)$ where $V = V_1 \cup V_2$ and $E = E_1 \cup E_2$ and is defined by the setting

$$(\gamma_1 \cup \gamma_2)(u) = \begin{cases} \gamma_1(u) & \text{if } u \in V_1 - V_2 \\ \gamma_2(u) & \text{if } u \in V_2 - V_1 \\ \gamma_1(u) \vee \gamma_2(u) & \text{if } u \in V_1 \cap V_2 \end{cases}$$

and

$$(\lambda_1 \cup \lambda_2)(uv) = \begin{cases} \lambda_1(uv) & \text{if } uv \in E_1 - E_2 \\ \gamma_2(uv) & \text{if } uv \in E_2 - E_1 \\ \gamma_1(uv) \vee \gamma_2(uv) & \text{if } uv \in E_1 \cap E_2 \end{cases}$$

2.9.2 Join of Two Fuzzy Graphs

Let us consider that $G_1 : (\gamma_1, \lambda_1)$ and $G_2 : (\gamma_2, \lambda_2)$ are two fuzzy graphs with the underlying classical graphs $G_1^* : (V_1, E_1)$, and $G_2^* : (V_2, E_2)$ where E′is the set of all. arcs joining the nodes of V_1 and V_2 with the assumption that $V_1 \cap V_2 = \emptyset$ and let $G^* = G_1^* + G_2^* = \left(V_1 \cup V_2, E_1 \cup E_2 \cup E'\right)$ be the join of G_1^* and G_2^*. Then the join of the fuzzy graphs G_1 and G_2 is a fuzzy graph denoted by the symbol $G = G_1 + G_2 : (\gamma_1 + \gamma_2, \lambda_1 + \lambda_2)$ on $G^* : (V, E)$ where $V = V_1 + V_2$ and $E = E_1 + E_2 + E'$ and is defined by the setting

$$(\gamma_1 + \gamma_2)(u) = (\gamma_1 \cup \gamma_2)(u) \text{ for all } u \in V_1 \cup V_2$$

and

$$(\lambda_1 + \lambda_2)(uv) = \begin{cases} (\lambda_1 \cup \lambda_2)(uv) \; if \;\; uv \in E_1 \cup E_2 \\ \gamma_1(u) \wedge \gamma_2(u) \quad\;\; if \;\; uv \in E' \end{cases}$$

2.9.3 Cartesian Product of Two Fuzzy Graphs

Let us suppose that $G_1 : (\gamma_1, \lambda_1)$ and $G_2 : (\gamma_2, \lambda_2)$ be two fuzzy graphs with the underlying classical graphs $G_1^* : (V_1, E_1)$, and $G_2^* : (V_2, E_2)$ with $V_1 \cap V_2 = \emptyset$. Then the Cartesian product of two fuzzy graphs G_1 and G_2 is a fuzzy graph denoted by the symbol $G = G_1 \times G_2 : (\gamma_1 \times \gamma_2, \lambda_1 \times \lambda_2)$ on $G^* : (V, E)$ where $V = V_1 \times V_2$ and

$$\begin{aligned} E =& \{(u_1, u_2)(v_1, v_2) : u_1 = v_1 \in V_1, u_2v_2 \in E_2\} \\ & \cup \{(u_1, u_2)(v_1, v_2) : u_2 = v_2 \in V_2, u_1v_1 \in E_1\} \end{aligned}$$

and is defined by the setting

$$(\gamma_1 \times \gamma_2)(u_1, u_2) = \gamma_1(u_1) \wedge \gamma_2(u_2) \text{ for all} (u_1, u_2) \in V \text{ and}$$

$$(\lambda_1 \times \lambda_2)(u_1, u_2)(v_1, v_2) = \begin{cases} \gamma_1(u_1) \wedge \lambda_2(u_2, v_2) \; if \;\; u_1 = v_1 \; and \;\; u_2v_2 \in E_2 \\ \gamma_2(u_2) \wedge \lambda_1(u_1, v_1) \; if \;\; u_2 = v_2 \; and \;\; u_1v_1 \in E_1 \end{cases}$$

2.9.4 Composition of Two Fuzzy Graphs

Let $G_1 : (\gamma_1, \lambda_1)$ and $G_2 : (\gamma_2, \lambda_2)$ be two fuzzy graphs with the underlying classical graphs $G_1^* : (V_1, E_1)$, and $G_2^* : (V_2, E_2)$. Then the composition of two fuzzy graphs G_1 and G_2 is a fuzzy graph denoted by the symbol $G = G_1 \circ G_2 : (\gamma_1 \circ \gamma_2, \lambda_1 \circ \lambda_2)$ on $G^* = G_1^* \circ G_2^* = (V, E)$ where $V = V_1 \times V_2$ and

$$\begin{aligned} E = \{(u, u_2)(u, v_2) : u \in V_1, u_2v_2 \in E_2\} \cup \{(u_i, w)(v_i, w) : w \in V_2, u_iv_i \in E_1\} \cup \\ \{(u_1, u_2)(v_1, v_2) : u_iv_i \in E_i, u_2 \neq v_2\} \end{aligned}$$

and is defined by the setting

$$(\gamma_1 \circ \gamma_2)(u_1, u_2) = \gamma_1(u_1) \wedge \gamma_2(u_2), \forall (u_1, u_2) \in V_1 \times V_2$$

$$(\lambda_1 \circ \lambda_2)((u, u_2)(u, v_2)) = \gamma_1(u_1) \wedge \lambda_2(u_2, v_2), \forall\, u \in V_1, u_2v_2 \in E_2$$

$$(\lambda_1 \circ \lambda_2)((u_1, w)(v_1, w)) = \gamma_2(w) \wedge \lambda_1(u_1, v_1), \forall\, w \in V_2, \forall\, u_1v_1 \in E_1$$

$$(\lambda_1 \circ \lambda_2)((u_1, u_2)(v_1, v_2)) = \gamma_2(u_2) \wedge \gamma_2(v_2) \wedge \lambda_1(u_1, v_1), \forall (u_1, u_2)(v_1, v_2) \in E - E'$$

where

$$E' = \{(u, u_2)(u, v_2) : u \in V_1, \forall u_2v_2 \in E_2\} \cup \{(u_1, w)(v_1, w) : w \in V_2, u_1v_1 \in E_1\}$$

2.10 Matrix Representation of Fuzzy Graph

Let $G = (\gamma, \lambda)$ be a fuzzy graph. Further, assume that λ is a fuzzy relation that is reflexive and symmetric and is completely determined by the fuzzy matrix denoted by M_λ, where

$$(M_\lambda)_{i,j} = \begin{cases} \lambda(u_i, u_j) & for\ i \neq j \\ \gamma_i & for\ i = j \end{cases}$$

If γ^* contains n elements, then M_λ is a square matrix of order n.

In other words, we can say that a fuzzy graph is an expression of fuzzy relation, and we can represent a fuzzy graph in terms of a fuzzy matrix.

Example 2.41 Let us assume that $G = (\gamma, \lambda)$ is a fuzzy graph, whose nodes and edges are defined as follows:

	u	v	w	x	y
γ	0.4	0.6	0.3	0.8	0.5

	uv	vw	wx	xy	wy
λ	0.3	0.2	0.1	0.4	0.2

The matrix representation of the fuzzy graph G is given by

$$M_\lambda = \begin{matrix} u \\ v \\ w \\ x \\ y \end{matrix} \begin{bmatrix} u & v & w & x & y \\ 0.4 & 0.3 & 0 & 0 & 0 \\ 0.3 & 0.6 & 0.2 & 0 & 0 \\ 0 & 0.2 & 0.3 & 0.1 & 0.2 \\ 0 & 0 & 0.1 & 0.8 & 0.5 \\ 0 & 0 & 0.2 & 0.4 & 0.5 \end{bmatrix}$$

Example 2.42 Let us assume that $G = (\gamma, \lambda)$ be a fuzzy graph, whose nodes and edges are defined as follows:

	u	v	w	x	y
γ	0.7	0.9	0.6	1	0.8

	uv	vw	wx	ux	vx	xy	wy
λ	0.5	0.4	0.3	0.6	0.9	0.7	0.8

The matrix representation of the fuzzy graph G is given by

$$M_\lambda = \begin{matrix} u \\ v \\ w \\ x \\ y \end{matrix} \begin{bmatrix} u & v & w & x & y \\ 0.7 & 0.5 & 0 & 0.6 & 0 \\ 0.5 & 0.9 & 0.4 & 0.9 & 0 \\ 0 & 0.4 & 0.6 & 0.3 & 0.8 \\ 0.6 & 0.9 & 0.3 & 1.0 & 0.7 \\ 0 & 0 & 0.8 & 0.7 & 0.8 \end{bmatrix}$$

Example 2.43 Express the following fuzzy matrix in the form of a fuzzy graph

$$\begin{bmatrix} & y_1 & y_2 \\ x_1 & 0.9 & 0.3 \\ x_2 & 0 & 0.4 \\ x_3 & 0.8 & 0.5 \end{bmatrix}.$$

2.11 α – Cut Of Fuzzy Graph

We have studied the α-cut of a fuzzy set and the α-cut of a fuzzy relation. The notion of the α—cut of a fuzzy graph can always be extended naturally as follows

Example 2.44 Let $X = \{u, v, w\}$. We know that a relation $R \subseteq X \times X$. Suppose that the relation R is defined as follows

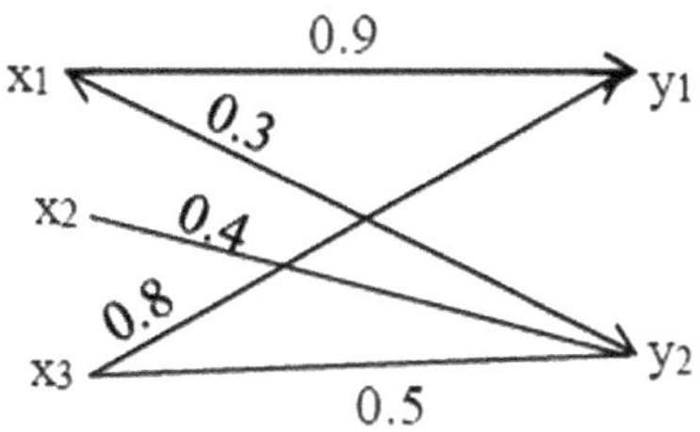

Fig. 2.75 Fuzzy matrix representation of Example 2.43

$$M_R = M_\lambda = \begin{bmatrix} & u & v & w \\ u & 0.9 & 0.5 & 0.0 \\ v & 0.5 & 1.0 & 0.9 \\ w & 0.0 & 0.0 & 1.0 \end{bmatrix}$$

Then, the level set of R is given by $\{0, 0.5, 0.9, 1\}$. Let the α-cut of the fuzzy graph be denoted by G_α. If $\alpha, \beta \in [0, 1]$, then we have the following relations between the α-cut of the fuzzy relation R and G_α, and is given by

$$\alpha \geq \beta \Rightarrow R_\alpha \subseteq R_\beta \text{ and } \alpha \geq \beta \Rightarrow G_\alpha \subseteq G_\beta$$

We have

$$\begin{bmatrix} M_{R_0} = & u & v & w \\ u & 0.9 & 0.5 & 0.0 \\ v & 0.5 & 1.0 & 0.9 \\ w & 0.0 & 0.0 & 1.0 \end{bmatrix}$$

$$\begin{bmatrix} M_{R_{0.5}} = & u & v & w \\ u & 1.0 & 1.0 & 0.0 \\ v & 1.0 & 1.0 & 1.0 \\ w & 0.0 & 0.0 & 1.0 \end{bmatrix}$$

$$\begin{bmatrix} M_{R_{0.9}} = & u & v & w \\ u & 1.0 & 0.0 & 0.0 \\ v & 0.0 & 1.0 & 1.0 \\ w & 0.0 & 0.0 & 1.0 \end{bmatrix}$$

$$\begin{bmatrix} M_{R_1} = & u & v & w \\ u & 0.0 & 0.0 & 0.0 \\ v & 0.0 & 1.0 & 0.0 \\ w & 0.0 & 0.0 & 1.0 \end{bmatrix}$$

2.12 Isomorphism of Fuzzy Graphs

Let $G_1 = (\gamma_1, \lambda_1)$, and $G_2 = (\gamma_2, \lambda_2)$ be any two fuzzy graphs with $\gamma_1^* = V_1$ *and* $\gamma_2^* = V_2$. Then an isomorphism between fuzzy graphs G_1 and G_2 is a bijective mapping $f : V_1 \to V_2$ satisfying the following conditions:

$$\gamma_1(u) = \gamma_2 f(u), \forall\, u \in V_1$$

$$\lambda_1(u, v) = \lambda_2(f(u), f(v)), \forall\, u, v \in V_1$$

And we denote the fact that G_1 is isomorphic to $G_2$by the symbol $\mathrm{G}_1 \approx G_2$.An isomorphism from G to itself is called an Automorphism of G.

2.13 Fuzzy Bridge

An arc (u, v) in a fuzzy graph $G = (\gamma,\ \lambda)$ is said to be a fuzzy bridge if the deletion of the edge uv, i.e., (u, v) reduces the strength of connectedness between some pair of nodes in G.

Thus, (u, v) is a fuzzy bridge if and only if there exist nodes x, y such that (u, v) is an edge of every strongest path from x *to* y.

In other words, let $G = (\gamma, \lambda)$ be a fuzzy graph and let u, v be any two distinct nodes of G. Let $G' = \left(\gamma, \lambda'\right)$ be the fuzzy subgraph of the fuzzy graph G obtained by deleting the edge (u, v), where.

$\lambda'(u, v) = 0;\ \lambda' = \lambda$ for all other pairs. Then, we say that (u, v) is a bridge in G if.

$\lambda'^{\infty}(u, v)\ < \lambda(u, v)$ for some u, v.

2.14 Fuzzy Cutnode

A node u in a fuzzy graph $G = (\gamma, \lambda)$ is called a cutnode if removing it reduces the strength of connectedness between some other pair of nodes.

Thus, u is a fuzzy cutnode if and only if nodes x and y exist distinct from u such that u is on the every strongest x, y path (Fig. 2.76).

Example 2.45 In the adjoining fuzzy graph $(u, w)(w, v)(v, x)$ and (x, y) are fuzzy bridges, and v, w, x are the fuzzy cutnodes of the fuzzy graph $G = (\gamma, \lambda)$.

Theorem 2.4 Show that the following statements are equivalent in a fuzzy subgraph $G = (\gamma, \lambda)$:

(i) (u, v) is a bridge.

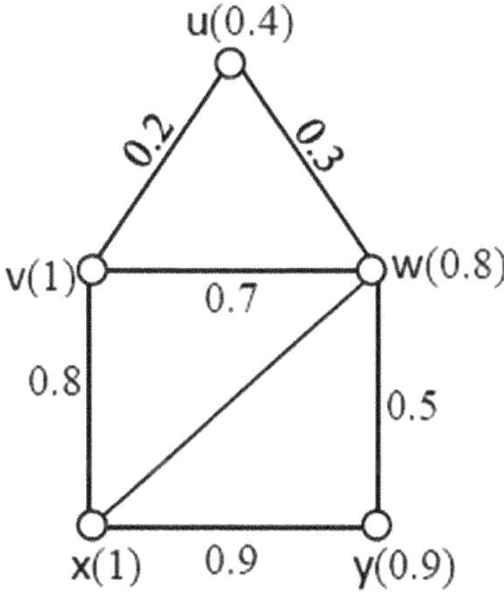

Fig. 2.76 Fuzzy bridge and fuzzy cutnodes

(ii) $\lambda'^{\infty}(u, v) < \lambda(u, v)$
(iii) (u, v) is not the weakest arc of any cycle.

Proof Let $G = (\gamma, \lambda)$ is a fuzzy graph.

(ii) $\Rightarrow$ (i): If (u, v) is not a bridge, then we must have $\lambda'^{\infty}(u, v) = \lambda^{\infty}(u, v) \geq \lambda(u, v)$. This is contrary to the notion. Thus, (ii) $\Rightarrow$ (i)

(i) $\Rightarrow$ (iii): Let (u, v) be the cycle's weakest arc; then any path p involving arc (u, v) can be converted into a path p' not involving (u, v) but at least as strong as p. Using the rest of the cycle as a path from u to v. Thus, (u, v) cannot be a fuzzy bridge, so that (i) $\Rightarrow$ (iii).

(iii) $\Rightarrow$ (ii): If $\lambda'^{\infty}(u, v) \geq \lambda(u, v)$, then there is a path from u to v, not involving (u, v), that has a strength greater than equal to $\lambda(u, v)$, and this path, together with uv, forms a cycle of the fuzzy graph G in which (u, v) is the weakest edge. Thus, (iii) $\Rightarrow$ (ii).

2.15 Block

A connected fuzzy graph $G = (\gamma, \lambda)$ is called non-separable or sometimes a block if it has no fuzzy cutnode. If between every two nodes u and v of the fuzzy graph G, the two strongest paths are disjoint (except for u and v themselves), G is a block. In his work, Rosenfeld has observed that blocks in fuzzy graphs may have fuzzy bridges (Fig. 2.78).

Example 2.46 In the following fuzzy graphs G_1 and G_2, we observe that G_1(Fig. 2.77) is a block without a fuzzy bridge and G_2 is a block with fuzzy bridges (u, v), and (w, x) (Fig. 2.78).

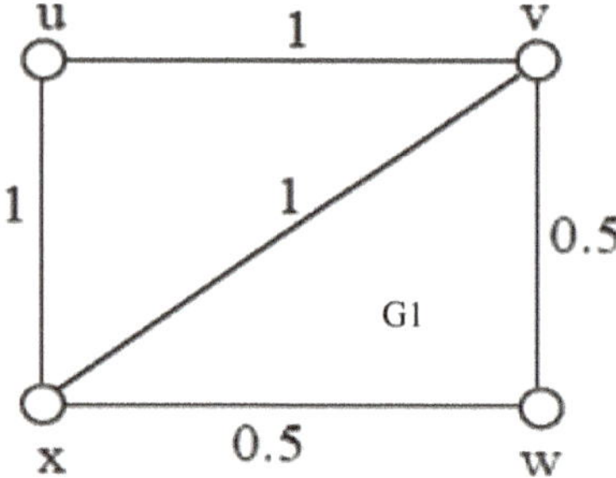

Fig. 2.77 Fuzzy graph G_1 representing block without fuzzy bridge

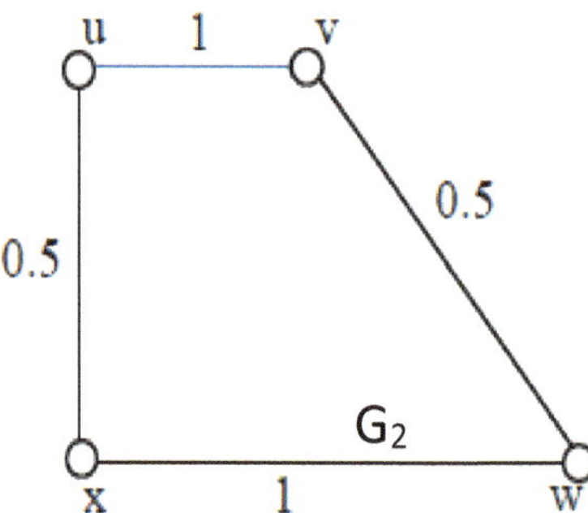

Fig. 2.78 Fuzzy Graph G_2 representing block with fuzzy bridge

Example 2.47 In the fuzzy graph, the edge (u, v) is a bridge. Its deletion reduces its strength of connectedness between u and v from 1 to 0.5. However, it can be easily verified that no node of this fuzzy graph is a cutnode.

2.16 Forest and Fuzzy Forest

A fuzzy graph $G = (\gamma, \lambda)$ is referred to as a **forest** if it has no cycles. In other words, a fuzzy graph having no cycles is called a forest. That is, it is an **acyclic fuzzy graph**. We may call a fuzzy graph (γ, λ) a forest if the graph consisting of its nonzero edges is a forest.

In general, a fuzzy graph (γ, λ) be referred to as a fuzzy forest if it has a fuzzy spanning subgraph $F = (\mu, \beta)$ which is a forest, where for all edges $(u, v) \notin F$ such that $\beta(u, v) = 0$, we have

$$\lambda(u, v) < \beta^{\infty}(u, v).$$

In other words, if $(u, v) \in G$ *but* $(u, v) \notin F$, there is a path in F between u and v whose strength is more significant than $\lambda(u, v)$. Hence a forest is a fuzzy forest (Figs. 2.79 and 2.80).

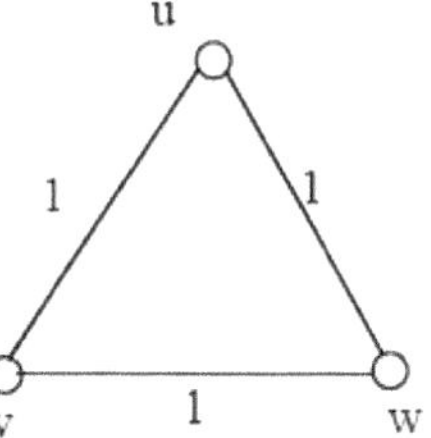

Fig. 2.79 Forest

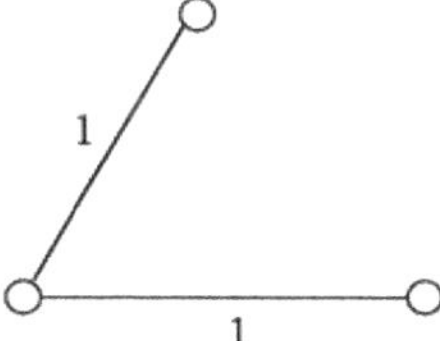

Fig. 2.80 Fuzzy Forest which is a subgraph of the fuzzy forest depicted in Fig. 2.79

If we replace $<$ by $\leq$ in the definition, then the fuzzy graph would be a fuzzy forest since it has subgraph such as (Figs. 2.80 and 2.82).

The following are examples of fuzzy forests (Figs. 2.81 and 2.85).

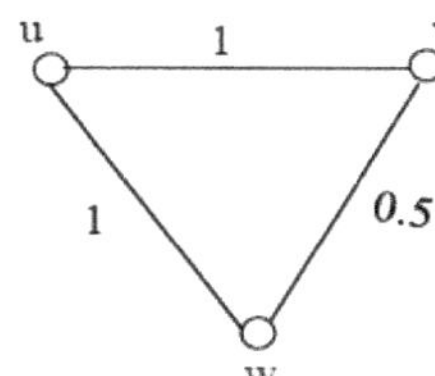

Fig. 2.81 Examples of fuzzy forests

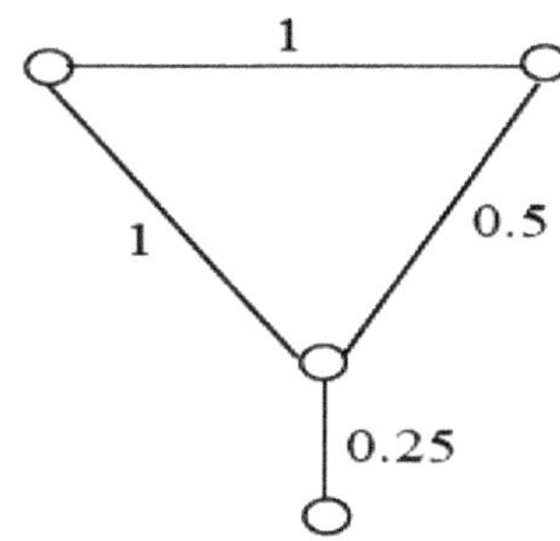

Fig. 2.82 Examples of fuzzy forests

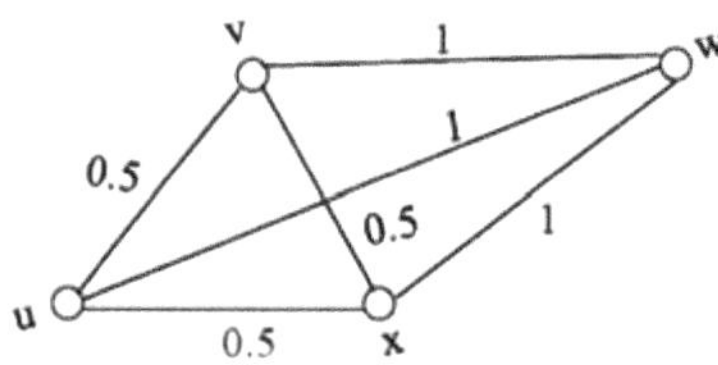

Fig. 2.83 Examples of fuzzy forests

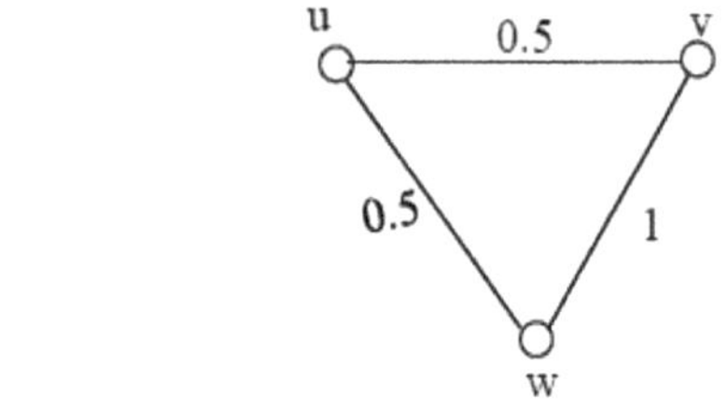

Fig. 2.84 Examples of fuzzy forests

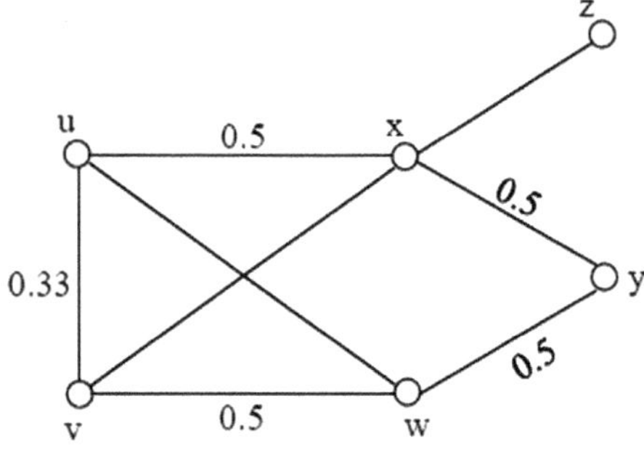

Fig. 2.85 Examples of fuzzy forests

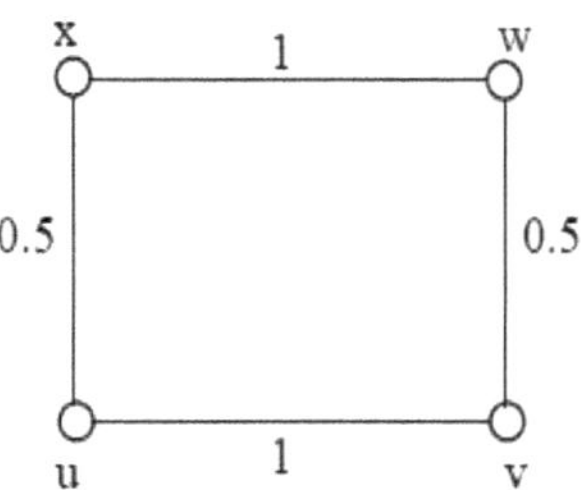

Fig. 2.86 Not a fuzzy forest

Example 2.48 Graphs that are not forests are represented in the following figures (Figs. 2.86 and 2.88).

2.17 Tree and Fuzzy Tree

A fuzzy forest that is connected is referred to as a tree. We may call a fuzzy graph (γ, λ) a tree if the graph consisting of its nonzero edges is a forest and connected too. A connected fuzzy forest is called a fuzzy tree.

Example 2.49 The examples of all fuzzy forests given above are all fuzzy trees (Figs. 2.87 and 2.88).

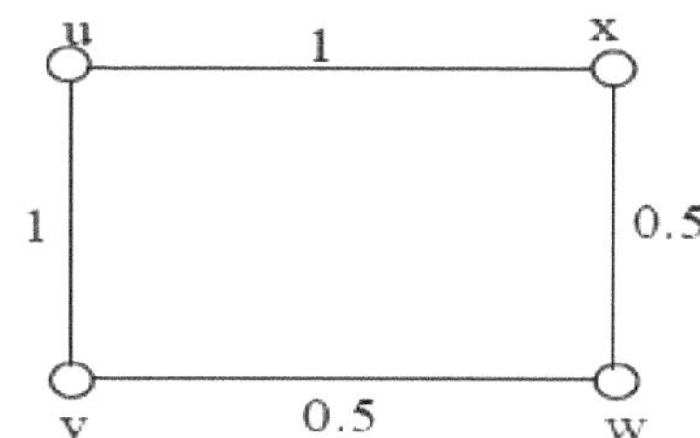

Fig. 2.87 Not a fuzzy forest

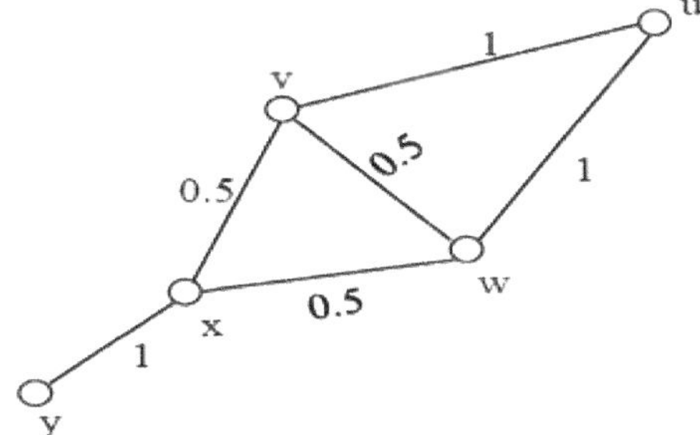

Fig. 2.88 Not a fuzzy forest

Theorem 2.5 Establish that G is a fuzzy forest if and only if, in any cycle of G, there is an arc (u, v) such that $\lambda(u, v) < \lambda'^{\infty}(u, v)$, where the prime denotes the deletion of the arc (u, v) from the graph G.

Proof Let $G = (\gamma, \lambda)$ is a fuzzy graph. Let G is a forest without cycles, and we have nothing to prove.

Let (u, v) be an edge of any cycle of G such that $\lambda'^{\infty}(u, v) > \lambda(u, v)$.

If we remove the edge (u, v) from the graph, the resulting fuzzy graph will satisfy the path property of the fuzzy forest. If there are still cycles in the fuzzy subgraph, we may repeat the above procedure. It is important to note that no previously deleted arc is stronger at each stage than the currently deleted arc.

Therefore, the path provided by the property of the theorem involves only edges that have not been deleted yet. When no cycles remain, the resulting fuzzy subgraph is a forest F. Suppose that (u, v) is not an edge of F. It means that (u, v) is one of the edges that have been deleted in the construction process of F, and there is a path from u to v that is stronger than $\lambda(u, v)$ and which does not contain (u, v) nor any of the edges deleted before it. If this path contains edges deleted later, it can detract around them by still longer edges. If any edges are deleted later, the path can be further detracted, and the process will continue. In the end, the path stabilizes with a path that contains entirely edges of F. And hence F is a fuzzy forest.

Conversely, let us assume that G is a fuzzy forest and let p be any cycle. Then some edge (u, v) of the cycle p is not in F. Thus, from the notion of a fuzzy forest, we have $\lambda(u, v) < \beta^{\infty}(u, v) \leq \lambda'^{\infty}(u, v)$. This proves the "only if" part.

Theorem 2.6 If there is at most one strongest path between any two nodes of the fuzzy graph G, then G must be a fuzzy forest.

Proof Let $G = (\gamma, \lambda)$ be a fuzzy graph. If possible, let us assume that G is not a fuzzy forest. Further, suppose that there is not at most one strongest path between the fuzzy graph $G's$ vertices, say u and v. Then, there must exist a path p in the fuzzy graph G such that for all edges (u, v) of the path, we have $\lambda(u, v) \geq \lambda'^{\infty}(u, v)$.

Thus, (u, v) itself forms the strongest path from u to v. If we choose (u, v) to be the weakest edge of the path, it follows that the rest of the path, p is also the strongest path from u to v. This is a contradiction, and hence G is a fuzzy forest.

Remarks 2.3 The converse of the above result is not true. That is, G can be a fuzzy forest and even have multiple vital paths between vertices. This is because the strength of a path is defined as the strength of its weakest edge. As long as this edge lies in F, there is minimal effect on the other edges. For example, the fuzzy graph G is a fuzzy forest. Here F consists of all edges except (u, v). The most robust paths between u and v have a strength of ¼ due to the edge (u, w). Both u, w, x, v and u, w, v are such paths, where the first one lies in F, but the second one does not lie.

Theorem 2.7 If G is a fuzzy forest, the arcs of F are just the bridges of G.

Proof Let $G = (\gamma, \lambda)$ be a fuzzy graph which is a fuzzy forest. Let (u, v) be an edge that is not in the fuzzy forest F. Then, certainly, it is not a fuzzy bridge, since

$(\lambda u, v) < \beta^{\infty}(u, v) \leq \lambda'^{\infty}(u, v)$. This proves the "if" part.

Conversely, let us assume that (u, v) is an edge in the fuzzy forest F. If the edge (u, v) is not a bridge, we would have a path p from $utov$, not involving (u, v), and is of strength $\geq (u, v)$.

This path must involve edges not in the fuzzy forest F. F is a fuzzy forest with no cycles. However, by the notion of the forest, any such edge (a_i, b_i) can be replaced by a path p_i in F whose strength is $> \lambda(a, b)$. Now p_i cannot involve (u, v) since all its arcs are strictly stronger than $\lambda(a, b) \geq \lambda(u, v)$. Thus by replacing each (a_i, b_i) with p_i , we can construct a path in the forest F from $utov$ that does not involve p_i, giving us a cycle in F. This leads to a contradiction. Hence, the edges of F are the bridges of G.

Note 1. G is a fuzzy forest, its spanning forest, and F is unique.

2.18 Introduction of Intuitionistic Fuzzy Graph

The concept of the intuitionistic fuzzy graph was propounded by *Atanassov & Shannon* in 1994. The intuitionistic fuzzy graph has broader applications in management science, medical science, computer science, engineering, etc. It helps in modeling real-time where the information inherited varies with different rent levels of precision. *Parvathi* and *Karunambigai* 2006 gave a new definition of the min–max intuitionistic fuzzy graph. Further, *Gani & Begum 2010 observed the properties of numerous kinds of degrees, sizes, and order of intuitionistic fuzzy graphs and gave definitions for complete and* regular fuzzy graphs. In 2011 *Karunambigai* defined constant intuitionistic fuzzy graphs and totally constant intuitionistic fuzzy graphs along with constant intuitionistic fuzzy graphs on a cycle. *Ramani and Kumar* 2011 proposed the concept of product intuitionistic fuzzy graph and their properties along with product partial intuitionistic fuzzy graphs and several results.

2.19 Fundamentals of Intuitionistic Fuzzy Graph

2.19.1 Intuitionistic Fuzzy Graph

An *intuitionistic fuzzy graph* (IFG) is of the form $G = (V, E)$, where (Fig. 2.89).

(i) $V = \{u_1, u_2, u_3, \ldots, u_n\}$ such that $\mu_A : [V \to 0, 1]$ and $\gamma_A : [V \to 0, 1]$ stand for the degree of membership and non-membership of the element $u_i \in V$, respectively, and $0 \leq \mu_A(u_i)+\gamma_A(u_i) \leq 1$, for every $u_i \in V(i = 1, 2, 3, \ldots, n)$.
(ii) $E \in V \times V$, where $\mu_B : V \times V \to [0, 1]$ and $\gamma_B : V \times V \to [0, 1]$ are such that

$$\mu_B(u_i, u_j) \leq min\{\mu_A(u_i), \mu_A(u_j)\}$$

$$\gamma_B(u_i, u_j) \leq min\{\gamma_A(u_i), \gamma_A(u_j)\}$$

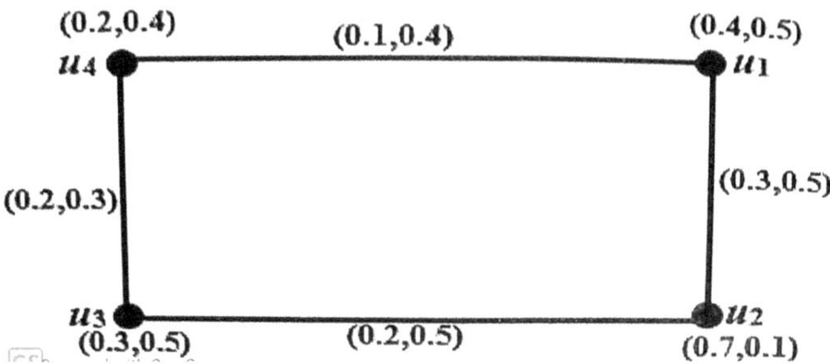

Fig. 2.89 Intuitionistic fuzzy graph

$$\text{and } 0 \le \mu_B(u_i, u_j) + \gamma_B(u_i, u_j) \le 1 \text{ for every} (u_i, u_j) \in E$$

Example 2.50 See Fig. 2.89.

2.19.2 Degree of a Vertex

Let $G = (V, E)$ be an intuitionistic fuzzy graph. Then the degree of a vertex v is denoted by the symbol $d(v)$ and is defined by

$$d(v) = (d_\mu(v), d_\gamma(v)) \text{where } d_\mu(v) = \sum_{u \neq v} \mu_B(u, v) \text{ and } d_\gamma(v) = \sum_{u \neq v} \gamma_B(u, v)$$

2.19.3 Order of the IFG

Let $G = (V, E)$ be an intuitionistic fuzzy graph. Then the order of the graph G is symbolically denoted by $O(G)$and is defined as

$$O(G) = (O_\mu(G), O_\gamma(G)) \text{where } O_\mu(G) = \sum_{u \in V} \mu_A(u) \text{ and } O_\gamma(G) = \sum_{u \in V} \gamma_A(u)$$

2.19.4 Size of the IFG

Let $G = (V, E)$ be an intuitionistic fuzzy graph. Then the size of the graph G is denoted by the symbol $S(G)$ and is defined as

$$S(G) = (S_\mu(G), S_\gamma(G)) \text{ where } S_\mu(G) = \sum_{u \neq v} \mu_B(u, v) \text{ and } S_\gamma(G) = \sum_{u \neq v} \gamma_B(u, v)$$

Example 2.51 Here, we have (Fig. 2.90)

$$O(G) = (O_\mu(G), O_\gamma(G)) = (1.6, 1.7) \text{ where } O_\mu(G) = 1.6 \text{ and } O_\gamma(G) = 1.7$$

$$S(G) = (S_\mu(G), S_\gamma(G)) = (1.0, 1.7) \text{ where } S_\mu(G) = 1.0 \text{ and } S_\gamma(G) = 1.7$$

$$d(a) = (d_\mu(a), d_\gamma(a)) = (0.4, 1.0) \text{ where } d_\mu(a) = 0.4 \text{ and } d_\gamma(a) = 1.0$$

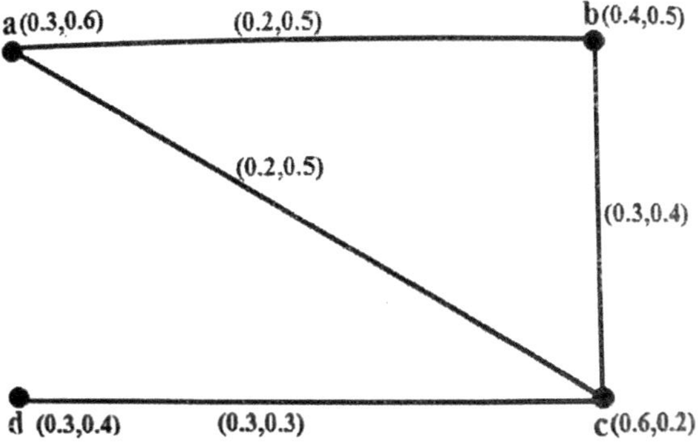

Fig. 2.90 Intuitionistic fuzzy graph

$$d(b) = \left(d_\mu(b), d_\gamma(b)\right) = (0.5, 0.9) \text{ where } d_\mu(b) = 0.5 \text{ and } d_\gamma(b) = 0.9$$

$$d(c) = \left(d_\mu(c), d_\gamma(c)\right) = (0.8, 1.2) \text{ where } d_\mu(c) = 0.8 \text{ and } d_\gamma(c) = 1.2$$

$$d(d) = \left(d_\mu(d), d_\gamma(d)\right) = (0.3, 0.3) \text{ where } d_\mu(d) = 0.3 \text{ and } d_\gamma(d) = 0.3$$

2.19.5 Strong Intuitionistic Fuzzy Graph

An intuitionistic fuzzy graph $G = (A, B)$ of the graph $G^* = (V, E)$ is said to be strong intuitionistic fuzzy graph if

$$\mu_B(u, v) = \min\{\mu_A(u), \mu_A(v)\}$$

$$\gamma_B(u, v) = \max\{\gamma_A(u), \gamma_A(v)\}, \text{ for all } (u, v) \in E$$

Example 2.52 See Fig. 2.91.

2.19.6 Complete Intuitionistic Fuzzy Graph

An intuitionistic fuzzy graph $G = (A, B)$ of the graph $G^* = (V, E)$ is said to be a complete intuitionistic fuzzy graph if (Fig. 2.92)

$$\mu_B(u, v) = \min\{\mu_A(u), \mu_A(v)\}$$

$$\gamma_B(u, v) = \max\{\gamma_A(u), \gamma_A(v)\}, \text{ for all edge } (u, v) \in E$$

Example 2.53 See Fig. 2.92.

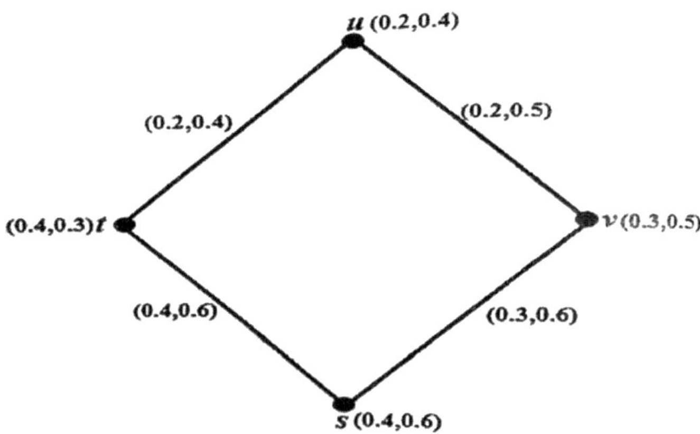

Fig. 2.91 Strong intuitionistic fuzzy graph

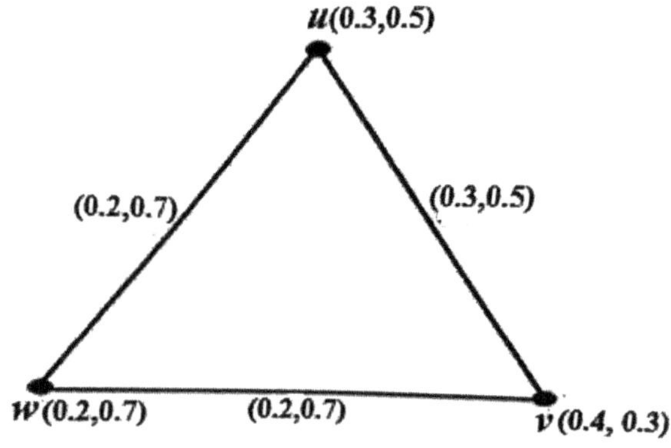

Fig. 2.92 Complete intuitionistic fuzzy graph

2.19.7 Path in an IFG

A path P in an intuitionistic fuzzy graph (IFG) G is defined as a sequence of distinct vertices $\{u_1, u_2, u_3, \ldots, u_n\}$ such that either one of the following conditions is satisfied:-

(i) $\mu_{B_{ij}} > 0$ and $\gamma_{B_{ij}} > 0$ for some i, j
(ii) $\mu_{B_{ij}} = 0$ and $\gamma_{B_{ij}} > 0$ for some i, j
(iii) $\mu_{B_{ij}} > 0$ and $\gamma_{B_{ij}} = 0$ for some i, j

2.19.8 Cycle in an IFG

A path P: $u_1, u_2, u_3, \ldots, u_n$ in an intuitionistic fuzzy graph (IFG) G is called a cycle if $u_1 = u_n$ and $n \geq 3$.

2.19.9 μ-Strength and γ-Strength of IFG

Let P: $u_1, u_2, u_3, \ldots, u_n$ be a path in an intuitionistic fuzzy graph (IFG) G. Then the μ-strength of the paths connecting any two vertices u_i, u_j is denoted by the symbol $\left(\mu_{B_{ij}}\right)^{\infty}$ and is defined as

$$\left(\mu_{B_{ij}}\right)^{\infty} = max\left\{\mu_B\left(u_i, u_j\right)\right\}$$

And the γ-strength of the paths connecting any two vertices u_i, u_j is denoted by the symbol $\left(\gamma_{B_{ij}}\right)^{\infty}$ and is defined as

$$\left(\gamma_{B_{ij}}\right)^{\infty} = min\left\{\gamma_B\left(u_i, u_j\right)\right\}$$

If the same edge possesses both the μ-strength and the γ-strength values, then it is the strength of the **strongest path** P and is denoted by the symbol S_P

$$S_P = \left(\left(\mu_{B_{ij}}\right)^{\infty}, \left(\gamma_{B_{ij}}\right)^{\infty}\right) \text{ for all i , j} = 1, 2, 3, \ldots, n.$$

2.19.10 Connected Intuitionistic Fuzzy Graph

An intuitionistic fuzzy graph (IFG) is said to be connected if any two vertices are joined by a path.

Exercise-2

1. Explain the notion of fuzzy graphs and a fuzzy subgraph of a graph with suitable examples.
2. What do you mean by a weighted graph? Could you give an example of it?
3. Define the terms adjacent nodes and adjacent edges in a fuzzy graph with proper examples of each.
4. Explain the terms fuzzy vertex set and the fuzzy edge set of a fuzzy graph $G = (\gamma, \lambda)$ with proper examples.
5. Give an example of a partial fuzzy subgraph.
6. Discus spanning subgraph with the help of an example.
7. Define the maximal partial fuzzy subgraph and could you give an example of it?
8. Discus-induced fuzzy subgraph with the help of an example.
9. What do you mean by a threshold fuzzy graph?
10. Give an example of an isolated vertex in a fuzzy graph.
11. What do you mean by the order and size of a fuzzy graph? Explain it with the help of examples.
12. Explain the concept of the degree of a node, minimum degree, maximum degree, and total degree of a node in a fuzzy graph.
13. Explain the concept of the degree of an edge, minimum degree, maximum degree, and the total degree of an edge in a fuzzy graph.
14. What do you mean by scalar cardinality in a fuzzy graph?
15. Distinguish between regular and irregular fuzzy graphs. Also, could you cite an example in support of your arguments?
16. Explain the complete fuzzy graph with the help of a proper example.
17. Explain the difference between totally irregular and strongly irregular fuzzy graphs. Could you give an example in support of your argument?
18. Establish the claim that strongly edges irregular fuzzy graphs need not be strongly edges totally irregular fuzzy graphs.
19. Show that if $G = (\gamma, \lambda)$ is a strongly edge irregular fuzzy graph, it implies a strongly edge totally irregular fuzzy graph.
20. Prove that a strongly edge totally irregular fuzzy graph implies a strongly edge irregular fuzzy graph.
21. If $G = (\gamma, \lambda)$ is both a strongly edge irregular fuzzy graph and a strongly edge totally irregular fuzzy graph, then it would not be a constant function.
22. If $G = (\gamma, \lambda)$ is a strongly edge irregular fuzzy graph, then G is neighborly edge irregular fuzzy graph. Prove this.
23. Show that a highly irregular fuzzy graph need not be a neighborly irregular.
24. Prove that a neighborly irregular fuzzy graph need not be highly irregular.
25. Define the terms μ-length and μ-distance in a fuzzy graph. Also, could you provide suitable examples for them?
26. Explain the notion of a path in a fuzzy graph. What do you mean by the strength of the path and the length of a path? Could you provide suitable examples for them?

27. Define the term cycle and fuzzy cycle in the fuzzy graph. Also, could you give proper examples for them?
28. Define tree, fuzzy tree, forest, and fuzzy forest and give examples.
29. Show that a complete fuzzy graph need not be a neighborly irregular fuzzy graph.
30. Define a connected fuzzy graph, and could you give an example?
31. Show that every complete fuzzy graph is a strong fuzzy graph but not conversely.
32. Give an example of a fuzzy forest having multiple strongest paths between nodes.
33. Define cut node and block in the fuzzy graph and give an example for each of them.
34. Show that (u, v) is a bridge $\Leftrightarrow (u, v)$ is not the weakest arc of any cycle.
35. Prove that $\delta(u, v)$ is a metric.
36. Give an example to establish that strongly edge irregular fuzzy graphs need not be strongly edge totally irregular.
37. Provide an example to show that a strong edge totally irregular fuzzy graph need not be a strong edge irregular fuzzy graph.
38. Explain the concept of intuitionistic fuzzy graphs and discuss its properties.
39. Define strong intuitionistic fuzzy graphs and complete intuitionistic fuzzy graphs and give examples of them.
40. Define the degree of a vertex, the order of the graph, and the size of the intuitionistic fuzzy graphs with suitable examples of them.

Objective-Type Questions and Answers on Graphs

1. A graph that contains only isolated vertex is called a

   (a) Regular graph
   (b) complete graph
   (c) null graph
   (d) pseudo graph

2. A graph in which all nodes are of equal degree is referred to as

   (a) Complete graph
   (b) simple graph
   (c) regular graph
   (d) directed graph

3. A path whose initial and terminal points are the same is called a

   (a) Simple path
   (b) open path
   (c) trivial path
   (d) closed path

4. The maximum number of arcs in a simple graph having n nodes is equal to
   (a) $\frac{n(n-1)}{2}$
   (b) $\frac{n(n+1)}{2}$
   (c) $n(n-1)$
   (d) n^2-1
5. A graph in which loops and multiple arcs are there, the graph is referred to as
   (a) Digraph
   (b) simple graph
   (c) multi-graph
   (d) pseudo graph
6. A node having degree 1 is called as
   (a) Vertex
   (b) isolated node
   (c) adjacent node
   (d) pendent vertex
7. If G is a tree with n vertices, then it has
   (a) $(n+1)$ edges
   (b) $(2n-1)$ edges
   (c) $(n-1)$ edges
   (d) n edges
8. The longest path in a tree is called as
   (a) Diameter of the tree
   (b) radius of the tree
   (c) eccentricity of the tree
   (d) ancestor
9. Distance from a vertex v of a graph G to the vertex farthest from v in G is known as
   (a) Diameter
   (b) center
   (c) radius
   (d) eccentricity
10. The number of branches in any spanning tree is referred to as
    (a) Rank
    (b) nullity
    (c) sibling
    (d) descendants

Answers

1. (c) **2.** (c) **3.** (d) **4.** (a) **5.** (d) **6.** (d) **7.** (c) **8.** (a) **9.** (d) **10.** (a)

Objective-Type Questions and Answers on Fuzzy Graphs

1. Any relation $R \subseteq S \times S = S^2$ on a set S can be regarded as defining a graph called as
   (a) Directed graphs
   (b) arc set S
   (c) node-set
   (d) none of these digraphs
2. A fuzzy relation $\lambda : S \times S \to [0, 1]$ is regarded as defining a
   (a) Graph
   (b) relation
   (c) weighted graph
   (d) fuzzy graph
3. The arc $(u, v) \in S \times S$ has weight $\lambda(u, v)$ belonging to
   (a) $\{0, 1\}$
   (b) $[0, 1]$
   (c) S
   (d) none of these
4. Any relation $R \subseteq S \times S$ on a set S can be regarded as defining a graph with
   (a) edge set R (b) arc set S (c) node-set R (d) none of these
5. A fuzzy relation $\lambda : S \times S \to [0, 1]$ is regarded as defining a fuzzy graph with
   (a) Edge set S
   (b) edge set $S \times S$
   (c) vertex set $S \times S$
   (d) none of these
6. A fuzzy graph is
   (a) a pair of functions
   (b) a graph
   (c) a fuzzy relation
   (d) none of these
7. In a fuzzy graph $G : (V, \gamma, \lambda)$
   (a) $\gamma : V \times V \to [0, 1]; \lambda : V \to [0, 1]$
   (b) $G : V \times V \to [0, 1]; \lambda : V \times V \to [0, 1]$
   (c) $\gamma : V \to [0, 1]; \lambda : V \to [0, 1]$
   (d) $\gamma : V \to [0, 1]; \lambda : V \times V \to [0, 1]$

8. In a fuzzy graph $G = (V, \gamma, \lambda)$
   (a) γ represent the membership value of an arc
   (b) λ represent the membership value of a node
   (c) γ represent the membership value of a node
   (d) none of these
9. The underlying graph of the fuzzy graph $G = (\gamma, \lambda)$ is denoted by
   (a) $G = (\gamma^*, \lambda^*)$
   (b) $G^* = (\gamma^*, \lambda^*)$
   (c) $G^* = (\gamma^*, \lambda)$
   (d) $G^* = (\gamma, \lambda^*)$
10. Our fuzzy relation involved in defining a fuzzy graph is
   (a) Antireflexive
   (b) antisymmetric
   (c) symmetric
   (d) none of these
11. Our fuzzy relation involved in defining fuzzy graph is
   (a) reflexive
   (b) antisymmetric
   (c) $\in$-reflexive
   (d) none of these
12. Our fuzzy relation involved in defining fuzzy graph is
   (a) reflexive
   (b) symmetric
   (c) transitive
   (d) none of these
13. All arcs are regarded as…pairs of nodes
   (a) Ordered
   (b) unordered
   (c) collection
   (d) none of these
14. The consecutive pairs (u_{i-1}, u_i) of the path are called as
   (a) Nodes
   (b) edges
   (c) span
   (d) none of these

15. A sequence of distinct nodes $x_0, x_1, x_2, \ldots, x_n$ for $1 \leq i \leq n$ and $n \geq 0$ in a fuzzy graph $G : (\gamma, \lambda)$ is called a path if

    (a) $\lambda(x_{i-1}, x_i) \geq 0$
    (b) $\lambda(x_{i-1}, x_i) \leq 0$
    (c) $\lambda(x_{i-1}, x_i) < 0$
    (d) $\lambda(x_{i-1}, x_i) > 0$

16. If u and v are nodes in a fuzzy graph and if they are connected by using a path, then the strength of the path is defined as

    (a) Strength of the strongest edge
    (b) Strength of the weakest edge
    (c) Strong arc
    (d) None of these

17. If u and v are nodes in a fuzzy graph $G : (\gamma, \lambda)$ and if they are connected using a path, then the strength of the path is defined as

    (a) $\bigvee_{i=1}^{n} \lambda(x_{i-1}, x_i)$
    (b) $\bigcup_{i=1}^{n} \lambda(x_{i-1}, x_i)$
    (c) $\bigwedge_{i=1}^{n} \lambda(x_{i-1}, x_i)$
    (d) $\bigcap_{i=1}^{n} \lambda(x_{i-1}, x_i)$

18. The length of a path p for $n > 0$ is called the number of … contained in the path.

    (a) Number of arcs
    (b) number of nodes
    (c) number of edges
    (d) none of these

19. The relation of connectedness between nodes in a fuzzy graph is a

    (a) Tolerance relation
    (b) fuzzy ordering relation
    (c) transitive relation
    (d) max–min transitive

20. A binary relation denoted by $R(S, S)$, or $R(S^2)$is a subset of $S \times S = S^2$ is referred to as

    (a) Undirected graph
    (b) digraph
    (c) directed graph
    (d) complete graph

21. According to the notion of the fuzzy graph, which pair of nodes is an arc?

    (a) (x_{i-1}, x_{i+2})
    (b) (x_{i+1}, x_i)
    (c) (x_i, x_{i+1})
    (d) (x_{i+1}, x_{i+3})

22. In a classical graph, each path is a … path with strength …

    (a) Strongest with strength 0
    (b) weakest with strength 1
    (c) Strongest with strength 1
    (d) weakest with strength 0

23. A fuzzy graph $G : (V, \gamma, \lambda)$ is connected if and only if

    (a) $\lambda^{\infty}(u, v) < 0 \; \forall \, u, v \in V$
    (b) $\lambda^{\infty}(u, v) \leq 0 \; \forall \, u, v \in V$
    (c) $\lambda^{\infty}(u, v) > 0 \; \forall \, u, v \in V$
    (d) $\lambda^{\infty}(u, v) \geq 0 \; \forall \, u, v \in V$

24 The degree of membership of a … arc is defined as the strength of the arc.

    (a) Weakest
    (b) strongest
    (c) connected
    (d) none of these

25. In a fuzzy graph $G : (V, \gamma, \lambda)$, if $S \subseteq V$, then the scalar cardinality of S is

(a) $\sum_{v \in S} \lambda(v)$

(b) $\sum_{v \in S} \gamma(v)$

(c) $\sum_{v \in S} \lambda(u, v)$

(d) none of these

26. A node v is said to be isolated in a fuzzy graph $G : (V, \gamma, \lambda)$ if

(a) $\gamma(u, v) = 0, \forall\, u = v$
(b) $\lambda(u, v) \neq 0, \forall\, u \neq v$
(c) $\gamma(u, v) \neq 0, \forall\, u = v$
(d) $\lambda(u, v) = 0, \forall\, u \neq v$

27. The order of a fuzzy graph $G : (V, \gamma, \lambda)$ is denoted by o(G) and is given by

(a) $\sum_{v \in V} \lambda(v)$

(b) $\sum_{v \in V} \gamma(v)$

(c) $\sum_{(u, v) \in V \times V} \lambda(u, v)$

(d) none of these

28. The size q of a fuzzy graph $G : (V, \gamma, \lambda)$ is denoted by S(G) and is given by

(a) $\sum_{v \in V} \lambda(v)$

(b) $\sum_{v \in V} \gamma(v)$

(c) $\sum_{(u, v) \in V \times V} \lambda(u, v)$

(d) none of these

29. Degree of a node v in a fuzzy graph G is denoted by $d(v)$ and is given by

(a) $\sum_{u \neq v} \lambda(u, v)$

(b) $\sum_{v \in V} \lambda(v)$

(c) $\sum_{v \in V} \gamma(v)$
(d) none of these

30. If each vertex has the same degree k in a fuzzy graph, then G is said to be
    (a) k-Complete fuzzy graph
    (b) k-spanning graph
    (c) k-totally regular
    (d) k-regular fuzzy graph

31. $\delta(x, y)$ represents a metric in a
    (a) Undirected graphs
    (b) digraph
    (c) fuzzy cycle
    (d) none of these

32. A path p in a fuzzy graph $G : (\gamma, \lambda)$ is called a cycle if $v_0 = v_n$ and n
    (a) $n > 3$
    (b) $n \leq 3$
    (c) $n <$
    (d) $n \geq 3$

33. A fuzzy cycle contains more than one
    (a) Weakest arc
    (b) strong arc
    (c) bridge
    (d) none of these

34. In a fuzzy graph $G = (\gamma, \lambda)$, the strongest path joining any two vertices u and v is that which is the path that has the strength
    (a) $\gamma^{\infty}(u, v)$
    (b) $\lambda^{\infty}(u, v)$
    (c) $\gamma(u, v)$
    (d) none of these

35. In a fuzzy graph $G = (\gamma, \lambda)$, the minimum degree of G is denoted by
    (a) $\Delta(G)$
    (b) $d(G)$
    (c) $\delta(G)$
    (d) none of these

36. If u and v are two nodes of a fuzzy graph G, then the degree of the arc(u, v) is given by

(a) $d_G(u) + d_G(v) - 2\lambda(u, v)$
(b) $d_G(u) + d_G(v) - \lambda(u, v)$
(c) $d_G(u) + d_G(v) + 2\lambda(u, v)$
(d) $d_G(u) + d_G(v) + \lambda(u, v)$

37. If the deletion of the arc (u, v) reduces the strength of connectedness between some pair of nodes, then the arc is a

(a) Cutnode
(b) block
(c) bridge
(d) none of these

38. If $n = 0$, then $\ell(p)$ is equal to

(a) 1
(b) ∞
(c) 0
(d) none of these

39. Every complete fuzzy graph is a

(a) Connected fuzzy graph
(b) strong fuzzy graph
(c) regular fuzzy graph
(d) none of these

40. If there is at most one strongest path between any two vertices of the graph G, then G is a

(a) Fuzzy tree
(b) forest
(c) tree
(d) fuzzy forest

Answers

(a & d) 2. (c & d) 3. (b) 4. (a) 5. (b) 6. (a, b & c) 7. (d) 8. (c) 9. (b) 10. (c) 11. (a) 12. (a, b & c) 13. (b) 14. (b) 15. (d) 16. (b) 17. (c & d) 18. (b) 19. (c) 20. (b & c) 21. (c) 22. (d) 23. (c) 24. (a) 25. (b) 26. (d) 27. (b) 28. (c) 29. (a) 30. (d) 31. (a) 32. (d) 33. (a) 34. (b) 35. (c) 36. (a) 37. (c) 38. (c) 39. (b) 40. (d)

True/False Type Statements and Answers

1. Fuzzy graph is an expression of a fuzzy relation.
2. Usually, fuzzy graphs are frequently expressed as a fuzzy matrix.
3. Let $R = \{[(x, y), R(x, y)] : (x, y) \in X \times Y\}$ be a fuzzy relation on fuzzy sets A and B. The degrees of membership of the elements of the related fuzzy sets define the "strength" in the respective edges of the graph.

4. Let $R(x, y) \leq \{[(x, y), R(x, y)] : (x, y) \in X \times Y\}$ is a fuzzy relation on fuzzy sets A and B. The degrees of membership of the corresponding pairs in the relation define the "flows" or "capacities" of the graph.
5. The flows in the edges of the graph can exceed the flows in the respective nodes (pair of edges).
6. Let E be the classical set of nodes. A fuzzy graph is then defined as $\mathrm{G}(\mathrm{x_i}, \mathrm{x_j}) = \{((\mathrm{x_i}, \mathrm{x_j}), \mathrm{G}(\mathrm{x_i}, \mathrm{x_j})) : (\mathrm{x_i}, \mathrm{x_j}) \in \mathrm{E} \times \mathrm{E}\}$
7. The fuzzy relation $R(x, y) \leq \min\{A(x), B(y)\}, \forall(x, y) \in X \times Y$ always represents a *weighted graph* or fuzzy graph, where the *arc* $(x, y) \in X \times Y$ has *weight* $R(x, y) \in [0, 1]$, where A and B are fuzzy sets on X and Y, respectively.
8. Elements of the fuzzy relation R are called the *"arcs"* of the fuzzy graph depicted by this fuzzy relation.
9. The *arc* $(x, y) \in X \times Y$ has *weight* $R(x, y) \in [0, 1]$ in the fuzzy relation R.
10. Two distinct edges, λ_1 and λ_2, are said to be adjacent if they have a common end-vertex. That is if $\lambda_1 \cap \lambda_2 \neq \emptyset$.
11. An unordered pair of nodes are considered edges.
12. A fuzzy graph is a pair $G = (\gamma, \lambda)$ where γ is a fuzzy subset of V, and λ is asymmetric fuzzy relation on γ.
13. If $\mathrm{G}(\mathrm{x_i}, \mathrm{x_j}) \leq \mathrm{H}(\mathrm{x_i}, \mathrm{x_j}) \forall (\mathrm{x_i}, \mathrm{x_j}) \in \mathbb{N} \times \mathbb{N}$, then $H(\mathrm{x_i}, \mathrm{x_j})$ is called a fuzzy subgraph of $G(\mathrm{x_i}, \mathrm{x_j})$.
14. $H(\mathrm{x_I}, \mathrm{x_j})$ is called a spans graph of the fuzzy graph $G(\mathrm{x_I}, \mathrm{x_j})$ if the node sets of $H(\mathrm{x_I}, \mathrm{x_j})$ and $R(\mathrm{x_I}, \mathrm{x_j})$ are unequal.
15. If $H(\mathrm{x_I}, \mathrm{x_j})$ and $R(\mathrm{x_I}, \mathrm{x_j})$ the node sets differ only in their arc weights, we say that $H(\mathrm{x_I}, \mathrm{x_j})$ spans the graph $R(\mathrm{x_I}, \mathrm{x_j})$.
16. A sequence of distinct nodes, $x_0, x_1, \ldots, x_n$, such that $\forall(x_i, x_{i+1}), G(x_i, x_{i+1}) > 0$ is referred to as a path in a graph $\mathrm{G}(\mathrm{x_i}, \mathrm{x_j})$.
17. $\max\{G(x_i, x_{i+1})\}$ for all nodes contained in the path, is called the strength of the path.
18. For $n > 0$, the number of nodes contained in the path is termed the length of the path.
19. If in the path of a fuzzy graph $\mathrm{G}(\mathrm{x_i}, \mathrm{x_j})$, $x_0 = x_n$ for $n > 3$, the path is called circular.
20. Two nodes joined by a path are referred to as connected nodes.
21. A fuzzy graph without cycles is summoned as a forest.
22. A forest is not an acyclic fuzzy graph.
23. A connected fuzzy forest is convoked as a tree.
24 The relation of connectedness is always transitive.
25. In a path $p = x_0, x_1, x_2, x_3 \ldots, x_n$, the term $L(p) = \sum_{i=1}^{n} \frac{1}{\mu(x_i, x_{i+1})}$ is usually convened as μ-length.
26. The shortest path between two graph nodes is candidly called the distance between the nodes.

27. The μ-distance $d(x_i, x_j)$ between two nodes, x_i and x_j, is the smallest μ-length of any path from x_i to x_j, where $x_i, x_j \in G(x_i, x_j)$.
28. A fuzzy graph $G = (V, \gamma, \lambda)$ is a triple consisting of a nonempty set V together with a pair of functions γ and λ, where γ is called the fuzzy vertex set of G and λ is called the fuzzy edge set of G.
29. A fuzzy graph $H = (V, \mu, \beta)$ is said to be a partial fuzzy subgraph of $G = (V, \gamma, \lambda)$ if $\mu \subseteq \gamma$ *and* $\beta \subseteq \lambda$.
30. The symbol $\langle V^1 \rangle$ usually denotes the fuzzy subgraph induced by V_1.
31. A vertex x in a fuzzy graph $G = (\gamma, \lambda)$ is called an isolated vertex if $\lambda(x, y) \neq 0$ for all $x = y$.
32. The set of all neighbors of x is denoted by $N(x)$.
33. $p = \circ(G) = \begin{matrix} \sum \gamma(x) \\ x \in V \end{matrix}$ gives the order of the fuzzy graph $G = (\gamma, \lambda)$.
34. The size of the fuzzy graph $G = (\gamma, \lambda)$ is defined as $q = S(G) = \begin{matrix} \sum \lambda(x, y) \\ x, y \in V. \end{matrix}$.
35. The symbol $d(u)$ or $d_G(u)$ denotes the degree of a node "u" in a fuzzy graph G.
36. The symbol $\delta_u(G)$ denotes the minimum degree of a node $''u''$ in a fuzzy graph $G = (\gamma, \lambda)$ and is defined as $\delta_u(G) = \begin{matrix} max \\ u \in V \end{matrix} \{d(u)\}$.

37 The total degree of the vertex $'u'$ is denoted by $td_G(u)$ and is defined as $td_G(u) = d_G(u) - \gamma(u), \forall (u, v) \in V$.

38 The degree of an edge (u, v), in the fuzzy graph $G = (\gamma, \lambda)$ is denoted by $d_G(u, v)$ and is defined as $d_G(u, v) = d_G(u) + d_G(v) - 2\lambda(u, v)$.

39 nThe total degree of an edge (u, v) in the fuzzy graph $G = (\gamma, \lambda)$ is denoted by $td_G(u, v)$ and is defined as $td_G(u, v) = d_G(u) + d_G(v) - \lambda(u, v)$.

40. If each node has the same degree k in a fuzzy graph $G = (\gamma, \lambda)$, it is called a regular or $k-$ regular fuzzy graph.

Answers

1. (T); 2. (T); 3. (T); 4. (T); 5. (F); 6. (T); 7. (T); 8. (F); 9. (T); 10. (T); 11. (T); 12. (T); 13. (F); 14. (F); 15. (T); 16. (T); 17. (F); 18. (T); 19. (F); 20. (T); 21. (T); 22. (F); 23. (T); 24. (T); 25. (T); 26. (T); 27. (T); 28. (T); 29. (T); 30. (F); 31. (F); 32. (T); 33. (T); 34. (T); 35. (T); 36. (F); 37. (F); 38. (T); 39. (T); 40. (T).

Chapter 3
Fuzzy Measures, Possibility, and Necessity

This chapter introduces the notions of possibility theory, which is intimately attached to fuzzy set theory. It also incorporates belief and plausibility measures, basic probability assignment, and Dempster's Rule of Combination. It also covers possibility theory, necessity measures, and other related aspects.

3.1 Introduction

The theory of fuzzy sets furnishes us instinctive gratifying procedure of illustrating some uncertainty. Let us think about an adjudicator of a criminal proceeding who is not clear about the culpability or ignorance of the suspect. In this situation, the unpredictability, vagueness, or uncertainty appears divergent; the collection of upright people and the set of the person responsible for the offense are supposed to have a highly well-defined borderline. I think the matter of interest is not with the level up to which the appellant is blameable of the offense but with the level to which the proofs exhibit his belongingness in either the classical set of a culpable person or a virtuous person. The flawless proof would represent full membership in one and one of these two sets. Although, our proof could be more ideal. In general, some ambiguity is normally overcome. To portray such ambiguity or uncertainty, we usually allocate a value to each particular classical set to which the element in inquiry perhaps belong. Indeed, this value might represent the degree of confirmation, evidence, or certainty of the element's belongingness in the concerned set. The abovementioned representation of uncertainty is usually called a "*fuzzy measure*." It provides us with a tool to assess an acquainted possibility theory and will allow us to amplify the dissimilarity between the theory of fuzzy sets and probability theory.

The central concept of the theory of fuzzy measure is the fuzzy measure which was introduced independently by *Sugeno* in 1974 in the context of fuzzy measure. There are several classes of fuzzy measures, including plausibility or belief measures;

M. K. Singh, *Applied Fuzzy Mathematics*, Forum for Interdisciplinary Mathematics,
https://doi.org/10.1007/978-981-97-3257-9_3

possibility or necessity measures; and probability measures, a subset of classical measures.

3.2 Fuzzy Measures

Let X be the universal set and $\mathcal{P}(X)$ be a nonempty family of subsets of X. Then a function g from $\mathcal{P}(X)$ to [0, 1] that is $g : \mathcal{P}(X) \to [0, 1]$ or $g : 2^X \to [0, 1]$ is said to be a fuzzy measure on $(X, \mathcal{P}(X))$ if and only if the following conditions are satisfied:

(g_1) $g(\emptyset) = 0$ and $g(X) = 1$, known as ***boundary conditions***.

(g_2) For all $A, B \in \mathcal{P}(X)$, if $A \subseteq B$, then $g(A) \leq g(B)$, called as condition of ***monotonicity***.

(g_3) For any increasing sequence $A_1 \subset A_2 \subset A_3 \subset \ldots\ldots\ldots$ *in* $\mathcal{P}(X)$, if $\bigcup_{i=1}^{\infty} A_i \in \mathcal{P}(X)$, then, $\lim_{n\to\infty} g(A_i) = g\left(\bigcup_{i=1}^{\infty} A_i\right)$, this condition is perceived as ***continuity from below***.

(g_4) For any decreasing sequence $A_1 \supset A_2 \supset A_3 \supset \ldots\ldots\ldots$ *in* $\mathcal{P}(X)$, if $\bigcap_{i=1}^{\infty} A_i \in \mathcal{P}(X)$, then, $\lim_{n\to\infty} g(A_i) = g\left(\bigcap_{i=1}^{\infty} A_i\right)$, this condition is usually referred to as ***continuity from above***.

3.2.1 Illustrations of the Requirements

(i) From the boundary conditions, we inferred that despite our evidence, we know that the element under consideration certainly does not belong to the void or null set and certainly belongs to the universal set. As the null set does not contain any element, the element of our interest also does not belong to it. On the contrary, the universal set contains all elements under our thought processes, so it must contain our element of concern.

(ii) The condition (g_2) expresses that the evidence of the belongingness of an element in a collection must be as strong as the evidence that the element belongs to any sub-collection of that collection. When one knows with a high degree of confidence that the element is in our collection, then it certainly belongs to a more extensive collection, which contains the former collection in it as a member, with a high degree of certainty but in no case, it will be lower in value.

(iii) The conditions (g_3) and (g_4) are pertinent only for an unlimited universal set. Indeed, it can be discarded for the finite universal set. It expresses that for an unlimited nested subset of X, which converges to the set A, where $A = \bigcup_{i=1}^{\infty} A_i$ for an increasing sequence and $A = \bigcap_{i=1}^{\infty} A_i$ for the decreasing sequence, and in this situation, the sequence of numbers $g(A_1), g(A_2), g(A_3) \ldots\ldots$ should converge to the number $g(A)$ viz. g is needed to be a continuous function. These two conditions can be contemplated as essentials for ***consistency***. One can estimate the value of $g(A)$ in two distinct ways, first, by applying the function g to the limit of A_i for $i \to \infty$ or secondly, as the limit of $g(A_i)$ for $i \to \infty$, both giving the same value.

Remarks 3.1

(i) The functions that fulfill or appease conditions (g_1), (g_2) and either (g_3) or (g_4) are usually called as *semi-continuous fuzzy measures*. Semi-continuous fuzzy measures imply either continuous from below or continuous from above.

(ii) The notion of fuzzy measure can be generalized by extending the range of the function g from the closed interval $[0, 1]$ to the set of non-negative real numbers and by excluding the boundary condition $g(X) = 1$.

(iii) Fuzzy measures are regarded as generalizations of probability or crisp measures in the comprehensive sense.

(iv) The number $g(A)$ allocated to a set $A \in \mathcal{P}(X)$ by a fuzzy measure g conveys the *total feasible evidence* that a given element of X, whose depiction is inadequate in some regard, belongs to A.

(v) For any two sets A and $B \in \mathcal{P}(X)$, we have $A \cap B \subseteq A$ and $A \cap B \subseteq B$. Then, on account of the monotonicity of fuzzy measures implies that every fuzzy measure g fulfills the inequality

$$g(A \cap B) \leq min\left[g(A), G(B)\right]$$

Similarly, for any three sets A, B, and $A \cup B \in \mathcal{P}(X)$, we have $A \subseteq A \cup B$ and $B \subseteq A \cup B$. Then, on account of the monotonicity of fuzzy measures suggested that every fuzzy measure g fulfills the inequality $g(A \cup B) \geq max\left[g(A), g(B)\right]$

(vi) The function g describing fuzzy measure can be elucidated on fuzzy sets instead of classical sets.

3.3 Evidence Theory

This section deals with the relationships between probability theory and evidence theory; to a limited extent, it also deals with the relationship between a possibility theory founded on classical sets and fuzzy set theory. All such theories are related under an umbrella called labeled monotone measures. A *monotone measure* describes the vagueness in assigning an element, say x, to two or more classical sets. The main difference between a monotone measure and a fuzzy set on a universe of elements is that, in the former, the imprecision is in assigning an element to one of two or more classical sets. In the latter, the imprecision is in the prescription of the boundaries of a set.

There are particular kinds of monotone measures. A form associated with preconceived notions is known as a *belief measure,* and a form associated with possible or *plausible* information is termed a plausibility measure. Particular forms of belief measures and plausibility measures are called *certainty* and *possibility measures.*

Evidence theory has been used to display uncertainty in expert systems, especially in the vicinity of diagnostics. The theory of belief functions also referred to as evidence theory or Dempster–Shafer theory is a framework of reasoning with uncertainty connected to other frameworks such as possibility, probability, plausibility, etc. Evidence theory is planted on two dual non-additive measures, known as belief measures and plausibility measures.

3.4 Belief Measures

Let X be a finite universal set. A belief measure is a quantity denoted symbolically by "*Bel*" is a function from $\mathcal{P}(X)$to[0, 1]. That is,

$$Bel : \mathcal{P}(X) \to [0, 1]$$

such that the following requirements are satisfied:

$$(b_1)Bel(\emptyset) = 0, Bel(X) = 1$$

$$(b_2)\text{ For all } A_1, A_2, A_3, \ldots, A_n \in \mathcal{P}(X),\ 0 \le Bel(A) \le 1$$

$$Bel(A_1 \cup A_2 \cup A_3 \cup \ldots \cup A_n) \ge \sum_j Bel(A_j)$$

$$+ \sum_{j<k} Bel(A_j \cap A_k) \cdots + (-1)^{n+1} Bel(A_1 \cap A_2 \cap A_3 \cap \ldots \cap A_n)$$

for all possible families of subsets of X. Due to the inequality described above, belief measures are also known as ***superadditive***.

3.4.1 *Properties of Belief Measures*

(i) For every $A \in \mathcal{P}(X)$, $Bel(A)$ is elucidated as *"the degree of belief"* that a given element of X is contained in the set A.
(ii) Let us assume that the sets $A_1, A_2, A_3, \ldots, A_n \in \mathcal{P}(X)$ are pairwise disjoint; the inequality (b_2) needs that the degree of belief related to the union of the sets should not be smaller than the aggregate of the degrees of belief relating to respective sets.
(iii) It is the weaker version of the additive property of probability measures.
(iv) Probability measures are peculiar kinds of belief measures for which (b_2) always holds good.
(v) Let $A,\ B \in \mathcal{P}(X)$ and $A \subseteq B$. Let $C = A - B$. Then $A \cup C = B$ and $A \cap C = \phi$.

For $n = 2$, we have from condition (b_2) that

$$\begin{aligned} Bel(A \cup C) = BelB &\geq Bel(A) + Bel(C) - Bel(A \cap C) \\ &\geq Bel(A) + Bel(C) - Bel(\phi) \\ &\geq Bel(A) + Bel(C),\ \text{since } Bel(\phi) = 0 \end{aligned}$$

As a consequence $Bel(B) \geq Bel(A)$.

(vi) Let $A_1 = A$ and $A_2 = \overline{A}$ in condition (b_2) for $n = 2$, we have two disjoint sets A and $\overline{A}$. Then, we have from condition (b_2)

$$\begin{aligned} &Bel\big(A \cup \overline{A}\big) \geq Bel(A) + Bel\big(\overline{A}\big) - Bel\big(A \cap \overline{A}\big) + (-1)^{2+1} Bel\big(A \cap \overline{A}\big) \\ &\Rightarrow Bel(X) \geq Bel(A) + Bel\big(\overline{A}\big) - Bel(\phi) + (-1)^{2+1} Bel(\phi) \\ &\Rightarrow Bel(X) \geq Bel(A) + Bel\big(\overline{A}\big),\ \text{since } Bel(\phi) = 0 \\ &\Rightarrow 1 \geq Bel(A) + Bel\big(\overline{A}\big),\ \text{since } Bel(X) = 1 \end{aligned} \tag{3.1}$$

This states that a lack of belief in $x \in A$ does not imply a strong belief in $x \in \overline{A}$.

Note 1 A belief measure expresses the degree of support, or evidence, for a collection of elements defined by one or more of the classical sets existing on the power set of a universe.

3.5 Belief and Plausibility Measure

Accompanied by each belief measure, there is a plausibility measure, denoted by *Pl* and defined by the equation

$$Pl(A) = 1 - Bel\big(\overline{A}\big),\ \forall\ A \in \mathcal{P}(X) \tag{3.2}$$

The plausibility measure of the collection A is defined as the "complement of the Belief of the complement of A" and conversely.

Similarly, we will have

$$\begin{aligned} Pl(\overline{A}) &= 1 - Bel\left(\overline{\overline{A}}\right), \forall A \in \mathcal{P}(X) \\ &= 1 - Bel(A), \forall A \\ \therefore\ Bel(A) &= 1 - Pl(\overline{A}), \forall A \end{aligned} \tag{3.3}$$

Therefore, belief and plausibility measures are mutually dual (**Shafer**, 1976). However, plausibility measures can also be explicated without the support of belief measures.

Since belief measures are quantities that measure the degree of support for a collection of elements defined by one or more of the classical sets existent in a universe, it is quite possible that the belief measure of some set A plus the belief measure of $\overline{A}$ will not be equal to unity. The total belief, or evidence, for all elements or sets in a universe of discourse, is always equal to 1. The belief measure is a probability when the sum $BelA + Bel\overline{A}$ equals 1. That is, the evidence supporting the set A can be explained probabilistically. The difference between $BelA + Bel\overline{A}$ and *one* is usually called *ignorance.* That is,

$$ignorance = 1 - (BelA + Bel\overline{A}).$$

If ignorance equals 0, we have a case where the evidence can be described in terms of probability measures.

3.6 Plausibility Measure (Independent of the Notion of Belief Measure)

A plausibility measure, denoted symbolically by Pl, is defined as a function from $\mathcal{P}(X)$ to [0, 1], i.e., $Pl : \mathcal{P}(X) \to [0, 1]$ such that

$(p_1) Pl(\phi) = 0$ and $pl(X) = 1$

$$(p_2) Pl(A_1 \cap A_2 \cap A_3 \cap \ldots \cap A_n) \le \sum_{j} Pl(A_j) - \sum_{j<k} Pl(A_j \cup A_k) + \cdots$$
$$+ (-1)^{n+1} Pl(A_1 \cup A_2 \cup A_3 \cup \ldots \cup A_n)$$

for all possible families of subsets of X. Due to the inequality (p_2), plausibility measures are also referred to as "*subadditive.*" If the universal set X is infinite, the plausibility measures should be continuous from below.

Let $A_1 = A$ and $A_2 = \overline{A}$ in condition (p_2). Then for $n = 2$, we have

$$\begin{aligned} &Pl(A \cap \overline{A}) \le Pl(A) + Pl(\overline{A}) - Pl(A \cup \overline{A}) + (-1)^{2+1} Pl(A \cup \overline{A}) \\ &\Rightarrow Pl(\emptyset) \le Pl(A) + Pl(\overline{A}) - Pl(X) + (-1)^{2+1} Pl(X) \\ &\Rightarrow 0 \le Pl(A) + Pl(\overline{A}) - 1 - 1 \\ &\Rightarrow 1 \le Pl(A) + Pl(\overline{A}) - 1 \Rightarrow Pl(A) + Pl(\overline{A}) \ge 1 \end{aligned} \tag{3.4}$$

Note 2

(i) A form associated with plausible information is also known as the plausibility measure.
(ii) Particular belief and plausibility measure forms are usually called *certainty* and *possibility measures*, respectively.
(iii) A form associated with preconceived notions is known as a belief measure.

3.7 Basic Probability Assignment

The notion of belief and plausibility measures can be easily described by a function given by.

$m : \mathcal{P}(X) \to [0, 1]$, such that

$$(a_1)\ m(\phi) = 0 \text{ and}$$

$$(a_2) \quad \sum_{A \in P(x)} m(A) = 1$$

This measure is called a *basic evidence assignment (bea)* has been a function and is usually referred to as basic probability assignment (bpa). For every set $A \in \mathcal{P}(X)$, the value $m(A)$ enunciates the percentage to which all feasible and pertinent evidence assists the claim that a specific element of X, whose depiction of suitable qualities is inadequate, belongs to the set A. This value, $m(A)$, concerns set A only and not to any subsets of A. If there exists some extra evidence aiding the assertion that the element is in a subset B of A, it should be demonstrated by other value $m(B)$.

Remarks 3.2 Though (a_2) is similar to probability distribution functions, there is a basic difference between them because probability assignment is defined over $\mathcal{P}(X)$, whereas probability distribution functions are defined over X.

3.7.1 Some Attributes of Basic Probability Assignment

From the notion of probability assignment, we infer that

(i) For probability assignment, it is not mandatory that $m(X) = 1$.
(ii) It is not necessary that $A \subseteq B \Rightarrow m(A) \leq m(B)$ for the probability assignment.
(iii) There does not exist any relationship between $m(A)$ and $m(\overline{A})$.

So, from the above, we conclude that basic probability assignments are not fuzzy measures. Although, for all $A \in \mathcal{P}(X)$, it is possible to determine the belief measure and plausibility measure, when the following formulas gives the basic probability assignment, m.

$$Bel(A) = \sum_{B \subseteq A} m(B) \tag{3.5}$$

The association between $m(A)$ and $Bel(A)$ demonstrated by Eq. (3.5) has the following interpretation: although $m(A)$ specifies the degree of evidence or belief that the component under consideration belongs precisely to the set A, whereas $Bel(A)$ describes the complete evidence or belief that the component belongs to A as well as to the other subsets of A.

and

$$Pl(A) = \sum_{A \cap B \neq \phi} m(B) \tag{3.6}$$

respectively.

The plausibility measure $Pl(A)$ given by Eq. (3.6) illustrates the complete evidence or belief that the component under consideration belongs to set A or any of its subsets. The supplementary evidence or belief attached with sets also conjoins with A. Hence, for all $A \in \mathcal{P}(X)$, we have on combining Eqs. (3.1) and (3.4), it can be shown that

$$Pl(A) \geq Bel(A) \tag{3.7}$$

Equation (3.7) states that for whatever evidence supports set A, its plausibility measure is always at least as great as its belief measure.

If a belief measure $Bel(A)$ is given, then for all $A \in \mathcal{P}(X)$, the corresponding probability assignment m can be determined by the formula given by

$$m(A) = \sum_{B \subseteq A} (-1)^{|A-B|} Bel(B) \tag{3.8}$$

By the account of Eqs. (3.5), (3.6), and (3.8), each of the three functions m, Bel, and Pl is adequate to find out the other two.

3.7.2 Focal Element

If X is the universal set, each set $A \in \mathcal{P}(X)$, for which we have $m(A) > 0$, is commonly termed as the focal element of m. Focal elements are subsets of X upon which the feasible evidence is spotlighted.

3.7.3 Body of Evidence

Let the universal set X is finite. Then, m can be described completely by a menu of its focal elements A with the corresponding value $m(A)$. In such a situation, the pair $(\mathcal{F}, m)$, where $\mathcal{F}$ and m stand for a set of focal elements and the associated basic assignment, is frequently labeled as a body of evidence.

3.7.4 Total Ignorance

Let $m : \mathcal{P}(X) \to [0, 1]$. That is, m is a basic probability assignment, then the total ignorance can be enunciating in terms of probability assignment by

$$m(X) = 1 \text{ and } m(A) = 0.$$

We can express the facts described above by saying that it is well-known that the element is in the universal set, but there is no evidence about its whereabouts in any subset of X.

From Eq. (3.8), we have $m(B) = 0$ as $B \subseteq A \Rightarrow m(A) = 0$ and consequently, we will get $Bel(A) = 0$ and $Bel(X) = 1$ for all $A \neq X$. Thus, the expression of total ignorance concerning belief measure is precisely the same, i.e., $Bel(X) = 1$ and $Bel(A) = 0 \; \forall \; A \neq X$. However the expression of total ignorance in terms of associated plausibility measure is distinct as $Pl(\phi) = 0$ and $Pl(A) = 1$ for all $A \neq \phi$. (by Eq. 3.2)

3.8 Dempster's Rule of Combination

Let us assume that X is the universal set and $\mathcal{P}(X)$ is the power set. Further, suppose that m_1 and m_2 be two basic probability assignments on $\mathcal{P}(X)$, providing the evidence in some circumstances given by independent experts. Then, the joint basic probability assignment is usually denoted by $m_{1\cdot 2}$ and is demonstrated by the formula given by

$$m_{1\cdot 2}(A) = \frac{\sum\limits_{B \cap C = A} m_1(B) \cdot m_2(C)}{1 - K} \quad \forall A \neq \phi \text{ and } m_{1\cdot 2}(\phi) = 0 \tag{3.9a}$$

where

$$K = \sum_{B \cap C = \phi} m_1(B) \cdot m_2(C) \tag{3.9b}$$

The aforesaid formula represented by Eq. (3.9a) is often called *Dempster's rule of combination.*

From the above rule, we infer that the degree of evidence $m_1(B)$ and $m_2(C)$ obtained from the first and second experts, respectively, that spotlight on set $B \in \mathcal{P}(X)$ and $C \in \mathcal{P}(X)$ are combined by considering the product $m_1(B) \cdot m_2(C)$ that spotlights on $B \cap C$, the process adopted in calculating $m_{1 \cdot 2}(A)$ is precisely matching with the joint probability distribution for independent marginal distributions.

Note 3 The sum of products $m_1(B) \cdot m_2(C)$ for all focal elements B of m_1 and all focal elements C of m_2 such that $B \cap C = \phi$ is equal to $1 - K$ for acquiring the normalized basic assignment $m_{1 \cdot 2}$, we should divide each of these products by $1 - K$.

3.9 Possibility Theory

The name theory of possibility was coined by *Zadeh* (in 1978) and was inspired by Gaines and *Kohout* (1975). In *Zadeh's* view, possibility distributions provided graded semantics to natural language statements.

An offshoot of evidence theory that concerns solitary with bodies of evidence having nested focal elements is usually referred to as possibility theory. In possibility theory, distinctive complements of belief and plausibility measures are labeled as necessity and possibility measures sequentially. Suppose we have a collection of some or all of the subsets on the power set of a universe of discourse, which has the attributes $A_1 \subset A_2 \subset A_3 \subset \cdots \subset A_n$. With this attribute, these sets are said to be *nested*, according to **Shafer**, 1976. When the elements of a set or universe have pieces of evidence nested, we say that belief measures ($Bel(A_i)$), and the plausibility measures $Pl(A_i)$, represents a **consonant body of evidence**. By the word *consonant*, we signify that the evidence allocated to the numerous elements of the set or subsets on the universe of discourse does not dispute; the evidence is at liberty of dissension or dissonance.

The condition that focal elements be nested or consonant curbs belief and plausibility measures in the course of action described by the subsequent results by (**Klir and Folger**, 1988).

Theorem 3.1 Given that $(\mathcal{F}, m)$ be a finite, nested body of evidence. Then, for all $A, B \in \mathcal{P}(X)$, the associated belief and plausibility measures have the following fascinating properties:

(i) $Bel(A \cap B) = \min\{Bel(A), Bel(B)\}$
(ii) $Pl(A \cup B) = \max\{Bel(A), Bel(B)\}$

Proof

(i) According to our assumption, the focal elements in $\mathcal{F}$ are nested. Therefore, subsets of $\mathcal{F}$ may be linearly ordered by the subset relation. Suppose that $\mathcal{F} = \{A_1, A_2, A_3, \ldots, A_n\}$ and are such that $A_i \subseteq A_j$ for all $i < j$. Let us choose two arbitrary subsets A and B of X. Le ti_1 be the greatest integer i such that $A_i \subseteq A$ and let i_2 be the greatest integer i such that $A_i \subseteq B$.

Then, $A_i \subseteq A$ and $A_i \subseteq B$ if and only if, $i \leq i_1$ and $i \leq i_2$ respectively. Furthermore, $A_i \subseteq A \cap B$ if and only if $i \leq \min\{i_1, i_2\}$. In consequence

$$
\begin{aligned}
Bel(A \cap B) &= \bigcap_{i=1}^{\min\{i_1, i_2\}} m(A_i) \\
&= \min\left[\bigcap_{i=1}^{i_1} m(A_i), \ \bigcap_{i=1}^{i_2} m(A_i)\right] \\
&= \min\{Bel(A), Bel(B)\}
\end{aligned}
$$

It states that the belief measure of the intersection of two sets is the smaller of the belief measures of the two sets.

(b) Let us suppose that (i) preserves. We know that for each belief measure *Bel*, there is associated a plausibility measure *Pl* and is related to the equation

$$
\begin{aligned}
Pl(A) &= 1 - Bel(\overline{A}), \forall A \in \mathcal{P}(X) \\
\therefore Pl(A \cup B) &= 1 - Bel(\overline{A \cup B}), \forall A, B \in \mathcal{P}(X) \\
&= 1 - Bel(\overline{A} \cap \overline{B}) \\
&= 1 - min\{Bel(\overline{A}), Bel(\overline{B})\}, \text{ using (i)} \\
&= \max\{1 - Bel(\overline{A}), 1 - Bel(\overline{B})\} \\
&= \max\{Pl(A), Pl(B), \forall A, B \in \mathcal{P}(X)
\end{aligned}
$$

It states that the plausibility measure of the union of two sets is the larger of the plausibility measure of the two sets.

The theory of possibility is a mathematical theory for dealing with certain forms of uncertainty and is an alternative to the theory of probability. It uses measures of possibility and necessity between 0 and 1, ranging from possible to impossible and necessary to unnecessary. It was propounded by *Zadeh* in 1978. *Dubois and Prade* further contributed a lot to its development.

Let us assume that the symbols "Nec and Pos" represent necessity and possibility measures. In some literature, *consonant belief* and *plausibility measures* are referred to as *necessity* (also denoted by η and *possibility*, denoted by π) Then, $\forall\ A, B \in \mathcal{P}(X)$ by Theorem 3.1, we will have the following relations of possibility theory.

$$\begin{aligned} Nec(A \cap B) &= \min\{Nec(A), Nec(B)\} \\ i.e., \quad \eta(A \cap B) &= \min\{\eta(A), \eta(B)\} \end{aligned} \tag{3.10}$$

$$\begin{aligned} Pos(A \cup B) &= \max\{Pos(A), Pos(B)\} \\ i.e., \quad \pi(A \cup B) &= \max\{\pi(A), \pi(B)\} \end{aligned} \tag{3.11}$$

If $A \subset B$, then $\pi(A) \leq \pi(B)$.

Now it is possible to consider the concept of necessity and possibility measures for arbitrary universal sets (of course, not finite).

3.10 Necessity Measure

Let X be the universal set and $\mathcal{P}(X)$ be a nonempty family of subsets of X. Then a fuzzy measure '*Nec*' on $\langle X, \mathcal{P}(X)\rangle$ be called a necessity measure if and only if

$$Nec\left(\bigcap_{k \in K} A_k\right) = \inf_{k \in K} Nec(A_k)$$

where K is an arbitrary index set and for any family $\{A_k : k \in K\} \in \mathcal{P}(X)$ and is such that $\bigcap_{k \in K} A_k \in \mathcal{P}(X)$.

3.11 Possibility Measure

Let Pos stands for a fuzzy measure on $\langle X, \mathcal{P}(X)\rangle$. Then, '$Pos$' be termed as a possibility measure if and only if

$$Pos\left(\bigcup_{k \in K} A_k\right) = \sup_{k \in K} Pos(A_k)$$

where K is an arbitrary index set and for any family $\{A_k : k \in K\} \in \mathcal{P}(X)$ and is such that $\bigcup_{k \in K} A_k \in \mathcal{P}(X)$.

Because necessity measures are distinctive belief measures and possibility measures are distinctive plausibility measures, they appease properties (3.1), (3.2), (3.3), and (3.4). In consequence, we will have

$$Nec(A) + Nec(\overline{A}) \leq 1 \tag{3.12}$$

$$Pos(A) + Pos(\overline{A}) \geq 1 \tag{3.13}$$

$$Nec(A) = 1 - Pos(\overline{A}) \tag{3.14}$$

From Eqs. (3.10) and (3.11), we at once get

$$Nec(A \cap \overline{A}) = min\{Nec(A), Nec(\overline{A})\} \Rightarrow min\{Nec(A), Nec(\overline{A})\} = Nec(\phi) = 0$$

As necessity measures are special belief measures, therefore $Nec(\phi) = Bel(\phi) = 0$

$$\therefore\ min\{Nec(A), Nec(\overline{A})\} = 0 \tag{3.15}$$

$$Pos(A \cup \overline{A}) = max\{Pos(A), Pos(\overline{A})\} \Rightarrow max\{Pos(A), Pos(\overline{A})\} = Pos(X) = 1$$

As possibility measures are special plausibility measures, therefore $Pos(X) = Pl(X) = 1$

$$\therefore\ max\{Pos(A),\ Pos(\overline{A})\} = 1 \tag{3.16}$$

Theorem 3.2 To show that for all $A \in \mathcal{P}(X)$, any necessity measure and the associated possibility measure fulfill the ensuing implications:

(i) $Nec(A) > 0 \Rightarrow Pos(A) = 1$;
(ii) $Pos(A) < 1 \Rightarrow Nes(A) = 0$.

Proof

(i) For some $A \in \mathcal{P}(X)$, Nec and Pos stands for necessity measure and the associated possibility measure. Suppose that $Nec(A) > 0$.

From Eq. (3.15), we have that $min\{Nec(A), Nec(\overline{A})\} = 0$ and by assumption $Nec(A) > 0$. Therefore, it follows that

$$Nec(\overline{A}) = 0 \tag{3.17}$$

Again, from Eq. (3.14), we have $Pos(\overline{A}) = 1 - Nec(A)$.
On replacing A by $\overline{A}$ in the above result, we get.
$Pos(\overline{\overline{A}}) = 1 - Nec(\overline{A}) \Rightarrow Pos(A) = 1 - 0$, using (3.17)

$$\therefore\ Pos(A) = 1$$

(ii) It is given that $Pos(A) < 1$ for some $A \in \mathcal{P}(X)$. From Eq. (3.16), we have $max\{Pos(A), Pos(\overline{A})\} = 1$. Therefore, it follows that $Pos(\overline{A}) = 1$. Again, from Eq. (3.14), we have $Nec(A) = 1 - Pos(\overline{A}) \Rightarrow Nec(A) = 1 - 1 = 0$.

3.11.1 Possibility Distribution Function

Let Pos be the possibility measure on $\mathcal{P}(X)$; then a function $r : X \to [0, 1]$ be termed as a possibility distribution function associated with $'Pos'$ if

$$r(x) = Pos(\{x\}) \forall x \in X \tag{3.18}$$

Now, we'll talk about a Paramount result proposed by **Klir and Folger**, 1988.

Theorem 3.3 Let X be a finite universal set and $\mathcal{P}(X)$ be a power set on it. Then, every possibility measure on $\mathcal{P}(X)$ can be uniquely determined by a possibility distribution function $r : X \to [0, 1]$ given by.

$$Pos(A) = \max_{x \in A} r(x) \forall A \in \mathcal{P}(X) \tag{3.19}$$

Proof We shall demonstrate this result by using the mathematical induction principle on set A's cardinality. Assume that $|A| = 1$. Then, we have.

$$|A| = \{x\}, \text{ where } x \in X$$

and hence the result is true for $n = 1$.

We now presume that the result (3.18) is true for $|A| = n - 1$, and let $A = \{x_1, x_2, x_3, \ldots, x_n\}$. We know that $Pos(A \cup B) = max\{Pos(A), Pos(B)\}$. Therefore,

$$\begin{aligned}
Pos(A) &= Pos\{x_1, x_2, x_3, \ldots\ldots, x_{n-1}, x_n\} \\
&= max\{Pos(\{x_1, x_2, x_3, \ldots\ldots, x_{n-1}\}), Pos(\{x_n\})\} \\
&= max\big[max\{Pos(\{x_1, x_2, x_3, \ldots\ldots, x_{n-2}\}), Pos(\{x_{n-1}\}), Pos(\{x_n\})\}\big] \\
&- - - - - - - - - - - - - \\
&= max\{Pos(\{x_1\}), Pos(\{x_2\}), Pos(\{x_3\}), \ldots\ldots., Pos(\{x_{n-1}\}), Pos(\{x_n\})\} \\
&= \max_{x \in A} r(x)
\end{aligned}$$

It follows that the result is valid for $|A| = n$.

Therefore, by the principle of mathematical induction, the result is valid for all coming values of n.

Note 4 In case the universal set X is not finite, then Eq. (3.19) will take the form

$$Pos(A) = \sup_{x \in A} r(x) \, \forall \, A \in \mathcal{P}(X)$$

3.11.2 Possibility Distribution

Let $X = \{x_1, x_2, x_3, \ldots, x_{n-1}, x_n\}$ be a finite universal set, and r be a possibility distribution function defined on X. Then, the n-tuple $r = \{r_1, r_2, r_3, \ldots, r_n\}$ where $r_i = r(x_i) \forall \, x_i \in X$ is labeled as the possibility distribution associated with the function r. The number of components present in the possibility distribution function r is usually referred to as its **length**.

3.12 Lattice of Possibility Distributions

For ordering possibility distribution, it is appropriate to do it in such a way that $r_i > r_j$ when $i < j$. Let the set of all ordered possibility distribution of length n be denoted by $n_{\mathcal{R}}$. Further, we suppose that $\mathcal{R} = \bigcup_{n \in N} n_{\mathcal{R}}$.

Assume that we have two possibility distributions, 1_r and $2_r \in n_{\mathcal{R}}$ for some $n \in \mathbb{N}$ and are such that

$$1_r = \langle 1_{r_1}, 1_{r_2}, \ldots, 1_{r_n} \rangle \in n_{\mathcal{R}}$$

and

$$2_r = 2_{r_1}, 2_{r_2}, \ldots, 2_{r_n} \in n_{\mathcal{R}}$$

Let us define ordering in $n_{\mathcal{R}}$ by setting

$$1_r \leq 2_r \text{ if and only if } 1_{r_i} \leq 2_{r_i} \forall \, i \in \mathbb{N}_n$$

Obviously, this ordering is reflexive, antisymmetric, and transitive and, consequently, it is a partial ordering on $n_{\mathcal{R}}$. This ordering on $n_{\mathcal{R}}$ forms a lattice whose meet ($\wedge$) and join ($\vee$) are designated individually by

$$i_r \vee j_r = \langle max(i_{r_1}, j_{r_1}), \; max(i_{r_2}, j_{r_2}), \ldots, max(i_{r_n}, j_{r_n}) \rangle$$

and

$$i_r \wedge j_r = \langle min(i_{r_1}, j_{r_1}),\ min(i_{r_2}, j_{r_2}), \ldots, min(i_{r_n}, j_{r_n}) \rangle$$

for all $i_r, j_r \in n_{\mathcal{R}}$. Then, for each $n \in \mathbb{N}$, $\langle n_{\mathcal{R}}, \leq \rangle$ is usually called a *lattice of possibility distributions of length n*.

According to **Klir and Folger**, every possibility measure can be described by the n-tuple, denoted as a basic distribution

$$m = (\mu_1, \mu_2, \mu_3, \ldots, \mu_n) \tag{3.20}$$

where

$$\sum_{i=1}^{n} \mu_i = 1 \tag{3.21}$$

$\forall \mu_i \in [0, 1]$ and $\mu_i = m(A_i)$. Obviously, sets A_i are nested as is compulsory for all consonant bodies of evidence. From Eq. (3.6) and the relations

$$r_i = r(x_i) = \pi(x_i) = Pl(x_i) \tag{3.22}$$

it can be shown (**Klir and Folger**) that

$$r_i = \sum_{k=i}^{n} \mu_k = \sum_{k=i}^{n} m(A_k) \tag{3.23}$$

Or in an iterative form

$$\mu_i = r_i - r_{i+1}, \tag{3.24}$$

where $r_{n+1} = 0$, by protocol.

Evidently, on putting $i = 1, 2, 3, \ldots, n$ in Eq. (3.24) and adding, we get

$$\begin{aligned}\mu_1 + \mu_2 + \mu_3 + \cdots + \mu_n &= r_1 - r_2 + r_2 - r_3 + r_3 - r_4 + \cdots + r_n - r_{n+1} \\ &= r_1 - r_{n+1} = r_1 - 0 = r_1\end{aligned}$$

i.e., $r_1 = \mu_1 + \mu_2 + \mu_3 + \cdots + \mu_n$

Thus, Eq. (3.24) gives rise to a set of equations of the form

$$\begin{aligned} r_1 &= \mu_1 + \mu_2 + \mu_3 + \cdots + \mu_n, \\ r_2 &= \quad\ \mu_2 + \mu_3 + \cdots + \mu_n, \\ r_3 &= \quad\quad\ \ \mu_3 + \cdots + \mu_n, \\ &= \ldots\ldots\ldots\ldots\ldots\ldots \end{aligned}$$

$$r_n = \qquad \mu_n.$$

The nesting of physical elements is a crucial physical characteristic of a body of evidence. In this light, let us contemplate the ensuing exemplification of physical nesting.

Example 3.1 Let seven nodes in a computer network X be designated a_1 to a_7 and illustrated by boxes. Out of these nodes, one is creating some difficulty. The company wishes to know about the node which is causing the difficulty in communications from the network expert. The expert aggregates these nodes into sets, as the associated table displays. Let the expert's primary distribution is described in the third column, and the possibility distribution is displayed in the last column and is calculated from Eq. (3.23).

Solution: The substantial importance of this nesting can be depicted as follows: In the opinion of the network experts, node a_1 is instigating the difficulty, because this node has up-to-date hardware and is an experimental CPU. Due to this very reason, the expert has put down the highest attention on set A_1. The following set with nonzero belief (supporting evidence) is A_2, which consists of the union a_1 or a_2; the expert has the meager belief that node a_1 or a_2 is creating the hindrance. Furthermore, node a_2 also has new hardware, but it has a credible CPU. The succeeding set, which has a nonzero belief is that A_4 has nodes a_1 or a_2 or a_3 or a_4. The expert has little belief that this set causes the hindrance. The expert thinks that a_3 or a_4 has credible hardware and CPUs. The adjacent set with a nonzero belief is A_6 having nodes a_1 or a_2 or a_3 or a_4 or a_5 or a_6. The expert has a moderately more vigorous belief that this particular set is creating hindrance compared to A_4 but still less than the foremost set A_1. The rationale is that two brand new programmers are utilizing these nodes for testing communications software. The decider set with evidence is the union of all seven nodes. The expert has the modest belief that this set is the cause of hindrance because node a_7 is generally disgusted Table 3.1.

Here, we notice that the first element of any ordered possibility distribution ρ_1 always equals unity; that is, $\rho_1 = 1$. This truth is endorsed by Eq. (3.21). The lowest possibility distribution of length n has the pattern $\rho = (1, 0, 0, 0, \ldots, 0)$, where there are $(n-1)$ zeros after a value of unity in the distribution. The connected primary

Table 3.1 Expert aggregates of focal elements

Set A	Aggregation of Focal Elements	$\mu_n = \mu(A_n)$	ρ_i
A_1	a_1	0.4	1
A_2	$a_1 \cup a_2$	0.2	0.6
A_3	$a_1 \cup a_2 \cup a_3$	0	0.4
A_4	$a_1 \cup a_2 \cup a_3 \cup a_4$	0.1	0.4
A_5	$a_1 \cup a_2 \cup a_3 \cup a_4 \cup a_5$	0	0.3
A_6	$a_1 \cup a_2 \cup a_3 \cup a_4 \cup a_5 \cup a_6$	0.2	0.3
A_7	$a_1 \cup a_2 \cup a_3 \cup a_4 \cup a_5 \cup a_6 \cup a_7$	0.1	0.1

distribution may be of the pattern $m = (1, 0, 0, 0, \ldots, 0)$. In this situation, there may be only one focal element with evidence, and it will have all the evidence. This state of affairs displays *perfect evidence*. In this state of affairs, no uncertainty is involved at all (Fig. 3.1).

On the other hand, the most significant possibility distribution of length n has the pattern $\rho = (1, 1, 1, 1, \ldots, 1)$, wherein the distribution of all values is unity. The connected primary distribution may be of the pattern $m = (0, 0, 0, 0, \ldots, 1)$. In this situation, all the evidence is on the focal element consisting of the whole universe. That is, $A_n = a_1 \cup a_2 \cup a_3 \cup a_4 \cup a_5 \cup a_6 \cup \ldots \cup a_n$. As a consequence, we know nothing regarding any particular focal element in the universe, excluding the universal set. This state of affairs is usually referred to as *total ignorance*. Primarily, the immense the possibility distribution, the less specific be the evidence and the more oblivious for drawing any inference.

Because possibility measures are some kind of plausibility measures and necessity measures are variants of belief measures, it is possible to connect possibility measures and necessity measures to probability measures. From the equation $Pl(A) = p(A) = Bel(A) \rightarrow p(A) + p(\overline{A}) = 1$, we infer that when all the evidence in a universe dwells entirely on the singletons of the universe, the belief and plausibility measures reduce to probability measures. In the same footsteps, it can be exhibited that the plausibility measures move toward the probability measures from an upper bound and that the belief measures move toward the probability measures from a lower bound. According to **Yager and Filev** (1994), the result is a range around the probability measure, that is,

$$Bel(A) \leq p(A) \leq Pl(A)$$

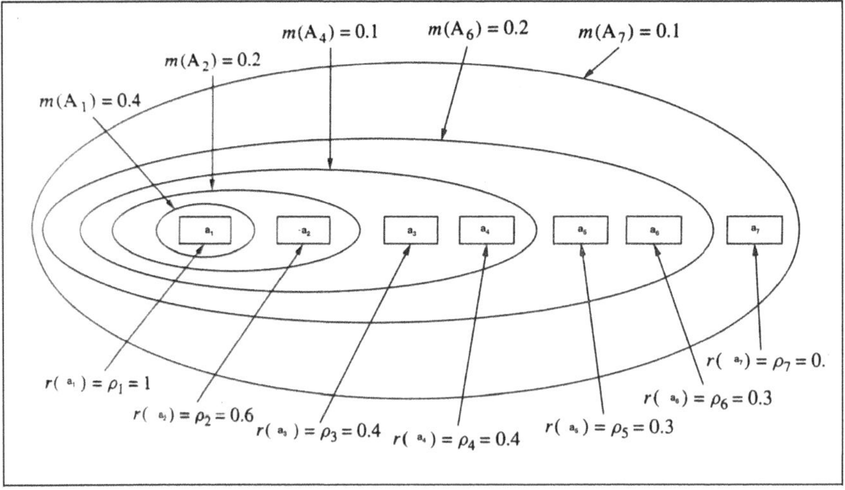

Fig. 3.1 Significance of nesting of nodes diagram

The graphical elucidation of the evidence theory evolved from *Dempster's and Shafer's* is furnished by *Zadeh* (1986) in what is known as the **ball-box analogy**. The following problem is due to *Zadeh*.

Example 3.2 Let the President of the country C believe that a submarine-launched cruise missile M is in the continental shelf (territorial water) of C. The President convened n experts say $E_1, E_2, E_3, \ldots, E_n$ in this field to offer their opinions on the position of the submarine-launched cruise missile to him. All the n experts estimated the position of the submarine-launched cruise missile M. Let the feasible positions be $P_1, P_2, \ldots, P_m, \ldots, P_n$, where $m \leq n$. Evidently,$P_i \subseteq C$, that is, the feasible positions are the subsets of the territorial waters C. Be a little bit more clear, let the experts $E_1, E_2, E_3 \ldots, E_n$ opine that M is in $P_1, P_2, \ldots, P_m$, and the experts $E_{m+1}, E_{m+2}, \ldots, E_{n-1}, E_n$ opines that M is not located in the territorial waters of C. That is $P_{m+1} = P_{m+2} = \cdots = P_{n-1} = P_n = \phi$. It means that there are $(m - n)$ experts who think that M is not situated in the territorial waters. Now, the President of the country queries, "Is M in a subset of S of our territorial waters?" The adjoining figure depicts the possible location regions P_i, and region $S's$ inquiry positions of interest (Fig. 3.2).

Let us assume that the i th expert suggested that S_i is the feasible position. Now, we have the following two rules:

(i) $S_i \subset S \Rightarrow$ that it is certain that $M \in S$.
(ii) $S_i \cap S \neq \phi \Rightarrow$ that it is possible that $M \in S$.

Here, we observe that certitude is incorporated in the set of possibilities; certitude inferred possibility. Additionally, we assume that the President accumulates the assessments of his experts by averaging. As a consequence, let k out of n experts choice for rule number (i), then the average certitude $= {}^{k}/_{n}$; however, if l out of n (where $l \geq k$) experts choice for rule number (ii), then the average possibility $= {}^{l}/_{n}$. Eventually, suppose the intelligence of those experts who think there is no submarine in any place in the territorial waters is neglected. In that case, the average certitude and

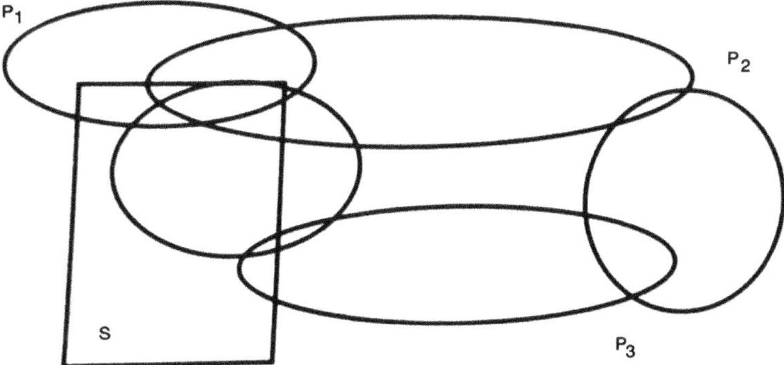

Fig. 3.2 Territorial waters

possibility will be $^k/_m$, and $^l/_m$ respectively (where $m \leq n$). Neglecting the opinion of those experts whose possible location regions P_i is the empty set conforms to the normalization equation (3.9b) in Dempster's and Shafer's evidence theory. However, Dempster's rule of combination may cause unreasonable outcomes due to the normalization issue. According to Zadeh (1984), normalization hurls out evidence that states firmly that the object under contemplation does not exist, that is, void set ϕ or core set.

3.13 Possibility Distributions as Fuzzy Sets

So far, we have seen that nested belief structures are usually referred to as consonants. One of the underlying properties of consonant belief structures is that their plausibility measures are nothing but possibility measures. According to **Dubois and Prade** (1988), possibility measures are equivalent to fuzzy sets. In this equivalence, the membership degree of an element x conforms to the plausibility of the singleton comprising that element, x. In other words, a consonant belief structure is identical to a fuzzy set F of X, where $F(x) = pl(\{x\})$.

Next, we will see a problem in which equating consonant belief structures with fuzzy sets is equivalent to combining two consonant belief functions by Dempster's rule of combination does not lead to a consonant result. It was shown by **Yager** in 1993. So Dempster's rule is a conjunction operation, the intersection of two fuzzy sets elucidated as consonant belief structures is not a sound fuzzy set (i.e., a consonant structure). The following example is due to *Yager*.

Example 3.3 Let us consider a universe of discourse consisting of five singletons, given as

$$U = \{a_1, a_2, a_3, a_4, a_5\}$$

Suppose that two experts, E_1 and E_2, furnish the evidence Table 3.2.

The above-appearing table contributes to the experts' degrees of belief regarding particular subsets of the universe of discourse, U. Let us assume that $X_1 = \{a_1, a_2, a_3\}, X_2 = U = \{a_1, a_2, a_3, a_4, a_5\}, Y_1 = \{a_3, a_4, a_5\}$, and $Y_2 = U = \{a_1, a_2, a_3, a_4, a_5\}$ be the focal elements; that is, subsets of U. Let $m_1(X_i)$and $m_2(Y_i)$ are the basic probability assignments.

Table 3.2 Degrees of belief regarding particular subsets of the universe of discourse

Focal elements	Expert 1,$m_1(X_i)$	Expert 2,$m_2(Y_i)$
$X_1 = \{a_1, a_2, a_3\}$	0.8	
$X_2 = U = \{a_1, a_2, a_3, a_4, a_5\}$	0.4	
$Y_1 = \{a_3, a_4, a_5\}$		0.9
$Y_2 = U = \{a_1, a_2, a_3, a_4, a_5\}$		0.3

Since $X_1 \subset X_2$ and $Y_1 \subset Y_2$, we have two consonant (nested) belief structures represented by X and Y. Under Dempster's rule of combination and on making use of Eqs. (3.9a) and (3.9b), we have for any set Z on the universe of discourse U as

$$m_{1\cdot 2}(Z) = \frac{1}{1-K} \sum_{X_i \cap Y_j = Z} m_1(X_i) \cdot m_2(Y_i)$$

and

$$K = \sum_{X_i \cap Y_j = \phi} m_1(X_i) \cdot m_2(Y_i)$$

Because there are two focal elements in each of the two experts' belief structures, there will have $2^2 = 4$(four) belief structures in the event of combined evidence. Let it be denoted by m. Here, we observe that for these data, we will get a value of $k = 0$, because no intersection exists between the focal elements of X and Y that result in the empty set ϕ. For instance, the intersection between X_1 and Y_1 is the singleton a_3 *i.e.*, $\{a_3\}$. Thus, we have

$$Z_1 = X_1 \cap Y_1 = \{a_3\} \quad \Rightarrow m(Z_1) = 0.8 \times 0.9 = 0.72$$

$$Z_2 = X_1 \cap Y_2 = \{a_1, a_2, a_3\} \quad \Rightarrow m(Z_2) = 0.8 \times 0.3 = 0.24$$

$$Z_3 = X_2 \cap Y_1 = \{a_3, a_4, a_5\} \quad \Rightarrow m(Z_3) = 0.4 \times 0.9 = 0.36$$

$$Z_4 = X_2 \cap Y_2 = \{a_1, a_2, a_3, a_4, a_5\} = U \Rightarrow m(Z_4) = 0.4 \times 0.2 = 0.08$$

For the focal elements Z_i, we observe that $Z_1 \subset Z_2 \subset Z_4$ and $Z_1 \subset Z_3 \subset Z_4$. However, we do not have $Z_2 \subset Z_3$ or $Z_3 \subset Z_2$. It follows that the integrated case is not consonant. In other words, the integrated case is partially nested.

Everyone in this society wanted to be an artist of any genre at some point in their life; however, walking in the shoes of *Picasso* and *Monet* is not everyone's cup of team. Sadly, not every artist becomes famous, and very few who can make a name for themselves are often forgotten in the sand of time. The following example concerns an illustrious painter *Raja Ravi Varma,* who made India proud due to his skill and talent. He is also regarded as the *"Father of Indian Modern Art."*

Example 3.4 An ancient painting has come across that entirely coincides with the paintings of *Raja Ravi Varma.* This finding causes several questions concerning the dignity of the painting.

Undertake the succeeding three questions:

(i) Whether the painting that comes across is an authentic painting by *Raja Ravi Varma?*

(ii) Whether the painting that comes across is a material of the painting by *Raja Ravi Varma's* followers?
(iii) Whether the painting that comes across is a replicate?

Let U be the set of all paintings that is, our universe of discourse-which consists of the set of all paintings by *Raja Ravi Varma,* the collection of all paintings by the followers of *Raja Ravi Varma,* and the collection of all replicates of *Raja Ravi Varma* paintings, respectively. Let V, F, and R represent the subsets of the universal set U and represent the set of all paintings by *Raja Ravi Varma*, the set of all paintings by his followers, and the set of all replicas.

If a team of two specialists carefully looks over the paintings and supply us with essential assignments m_1 and m_2 provided in the adjoining table that would be great. These are the degrees of evidence which each specialist presumed by the inspection and, simultaneously, which support the numerous claims that the concerned painting's connection to one of the four sets of our interest. For instance, $m_1(V \cup F) = \cdot 15$ is the degree of evidence supplied by the first specialist that the painting of interest was made by *Raja Ravi Varma* himself or that the painting was made by one of his followers? For calculating the total evidence, Bel_1 and Bel_2 in each set, we can use Eq. (3.5) specified in the following Table 3.3.

We will get the joint primary assignment $m_{1.2}$ on administering Dempster's rule to m_1 and m_2.

Which has also been displayed in the above Table 3.3. First we must determine the normalization factor $1 - K$ to determine the values of $m_{1.2}$. On applying

$$K = \sum_{V \cap F = \phi} m_1(V) \cdot m_2(F)$$

$$\Rightarrow K = m_1(V) \cdot m_2(F) + m_1(V) \cdot m_2(R) + m_1(R) \cdot m_2(F \cup R) + m_1(F) \cdot m_2(V) + m_1(F) \cdot m_2(R) + m_1(F) \cdot m_2(V \cup R)$$

Table 3.3 Combination of degrees of evidence from two independent specialist

	Specialist-1		Specialist-2		Combined evidence	
Focal Elements	m_1	Bel_1	m_2	Bel_2	$m_{1\cdot 2}$	$Bel_{1\cdot 2}$
V	0.05	0.05	0.15	0.15	0.21	0.21
F	0	0	0	0	0.01	0.01
R	0.05	0.05	0.05	0.05	0.09	0.09
$V \cup F$	0.15	0.2	0.05	0.2	0.12	0.34
$V \cup R$	0.1	0.2	0.2	0.4	0.2	0.5
$F \cup R$	0.05	0.1	0.05	0.1	0.06	0.16
$V \cup F \cup R$	0.6	1	0.5	1	0.31	1

$$+ m_1(R) \cdot m_2(V) + m_1(R) \cdot m_2(F) + m_1(R) \cdot m_2(V \cup F)$$
$$+ m_1(V \cup F) \cdot m_2(R) + m_1(V \cup R) \cdot m_2(F) + m_1(F \cup R) \cdot m_2(V)$$

$$= 0.05 \times 0 + 0.05 \times 0.05 + 0.05 \times 0.05 + 0 \times 0.15$$
$$+ 0 \times 0.15 + 0 \times 0.2 + 0.05 \times 0.15 + 0.05 \times 0 + 0.05 \times 0.05$$
$$+ 0.15 \times 0.05 + 0.1 \times 0 + 0.05 \times 0.15$$

$$= 0.0025 + 0.0025 + 0.0075 + 0.0025$$
$$+ 0.0075 + 0.0075 = 0.0300 = 0.03$$

Thus, the normalization factor $1 - K = .97$. Now, the values of $m_{1.2}$ can be determined by

$$m_{1 \cdot 2}(V) = \frac{1}{1-K} \sum_{V \cap F = Z} m_1(V) \cdot m_2(F)$$

$$m_{1 \cdot 2}(V) = \frac{1}{0.97} [m_1(V) \cdot m_2(V) + m_1(V) \cdot m_2(V \cup F)$$
$$+ m_1(V) \cdot m_2(V \cup R) + m_1(V)$$
$$\cdot m_2(V \cup F \cup R) + m_1(V \cup F) \cdot m_2(V) + m_1(V \cup F)$$
$$\cdot m_2(V \cup R) + m_1(V \cup R) \cdot m_2(V) + m_1(V \cup R) \cdot m_2(V \cup F)$$
$$+ m_1(V \cup F \cup R) \cdot m_2(V)]$$
$$= \frac{1}{0.97} [0.05 \times 0.15 + 0.05 \times 0.05 + 0.05 \times 0.2 + 0.05 \times 0.5$$
$$+ 0.15 \times 0.15 + 0.15 \times 0.2 + 0.1 \times 0.15 + 0.1 \times 0.05 + 0.6 \times 0.15]$$
$$= \frac{1}{0.97} [0.0075 + 0.0025 + 0.01 + 0.025 + 0.0225 + 0.03$$
$$+ 0.015 + 0.005 + 0.09]$$
$$= \frac{1}{0.97} \times 0.2075 = 0.2139 = 0.21$$

$$m_{1 \cdot 2}(F) = \frac{1}{0.97} [m_1(F) \cdot m_2(F) + m_1(F) \cdot m_2(V \cup F)$$
$$+ m_1(F) \cdot m_2(F \cup R) + m_1(F)$$
$$\cdot m_2(V \cup F \cup R) + m_1(V \cup F) \cdot m_2(F)$$
$$+ m_1(V \cup F) \cdot m_2(F \cup R) + m_1(F \cup R)$$
$$\cdot m_2(F) + m_1(F \cup R) \cdot m_2(V \cup F) + m_1(V \cup F \cup R) \cdot m_2(F)]$$
$$= \frac{1}{0.97} [0 \times 0 + 0 \times 0.15 + 0 \times 0.05 + 0 \times 0.5 + 0.15$$
$$\times 0.05 + 0.05 \times 0 + 0.05 \times 0.15 + 0.05 \times 0.05 + 0.6 \times 0$$

$$= \frac{1}{0.97} \times [0.0075 + 0.0075 + 0.0025]$$
$$= \frac{1}{0.97} \times 0.0175 = 0.0180 = 0.01$$

$$m_{1\cdot 2}(V \cup R) = \frac{1}{0.97}[m_1(V \cup R) \cdot m_2(V \cup R) + m_1(V \cup R) \cdot m_2(V \cup F \cup R)$$
$$+ m_1(V \cup F \cup R) \cdot m_2(V \cup R)]$$
$$= \frac{1}{0.97}[0.1 \times 0.2 + 0.1 \times 0.5 + 0.6 \times 0.2]$$
$$= \frac{1}{0.97} \times 0.19 = 0.1958 = 0.2$$

$$m_{1\cdot 2}(V \cup F \cup R) = \frac{1}{0.97}[m_1(V \cup F \cup R) \cdot m_2(V \cup F \cup R)]$$
$$= \frac{1}{0.97}[0.6 \times 0.5] = \frac{1}{0.97} \times 0.3 = 0.309$$
$$= 0.31$$

Moving in the same footsteps, one can calculate the remaining focal elements $R, V \cup F$, and $F \cup R$

3.14 Degree of Confirmation and Disconfirmation

Possibility theory is constructed on two dual functions: necessity measure,$Nec(\eta)$, and possibility measure, $Pos(\pi)$. The two functions Nec and Pos, whose range is [0, 1], can be transformed into a single function,C, whose range is $[-1, 1]$. For one and all $A \in \mathcal{P}(X)$, the function C is defined (**Klir and Yuan**, 1995) in the equation:

$$C(A) = Nec(A) + Pos(A - 1$$
$$= \eta(A) + \pi(A) - 1$$

On the contrary, for each $A \in \mathcal{P}(X)$

$$\eta(A) = Nec(A) = \begin{cases} 0 \text{ when } C(A) \leq 0 \\ C(A) \text{ when } C(A) > 0 \end{cases}$$

and

$$\pi(A) = Nec(A) = \begin{cases} C(A) + 1 \quad \text{when } C(A) \leq 0 \\ 1 \quad \text{when } C(A) > 0 \end{cases}$$

Positive values of $C(A)$ specify the *degree of confirmation* of A by the available evidence, whereas negative values of $C(A)$ convey the *degree of disconfirmation* by the evidence.

3.15 Possibility Theory and Fuzzy Sets

In this section, we will observe that possibility measures are openly associated with fuzzy sets under the associated possibility distribution functions. To illustrate this association, let us assume that $\mathcal{V}$ represents a variable that takes values in the universal set V and let the equation $\mathcal{V} = v$ where $v \in V$ is used to express the value of $\mathcal{V}$ is v.

Let F is a fuzzy set on V that enunciates an elastic constraint on values that may be earmarked to $\mathcal{V}$. Then, for a specific value $v \in \mathcal{V}$, $F(v)$ is elucidated as the degree of compatibility of v with the notion described by F. Further, for the given proposition $''\ \mathcal{V}$ is $F\ ''$ based upon F, it is relevant to expound $F(v)$ as the *degree of possibility* that $\mathcal{V} = v$. That is, for a given fuzzy set F on V and the proposition $''\ \mathcal{V}$ is $F\ ''$ the possibility $r_F(v)$ of $\mathcal{V} = v$ for each $v \in V$ is numerically equivalent to the degree $F(v)$ to which v belongs to F. Conventionally, we have

$$r_F(v) = F(v)\ \forall\ v \in V \tag{3.25}$$

The function $r_F : V \to [0, 1]$ defined by the Eq. (3.25) is a possibility distribution function on V.

3.16 Associated Possibility Measure

For a given possibility distribution function r_F on V, the associated possibility measure Pos_F is defined for all $A \in \mathcal{P}(V)$ by the relation

$$Pos_F(A) = \underset{v \in A}{Sup}\ r_F(v)$$

This is equivalent to Eq. (3.18), i.e., $\underset{x \in A}{Sup}\ r(x)$. This measure communicates the uncertainty concerning the genuine value of the variable $\mathcal{V}$ under imperfect information provided in the form of the proposition $''\ \mathcal{V}$ is $F\ ''$ The associated necessity measure for normal fuzzy sets is defined for all $A \in \mathcal{P}(V)$ by the relation given by

$$Nes_F(A) = 1 - Pos_F\left(\overline{A}\right)$$

This is equivalent to Eq. (3.14) $Nec(A) = 1 - Pos\left(\overline{A}\right)$

3.17 Possibility Theory Vis-À-Vis Probability Theory

One common question about possibility distribution is its relationship with probability distribution. The subject is too complex for a comprehensive discussion; it is essential to discuss it more. The easiest way to acknowledge the relationship between possibility distributions and probability distribution is to compare interval values with probability. As we know, an interval-valued assignment constrains the possible value of a variable without indicating the likelihood that the variable takes a specific value in the interval. Similarly, a possibility distribution states the degree of ease (i.e., possibility) for the variable to take a specific value without indicating the likelihood that the variable has such a value. Even though possibility and probability distribution are different, they are also related... if a value is impossible, it is improbable.

Imagine that you were an eyewitness of a chain-snatching event. When the police questioned you about the height of the snatcher, your answer may be one of the following:

(i) He is between four and a half feet and five and a half feet.
(ii) He is not tall.

Indeed, both of these answers impose a constraint on the feasible heights of the culprit. The first answer is a classical constraint, i.e., that the culprit can't be smaller than four and a half feet or higher than five and a half feet. The second answer imposes an ambiguous or inexplicit constraint on the culprit's height. On the contrary, the interval-value answer, the second answer, does not have a well-defined frontier between the feasible heights and the unfeasible ones. Preferably, it advocates that a fixed height is more feasible than another one.

Usually, when one assigns a fuzzy set, a variable whose value is not precisely known to him, the assignment introduces a vague constraint on the variable's value. Such a constraint is usually called *possibility distribution*, as it enumerates the degree of possibility for the variable to take a definite value. The constraint forced by an internal-value allocation is a peculiar type of possibility distribution. In other words, we can say that the conception of possibility distribution generalizes the concept of interval value to smooth the boundary between the possible and impossible values so that possibility becomes a matter of degree. We use the notation $\Pi(x)$ to represent the possibility distribution of a variable x.

Even though possibility and probability are distinct, they are related in some way. If a variable x can take a value x_i, it is also unlikely for x to have the value x_i, i.e.,

$$\Pi_X(x_i) = 0 \Rightarrow \mathrm{P}_X(x_i) = 0$$

In general, the possibility can deliver as an upper bound on probability: $\mathrm{P}_X(x_i) \leq \Pi_X(x_i)$.

Possibility measures the degree of ease for a variable to take a value, whereas probability measures the likelihood for a variable to take a value. Hence, both deal with two distinct types of uncertainty. Possibility theory operates imprecision, and probability theory operates likelihood of occurrence. This elemental dissimilarity leads to distinct mathematical properties of their distributions. It is a measure of occurrence; the probability distribution of a variable must add up to exactly 1. However, the possibility distribution is not subject to this restriction because a variable can have multiple possible values. Therefore, in our case, the chain-snatching culprit, we may have the following normal distributions with $\mu = 4.5$ ft., $\sigma = 0.2$ ft. to describe the probabilities of the culprit's height:

$$P(h) = \frac{1}{0.2\sqrt{2\pi}} e^{-\frac{1}{2}\left(\frac{h-4.5}{0.2}\right)^2}$$

We, however, know that the area under a normal distribution is 1.

3.17.1 Possibility Vis-À-Vis Probability Comparison

We know that the probability measure "Pro" is required to satisfy $Pro(A \cup B) = Pro(A) + Pro(B) - Pro(A \cap B)$, for all sets $A, B \in \mathcal{P}(X)$.

If $A \cap B = \phi \ \forall \ A, B \in \mathcal{P}(X)$, then the above equation reduces to

$$Pro(A \cup B) = Pro(A) + Pro(B)$$

This result is customarily called the *axiom of additivity of probability measures.* This axiom is more potent than the superadditivity axiom of belief measures given by (b_2). This infers that probability measures are exceptional kinds of belief measures Table 3.4.

Theorem 3.4 A belief measures *Bel* on a finite power set $\mathcal{P}(X)$ is a probability measure if and only if $m(\{x\}) = Bel(\{x\})$ and $m(A) = 0$ give associated basic probability assignment function m for all subsets of X that are not singletons.

Proof Let us assume that *Bel* is a probability measure. For the null set ϕ, the result is obviously accurate, since by the definition of basic probability assignment function $m(\phi) = 0$. So, now we assume that $A \neq \phi$ and let $A = \{x_1, x_2, x_3, \ldots, x_n\}$. Now on applying the additivity axiom of probability measure repeatedly on A, we will have.

Table 3.4 Comparison of possibility and probability distribution

Possibility distribution	Probability distribution
$0 < \pi(x) < 1$	$0 < P(x) < 1$
$\pi(x, y) = \pi(x) \vee \pi(y)$ if x and y are not interacting	$p(x, y) = p(x) * p(y)$ if x and y are independent, i.e.,$\int p(x)dx = 1$

$$\begin{aligned} Bel(A) &= Bel(\{x_1, x_2, x_3, \ldots, x_n\}) \\ &= Bel(\{x_1\}) + Bel(\{x_2, x_3, \ldots, x_n\}) \\ &= Bel(\{x_1\}) + Bel(\{x_2\}) + Bel(\{x_3, \ldots, x_n\}) \\ &= \ldots\ldots\ldots\ldots\ldots\ldots\ldots\ldots\ldots\ldots \\ &= Bel(\{x_1\}) + Bel(\{x_2\}) + Bel(\{x_3\}) + \cdots Bel(\{x_n\}) \end{aligned}$$

According to our assumption for all $x \in X$, we have $Bel(\{x\}) = m(\{x\})$ and

$$Bel(A) = \sum_{B \subseteq A} m(B)$$

Therefore, we have

$$Bel(A) = \sum_{i=1}^{n} m(\{x_i\})$$

It shows that *Bel* is defined concerning a basic assignment function *m* that addresses only singletons.

Conversely, let us suppose that a basic assignment function *m* is given such that

$$\sum_{x \in X}^{n} (\{x_i\}) = 1$$

Now, for any sets $A, B \in \mathcal{P}(X)$ such that $A \cap B = \phi$, *we have*

$$\begin{aligned} Bel(A) + Bel(B) &= \sum_{x \in A} (\{x\}) + \sum_{x \in B} (\{x\}) \\ &= \sum_{x \in A \cup B} (\{x\}) \\ &= Bel(A \cup B) \end{aligned}$$

As a consequence, *Bel* is a probability measure.

Thus, probability measures on finite sets are entirely characterized by a function $p : X \to [0, 1]$, such as $p(x) = m(\{x\})$, which is usually referred to as a *probability distribution function.*

Suppose $p = \langle p(x) : x \in X \rangle$, then p is a probability distribution function on X. When the basic probability assignment function addresses only singletons, we have

$$Bel(A) = Pl(A) = \sum_{x \in A} m(\{x\}) \forall A \in \mathcal{P}(X)$$

Which is (under the probability distribution function) equivalent to

$$Bel(A) = Pl(A) = \sum_{x \in A} p(x)$$

From this, we infer that the dual belief and plausibility measures merge under the notion of additivity axiom of probability measures. Therefore, it is appropriate to represent both by a sole symbol "*Pro.*" Then, we have

$$Pro(A) = \sum_{x \in A}^{n} m(\{x\}) \forall A \in \mathcal{P}(X)$$

3.18 Joint Probability Distribution

Let p_X and p_Y be the projections on X and Y, respectively. Then, the probability distribution function p defined on the Cartesian product of X and Y, i.e., on $X \times Y$, is called joint probability distribution, and p_X and p_Y are known as marginal probability distributions. The following formula elucidates the marginal probability distributions

$$p_X(x) = \sum_{y \in Y} p(x, y) \ \forall x \in X$$

and

$$p_Y(y) = \sum_{x \in X} p(x, y) \ \forall \ y \in Y$$

We will call the marginal probability distributions non-interactive about p if and only if.

$p(x, y) = p_X(x) \cdot p_Y(y)$ for all $x \in X$ and $y \in Y$.

3.19 Sugeno's Integral or Fuzzy Integral

Sugeno proposed the notion of fuzzy integral with respect to a fuzzy measure with the help of max and min operators in the early seventies. The theory of integration and measure plays a vital role in game theory, economics, and decision-making theory. Naturally, it has a significant role in several disciplines, including probability theory and analysis of recursive methods in dynamic macroeconomics. The concept of a fuzzy measure was introduced by **Choquet** in 1954. In this section, we wish to discuss Choquet's Integral and Sugeno's integral. The work originated from the problem, whose significance had been emphasized by **M. Brelot** and **H. Cartan**.

3.19.1 Boolean Algebra

Let X be a set. A collection $\mathcal{F}$ of subsets of X is called a Boolean algebra of sets if:

(i) If $A, B \in \mathcal{F} \Rightarrow A \cup B \in \mathcal{F}$
(ii) If $A \in \mathcal{F} \Rightarrow \overline{A} \in \mathcal{F}$

3.19.2 σ-Algebra on Set

A Boolean algebra is called a σ -algebra on X if every union of a sequence $\{A_i\}_{i=1}^{\infty}$ of members of $\mathcal{F}$ is a member of $\mathcal{F}$, i.e.,$\forall\, i = 1, 2, \ldots\ldots\ldots\ldots$

$$\bigcup_{i=1}^{\infty} A_i \in \mathcal{F}$$

For example, $\mathcal{P}$ and $\{\emptyset, X\}$ are Boolean algebras.

3.19.3 Additive Measure

Let $(X, \mathcal{F}, m))$ be a measure space. An additive measure on a σ-algebra $\mathcal{F}$, denoted by m, is a map $m : \mathcal{F} \to [0, \infty]$, such that:

(i) $m(\phi) = 0$ when $\phi \in \mathcal{F}$(vanishing at ϕ)
(ii) $\{E_n\} \subset \mathcal{F}$ a countable sequence of pairwise disjoint subsets of

$$\mathcal{F} \Rightarrow m\left(\bigcup_{i=1}^{\infty} E_i\right) = \sum_{i=1}^{n} m(E_i)$$

For example, the Lebesgue measure is an additive measure.

3.19.4 Fuzzy Measure

Let X be an arbitrary set and let $\mathcal{F}$ be a σ-algebra defined on X. A measure μ is said to be a fuzzy measure on a **measurable space** $m(X, \mathcal{F})$ if and only if the following conditions hold:

(i) $\mu(\phi) = 0$, $\mu(X) = 1$ when $\phi \in \mathcal{F}$(vanishing at ϕ).
(ii) For all $E \in \mathcal{F}$, $F \in \mathcal{F}$ and $E \subset F \Rightarrow \mu(E) \leq \mu(F)$

This axiom is called the axiom of **monotonicity**.

(iii) $E_n \subset \mathcal{F}, E_1, E_2, E_3, \ldots, \quad \bigcup_{i=1}^{\infty} E_i \in \mathcal{F}$

This axiom is called the axiom of **continuity from below**.

(d) $E_n \subset \mathcal{F}, E_1, E_2, E_3, \ldots\ldots\ldots\ldots\ldots\ldots.., \quad \bigcap_{i=1}^{\infty} E_i \in \mathcal{F}$

This axiom is called the axiom of **continuity from above**.

Further, we will say that μ is a lower (upper) semi-continuous fuzzy measure on $(X, \mathcal{F})$ if and only if it satisfies (i), (ii), and (iii) (or (i), (ii), and (iv)), or we can call both of them simply as semi-continuous measures and $(X, \mathcal{F}, \mu)$ is said to be a **fuzzy measure space**.

These measures help us to define and discuss the fuzzy integrals.

Fuzzy measures include probability measures, possibility measures, necessity measures, etc. The condition $\mu(X) = 1$ is only a matter of convention, and even it can be dropped.

3.19.5 Properties

For all $E, F \in \mathcal{F}$, we have the following properties:

(i) If $\mu(E \cup F) = \mu(E) + \mu(F)$ for all $E \cap F = \phi$, then μ is said to be additive.
(ii) If $\mu(E \cup F) \geq \mu(E) + \mu(F)$ for all $E \cap F = \phi$, then μ is said to be superadditive.
(iii) If $\mu(E \cup F) \leq \mu(E) + \mu(F)$ for all $E \cap F = \phi$, then μ is said to be subadditive.
(iv) If $\mu(E \cup F) + \mu(E \cap F) \geq \mu(E) + \mu(F)$, then μ is said to be supermodular.
(v) If $\mu(E \cup F) + \mu(E \cap F) \leq \mu(E) + \mu(F)$, then μ is said to be submodular.
(vi) If $|E| = |F| \Rightarrow \mu(E) = \mu(F)$, then μ is said to be symmetric.
(vii) If $\mu(E) = 0$ or $\mu(1)$, then μ is said to be Boolean.

3.19.6 Sugeno Integral or Fuzzy Integrals

Let $(X, \mathcal{F})$ be a measurable space, where $A \in \mathcal{F}$ and $f \in F$. Let $\mu : X \to [0, 1]$ be a fuzzy measure; the Sugeno Integral of f on A with respect to μ is defined by

$$\int_A f d\mu = \max_{\alpha \in [0,1]} [min(\alpha, \mu(A \cap F_\alpha))]$$

and is also known as fuzzy integral.

To create another type of integral over a fuzzy set A, we will have

$$\int_A f d\mu = \int_X \min(\mu_A(x), \mu_x) d\mu$$

where $\mu_A(x)$ is the membership function of the fuzzy set A.

Alternatively, let $(X, \mathcal{F}, \mu)$ be a fuzzy measure space,$A \in \mathcal{F}, f \in \{f : f \in \overline{M}, f \geq 0\}$ where

$\overline{M} = \{f : f : X \to (-\infty, \infty), \{x; f(x) \geq \alpha\} \in \mathcal{F}, \alpha \in [-\infty, \infty]\}$. The fuzzy integral of f on A.

with respect to μ is defined by

$$\int_A f d\mu \triangleq \sup_{\alpha \in [0,\infty]} [\alpha \wedge \mu(A \cap F_\alpha)]$$

where $F_\alpha \triangleq \{x : f(x) \geq \alpha\}, \alpha \in [0, \infty]$.

Example 3.5 Let $X = [0, 1]$ *and let* $\mathfrak{B}$ be the class of Borel set in X. Let $\mu = m^2$, where m is the Lebesgue measure, and $f(x) = \frac{1}{2}$. We have

$$F_\alpha = \{x : f(x) \geq \alpha\} = [2\alpha, 1]$$

Since $\theta \in \left[0, \frac{1}{2}\right)$, we need only to consider $\alpha \in \left[0, \frac{1}{2}\right)$. So we have

$$\int_A f d\mu = \max_{\alpha \in \left[0, \frac{1}{2}\right)} [min(\alpha, \mu(F_\alpha))]$$

$$= \max_{\alpha \in \left[0, \frac{1}{2}\right)} \left[min\left(\alpha, (1 - 2\alpha)^2\right)\right]$$

In the expression, $(1 - 2\alpha)^2$ is a decreasing continuous function of α when $\alpha \in \left[0, \frac{1}{2}\right)$. Hence the maximum will be attained at the point which is one of the solutions of the equation

$$\alpha = (1 - 2\alpha)^2 \Rightarrow \alpha = \frac{1}{4}$$

Therefore, we have

$$\int_A f d\mu = \frac{1}{4}$$

3.19.7 Properties of Sugeno Integrals

The following theorem gives the most elementary properties of the fuzzy integral.

(i) If $\mu(A) = 0$, then $\int_A f d\mu = 0$ for any $f \in F$
(ii) If $\int_A f d\mu = 0$, then $\mu(A \cap x : f(x) > 0) = 0$
(iii) If $f_1 \leq f_2$, then $\int_A f_1 d\mu \leq \int_A f_2 d\mu$
(iv) $\int_A f d\mu = \int_A f C_A d\mu$, where C_A is the characteristic function of A.
(v) $\int_A k d\mu = min\{a, \mu(A)\}$ for any constant $k \in [0, \infty)$.
(vi) $\int_A (f + k) d\mu \leq \int_A f d\mu + \int_A k d\mu$, for any constant $k \in [0, \infty)$.

3.20 Choquet's Integral

Let $(X, \mathcal{F})$ be a measurable space, where $A \in \mathcal{F}$ and $f \in F$. Let $\mu : X \to [0, 1]$ be a fuzzy measure and the fuzzy integral of f on A with regard to μ such that

$$\forall x \mathbb{R} : \{t | f(t) \geq x\} \in \mathcal{F}$$

Then, Choquet's integral is given by:

$$\int f d\mu = \int_{-\infty}^{0} (\mu(\{t | f(t) \geq x\}) - \mu(t)) dx + \int_{0}^{-\infty} (\mu(\{t | f(t) \geq x\}) dx)$$

Here, we observe that the integrals on the right-hand side are Riemann's Integrals.

3.20.1 Some Elementary Properties of Choquet's Integral

Followings are the most elementary properties of Choquet's Integral

(i) If $f \leq g$, *then* $\int f d\mu \leq \int g d\mu$
(ii) If $f, g : X \to \mathbb{R}$ are comonotone functions, that is, if for all $t, t' \in X$, it holds that

$$(f(t) - f(t'))(g(t) - g(t')) \geq 0, \text{ then } \int f d\mu + \int g d\mu + \int (f + g) d\mu$$

(iii) If μ is 2-alternative, then $\int f d\mu + \int g d\mu \geq \int (f+g) d\mu$
(iv) If μ is 2-monotone, then $\int f d\mu + \int g d\mu \leq \int (f+g) d\mu$

Exercise-3

1. Determine the basic assignment, possibility measure, and necessity measure for each of the following possibility distributions defined on $X = \{x_i : i \in \mathbb{N}_n\}$:

   (i) $1_r = \langle 1, .8, .8, .5, .2 \rangle$
   (ii) $2_r = \langle 1, 1, 1, 1, .7, .7, .7, .7 \rangle$
   (iii) $3_r = \langle 1, .9, .8, .6, .5, .3, .3 \rangle$
   (iv) $4_r = \langle 1, .5, .4, .3, .2, .1 \rangle$
   (v) $5_r = \langle 1, 1, .8, .8, .5, .5, .5, , 1 \rangle$

2. Show that the function Bel determined by $Bel(A) = \sum_{B|B \subseteq A} m(B)$ for any given basic assignment m is a belief measure.
3. Show that the function Pl determined by $Pl(A) = \sum_{B|A \cap B \neq \phi} m(B)$ for any given basic assignment m is a plausibility measure.
4. Let $X = \{a, b, c, d\}$. Given the basic assignment $m(\{a, b, c\}) = .5, m(\{a, b, d\}) = .2$

and $m(X) = .3$, determine the corresponding belief and plausibility measures.

5. Let $X = \{a, b, c, d\}$. Given the belief measure $Bel(\{b\}) = .1, Bel(\{a, b\}) = .2, Bel(\{b, c\}) = .3, Bel(\{b, d\}) = 1$, determine the corresponding basic assignment.
6. Calculate the possibility distribution, possibility measure, and necessity measure for each of the following basic distributions defined on $X = \{x_i : i \in \mathbb{N}_n\}$ for appropriate values of n:

   (i) $1_m = \{0, 0, .3, .2, 0, .4, .1\}$
   (ii) $2_m = \{.1, .1, .1, 0, .1, .2, .2, .2\}$
   (iii) $3_m = \{0, 0, 0, 0, .5, .5\}$
   (iv) $4_m = \{0, .2, 0, .2, 0, .3, 0, .3\}$
   (v) $5_m = \{.1, .2, .3, .4\}$
   (vi) $6_m = \{.4, .3, .2, .1\}$

7. Let a fuzzy set S be defined on $\mathbb{N}$ by $S = .4/1 + .7/2 + 1/3 + .8/4 + .5/5$ and

$A(x) = 0$ for all $x \notin \{1, 2, 3, 4, 5\}$. Determine $Nec(A)$ and $Pos(A)$ induced by S for all $A \in \mathcal{P}(\{1, 2, 3, 4, 5\})$.

8. Suppose that the basic probability assignments m_1 and m_2 on $X = \{a, b, c, d\}$, which are obtained from two independent sources, be defined as follows: $m_1(\{a, b\}) = .2, m_1(\{a, c\}) = .3, m_1(\{b, d\}) = .5, m_2(\{a, d\}) = .2, m_2(\{b, c\}) = .5, m_2(\{a, b, c\}) = .3$. Calculate the combined basic probability assignments $m_{1,2}$ by using the Dempster rule of combination.

9. By using $Pl(A) = 1 - Bel(\overline{A})$ and $m(A) = \sum_{B|B \subseteq A} (-1)^{|A-B|} Bel(B)$, derive a formula by which the basic assignment for a given plausibility measure can be determined.
10. Show that the function Pl determined by $Pl(A) = \sum_{B|A \cap B \neq \phi} m(B)$ for any given basic assignment m is a plausibility measure.

Objective-Type Questions and Answers

1. A monotone measure describes the vagueness or imprecision in the assignment of an element x to

   (a) One crisp set (b) two or more fuzzy sets
   (c) two or more classical sets (d) one fuzzy set

2. The plausibility measure of a set A is defined as

   (a) $1 - bel(A)$ (b) $1 + bel(A)$
   (c) $1 + bel(\overline{A})$ (d) $1 - bel(\overline{A})$

3. Let X be a universal set and M be the family of subsets of X. Then a mapping $g : M \rightarrow [0, 1]$ is said to be semi-continuous fuzzy measures if it satisfies the conditions of
(a) boundary (b) monotonicity
(c) continuity from below (d) continuity from above
4. For any set A, $bel(A) + bel(\overline{A})$ is
(a) $\geq 1 (b) \gg 1 (c) \ll 1 (d) \leq 1$
5. Belief measure and plausibility measures are

   a. mutually dual (b) complementary to each other
   b. are mappings from $P(X) \rightarrow [0, 1]$ (d) additive measures

6. For all $A, B \in P(X), pl(A \cup B)$ is equal to

   a. $min\{bel(A), bel(B)\}$ (b) $max\{bel(A), bel(B)\}$
   b. $min\{pl(A), pl(B)\}$ (d) $max\{pl(A), pl(B)\}$

7. For any set $A \in P(X), nec(A) > 0 \Rightarrow$

   a. $pos(A) > 1$ (b) $pos(A) < 1$ (c) $pos(A) = 0$ (d) $pos(A) = 1$

8. Let X be the universal set, and M be the family of subsets of X; then, a function $g : M \rightarrow [0, 1]$ is a fuzzy measure on (X, M). Then for all $A, B \in M$, $if\ A \subseteq B$, then
(a) $g(A) \geq g(B)$ (b) $g(A) < g(B)$ (c) $g(A) \leq g(B)$ (d) $g(A) > g(B)$
9. If g is a fuzzy measure on (X, M) and $A, B \in M$, then
(a) $g(A \cup B) = max\{g(A), g(B\}$ (b) $g(A \cup B) \leq max\{g(A), g(B\}$
(c) $g(A \cup B) \gg max\{g(A),\ g(B\}$(d) $g(A \cup B) \geq max\{g(A), g(B\}$
10. If g is a fuzzy measure on (X, M) and $A, B \in M$, then

(a) $g(A \cap B) = min\{g(A), g(B)\}$ (b) $g(A \cap B) \le min\{g(A), g(B)\}$
(c) $g(A \cap B) \nleq min\{g(A), g(B)\}$ (d) $g(A \cap B) \ge min\{g(A), g(B)\}$

11. For all $A, B \in P(X)$, $Bel(A \cap B)$ is equal to

    a. $min\{bel(A), bel(B)\}$ (b) $max\{bel(A), bel(B)\}$
    c. 1- $bel(\overline{A} \cap \overline{B})$ (d) none of these

12. If pl represents the plausibility measure, then, we have
(a) $pl(A) + pl(\overline{A}) < 0$ (b)$pl(A) + pl(\overline{A}) > 0$
(c) $pl(A) + pl(\overline{A}) \le 1$ (d) $pl(A) + pl(\overline{A}) \ge 1$

13. For all $A \in P(X)$, any necessity measure and the associated possibility measure satisfy the implication $Pos(A) < 1 \Rightarrow$
(a) $nec(A) > 0$ (b) $nec(A) < 0$ (c) $nec(A) = 0$ (d) $nec(A) \ge 0$

14. For all $A \in P(X)$, any necessity measure satisfies the relation
(a) $min\{nec(A), nec(\overline{A})\} \ge 0$ (b)$min\{nec(A), nec(\overline{A})\} \le 0$
(c) $min\{nec(A), nec(\overline{A})\} = 0$ (d) $min\{nec(A), nec(\overline{A})\} \ge 1$

15. For all $A \in P(X)$, any possibility measure satisfies the relation
(b) $max\{Pos(A), Pos(\overline{A})\} = 1$ (b)$max\{Pos(A), Pos(\overline{A})\} \le 0$
(c) $max\{Pos(A), Pos(\overline{A})\} = 0$ (d) $max\{Pos(A), Pos(\overline{A})\} \ge 1$

16. Probability measures on finite sets are represented by a function $p : X \to [0, 1]$ such that $p(x) =$

(a) $m(\{x\})$ (b) $m\{(x)\}$ (c) $\sum_{x \in A} m\{x\}$ (d) none of these.

17. Belief and plausibility measures can be characterized by a function $m : \mathcal{P}(X) \to [0, 1]$ such that
(a) $m(\phi) = 0$ (b) $\sum_{x \in \mathcal{P}(X)} m(A) = 1$
(c) $m(\phi) = 0$ and $\sum_{x \in \mathcal{P}(X)} m(A) = 1$ (d) none of these.

18. A possibility measure $\prod$ over a set of events S is a function from 2^S to[0, 1] such that
(a) $\prod(\phi) = 0$ (b)$\pi(S) = 1$
(c) If $A \subset B$, then $\pi(A) \le \pi(B)$ (d) $\pi(A \cup B) = max\{\pi(A), \pi(B)\}$

19. A fuzzy measure g over a set of events S is a function from S to [0, 1] such that
(a) $g(\phi) = 0$ (b) $m(S) = 1$(c) If $A \subset B$, then $P(A) \ge P(B)$ (d) none of these.

20. Since the set of axioms for the fuzzy measure is a subset of the axioms of probability and possibility measures, it is
(a) same as both of them (b) more general than both of them
(c) probability and possibility measures are more general than it (d) none of these.

Answers

1. (c) **2.** (d) **3.** (a, b, c, d) **4.** (d) **5.** (a & c) **6.** (b) **7.** (d) **8.** (c) **9.** (d) **10.** (b). **11**. (a) **12.** (d) **13**. (c) **14**. (c) **15**. (a) **16**. (a) **17.** (c) **18**. (a, b, c, d) **19**. (a, b, c) **20**. (b)

True/False Type Statements and Answers

1. $g(\phi) = 1$ and $g(X) = 0$ are the boundary requirements of fuzzy measure on the power set of $\mathcal{P}(X)$.
2. Belief measure is a function Bel : $\mathcal{P}(X) \rightarrow [0, 1]$.
3. Belief measures are superadditive.
4. For all $A \in \mathcal{P}(X)$, $\text{Bel}(A) + Bel(\overline{A}) \geq 1$.
5. Belief measure is a function such that $\text{Bel}(\phi) = 0$ and $Bel(X) = 1$.
6. Associated with each belief measure, there exists a plausibility measure Pl.
7. Belief measures, and plausibility measures are mutually dual.
8. The relation $Pl(A) = Bel(\overline{A}) - 1$ defines a plausibility measures.
9. A plausibility measure is not subadditive.
10. From the notion of basic probability assignment, we infer that it is not required $m(A) \leq m(B)$ when $A \subseteq B$.
11. There does not exist any relationship between $m(A)$ and $m(B)$ for basic probability assignment.
12. Total ignorance of basic probability assignment is expressed by $m(X) = 1$ and $m(A) = 0$ for $A \neq X$.
13. The expression of total ignorance in terms of the corresponding belief measures is given by $Bel(X) = 0$ and $Bel(A) = 1$ for $A \neq X$.
14. We have a relation for the plausibility measures $Pl(A) + Pl(\overline{A}) \geq 1$, for all $A \in \mathcal{P}(X)$.
15. A basic probability assignment function can conveniently characterize belief and plausibility measures.
16. For all $A \in \mathcal{P}(X)$, $Bel(A) \geq Pl(A)$.
17. For all $A, B \in \mathcal{P}(X)$, we have $Bel(A \cap B) = min[Bel(A), Bel(B)]$.
18. For all $A, B \in \mathcal{P}(X)$, we have $Pl(A \cup B) = max\left[Bel(\overline{A}), Bel(\overline{B})\right]$.
19. For each $A \in \mathcal{P}(X)$, the value $Q(A) = \sum_{A \subseteq B} m(B)$ represents the total portion of belief that can move freely to every point of A.
20. For basic probability assignment, it is not required that $m(X) = 1$.
21. Evidence theory is not based on two dual non-additive measures: belief measures and plausibility measures.
22. Fuzzy measures are generalizations of classical measures.
23. Fuzzy measures are not generalizations of probability measures.
24. From the monotonicity of fuzzy measures, it follows that every fuzzy measure g satisfies the inequality $g(A \cap B) \leq min\left[g(A), g(B)\right] \forall\ A, B, A \cap B \in \mathcal{C}$, a nonempty family of subsets of X.
25. The monotonicity of fuzzy measures does not imply that every fuzzy measure g satisfies the inequality $g(A \cup B) \geq max\left[g(A), g(B)\right] \forall\ A, B, A \cap B \in \mathcal{C}$, a nonempty family of subsets of X.
26. For all $A \in \mathcal{P}(X)$, $Bel(A) = 1 - Pl(\overline{A})$.
27. For all $A, B \in \mathcal{P}(X)$, $Nec(A \cap B) = max[Nec(A), Nec(B)]$.
28. For every $A, B \in \mathcal{P}(X)$, $Pos(A \cup B) \neq max[Pos(A), Pos(B)]$.

29. Possibility theory is based on the extreme values of fuzzy measures concerning set intersections and unions.
30. For all $A \in \mathcal{P}(X), Nec(A) + Nec(\overline{A}) \leq 1$.
31. For all $A \in \mathcal{P}(X)$, $Bel(A)$ is interpreted as the degree of belief that a given element of X belongs to the set A.
32. A plausibility measure is a function $Pl : \mathcal{P}(X) \rightarrow [0, 1]$ such that $Pl(\phi) = 1$ and $Pl(X) = 0$.
33. Evidence theory that deals only with bodies of evidence whose focal elements are nested is called possibility theory.
34. Necessity measures are not particular counterparts of belief measures.
35. Every set $A \in \mathcal{P}(X)$ for which $m(A) < 0$ is usually called a focal element of m.
36. The pair $(\mathcal{F}, m)$, where$\mathcal{F}$ denotes a set of focal elements and m, the associated basic assignment, is referred to as the body of evidence.
37. Possibility measures are particular counterparts of plausibility measures.
38. For all $A \in \mathcal{P}(X)$, Nec and Pos on $\mathcal{P}(X)$, satisfy the implication $Nec(A) > 0 \Rightarrow Pos(A) = 1$.
39. For all $A \in \mathcal{P}(X)$, Pos and Nec on $\mathcal{P}(X)$, satisfy the implication $Pos(A) < 1 \Rightarrow Nes(A) \neq 0$.
40. A lattice of possibility distribution of length n is denoted by $n_{\mathcal{R}}, \leq \forall\, n \in \mathbb{N}$.
41. The number of components in a possibility distribution is called its length.
42. The range of, Pos and Nec functions is $(0, 1)$.
43. The two functions, Pos and Nec, can be converted to a single combined function, C, whose range is $[-1, 1]$.
44. For each $A \in \mathcal{P}(X)$ the positive value of $C(A)$ represent the degree of disconfirmation of A by the evidence available.
45. A monotone measure describes the vagueness in the assignment of an element, say x to two or more crisp sets.
46. The difference between the sum of two quantities $Bel(A) + Bel(\overline{A})$ and 1 is called the ignorance.
47. When ignorance equals 0, we have the case when the probability measure cannot describe the evidence.
48. When the elements of a set, or universe, having evidence are nested, we say that the belief and plausibility measures represent a consonant body of evidence.
49. For each $A \in \mathcal{P}(X)$, the negative value of $C(A)$ represents the degree of confirmation of A by the evidence available.
50. For each $A \in \mathcal{P}(X)$, $\min\left[Nec(A), Nec(\overline{A})\right] = 0$.
51. Necessity measures are special kind of belief measures.
52. For each $A \in \mathcal{P}(X)$, $'Nec'$ and $'Pos'$ are related by the relation $Nec(A) = 1 + Pos(\overline{A})$.
53. For all $A \in \mathcal{P}(X), min\left[Nec(A), NEC(\overline{A})\right] = 0$
54. Given a possibility measure Pos on $\mathcal{P}(X)$, there exists a function $r : X \rightarrow [0, 1]$ such that $r(x) = Pos(\{x\}) \forall\, x \in \mathcal{P}(X)$ is called as possibility distribution function associated with Pos.
55. For all $A \in \mathcal{P}(X), max\left[Pos(A), Pos(\overline{A})\right] = 1$.

56. A function characterized by a function F for all $m : \mathcal{P}(X) \to [0, 1]$ such that $m(\phi) = 0$ and $\sum_{A \in \mathcal{P}(X)} m(A) = 1$, is called a basic probability assignment.
57. The value $m(A)$ expresses the proportion to which all available and relevant evidence supports the claim that a particular element of X, whose characterization in terms of relevant attributes is deficient, belongs to the set A.
58. Plausibility measures are subadditive and not continuous from below.
59. Basic assignments are not fuzzy measures.
60. Every possibility measure Pos on a finite power set $\mathcal{P}(X)$ is uniquely determined by a possibility distribution function $r : X \to [0, 1]$ by the formula $Pos(A) = \max_{x \in A} r(x), \forall A \in \mathcal{P}(X)$.

Answers

1. (F); 2. (T); 3. (T); 4. (F); 5. (T); 6. (T); 7. (T); 8. (F); 9. (F); 10. (T); 11. (T); 12. (T); 13. (F); 14. (T); 15. (T); 16. (F); 17. (T); 18. (F); 19. (T); 20. (T); 21. (T); 22. (T); 23. (F); 24. (T); 25. (T); 26. (T); 27. (T); 28. (F); 29. (T). 30. (T); 31. (T); 32. (F); 33. (T); 34. (F); 35. (F); 36. (T); 37. (T); 38. (T); 39. (F); 40. (T); 41. (T); 42. (F); 43. (T); 44. (F); 45. (T); 46. (T); 47. (F); 48. (T); 49. (F); 50. (T); 51. (T); 52. (F); 53. (); 54. (F); 55. (T); 56. (T); 57. (T); 58. (F); 59. (T); 60. (T).

Chapter 4
Compositional Rule of Inference

In the first section of this chapter, classical logic has been reviewed along with peculiar properties and the truth tables (a helpful device to deal with the truth values of compound propositions) for almost all laws of equivalent logical relations. It also presents other primary inference forms, three-valued logic, and many valued logic. The second section of this chapter deals with different facets of fuzzy logic: fuzzy propositions, quantifiers, linguistic hedges, and inference from fuzzy propositions.

4.1 Introduction

One of the main aims of logic is to provide rules and techniques by which one can determine whether a given argument is valid (correct). It is concerned with all kinds of reasoning, whether legal arguments, mathematical proofs, or conclusions in a scientific theory based on a set of hypotheses. The method of drawing conclusions in some manner is called logic. Because of the diversity of their application, these rules, called **rules of inference**, must be stated in general terms and must be independent of any particular argument or discipline involved. In logic, we are concerned with the forms of the arguments rather than with the arguments themselves. Like any other theory in science, the theory of inference is formulated in such a way that we should be able to decide about the validity of an argument by following the rules mechanically and independently of our feelings about the arguments.

4.2 Classical Logic

Logic is also known as classical logic, two-valued logic, Boolean logic, symbolic logic, etc. Propositional logic deals with statements (or propositions and hence the name propositional logic) that can be only either True or False but not both.

M. K. Singh, *Applied Fuzzy Mathematics*, Forum for Interdisciplinary Mathematics,
https://doi.org/10.1007/978-981-97-3257-9_4

Many proofs in mathematics and many algorithms in computer science use logical expressions such as:

"If p then q" or "If p_1 and p_2, then q_1 or q_2"

It is, therefore, necessary to know the cases in which these expressions are either True or False: what we refer to as the truth values of such expressions. We discuss these issues in the subsequent section of this chapter.

We will also investigate the truth values of quantified statements, which use the logical quantifier "for all" and "there exists."

4.2.1 Proposition

A proposition (or statement) is a declarative sentence that is either true or false but not both (and neither should it be!). Since this logic deals with propositions, the name propositional logic is also assigned. The terms True, False, and proposition are considered undefined concepts (which are intuitively clear).

True and False are usually denoted by the symbols T and F, respectively. It is also denoted by the symbols 1 and 0, respectively. This is also called two-valued logic since we allow only two possible truth values. Let us consider, for example, the following statements or sentences:

(i) Kolkata is in India.
(ii) $2 + 2 = 4$.
(iii) $3 + 3 = 5$.
(iv) Singapore is in England.
(v) $11 < 8$.
(vi) $x = 5$ is a solution of $x^2 = 25$.
(vii) Where are you going?
(viii) I wish you a happy new year.
(ix) Please wait for a while.
(x) Do your homework.
(xi) 11 is a prime number.
(xiii) May God bless you with good health and peace.

The statements (i)–(vi) and (xi) are propositions as they are either true or false. Moreover, (i), (ii), (vi), and (xi) are true, whereas (iii), (iv), and (v) are false. (vii) is (an inquiry), (viii) is a (wish), (ix) is (a request), (x) is (an instruction), and (xiii) is (a wish) are not propositions since they do not have a definite truth value (T) or false (F).

4.2.2 Fundamental Laws of Logic

Aristotle has provided three fundamental laws of logic to retain a statement beyond a doubt. These laws are:

(i) The Law of Excluded Middle: It expresses whether a statement is true or false. In symbolic notation, either p or not p must be true, there is no third or middle true proposition between them.
(ii) The Law of Non-Contradiction: It simultaneously expresses that no statement is true or false. In symbolic notation, for all propositions p, it is impossible for both p and not p to be true.
(iii) The Law or Principle of Identity: Under this law, the symbol used will have the same meaning throughout the specific problems. That is, $(\forall\, x)(x = x)$.

We begin by assuming that the object language contains a set of declarative sentences which cannot be further broken down into more straightforward sentences. These are the primary statements. Now, we will discuss the types of statements.

4.2.3 Simple Statements

The statements we consider initially are called simple statements, *atomic or* primitive *or primary statements*. Thus, the atomic statements are those which do not have connectives. All the declarative sentences to which it is possible to assign one and only one of the two possible truth values are called statements or simple statements.

4.2.4 Compound Statements

When a proposition is composed of sub-propositions and various connectives, it is called a compound proposition. Those statements containing one or more primary statements and some connectives are called *molecular or composite or compound statements*. The truth value of a compound statement depends only on the truth values of the statements being combined and on the types of connectives used.

Thus, a proposition is considered primitive if it cannot be broken down into more straightforward propositions, i.e., if it is not composite.

As an example, let P and Q be any two statements. Then, some of the compound statements formed by using P and Q are

$$P \vee Q, (P \vee Q) \wedge (\neg P), P \vee (\neg Q), \neg P, P \rightarrow Q, P \vee \neg P, P \wedge \neg P, (P \wedge Q) \vee (\neg P).$$

The compound statements cited above are statement formulas derived from the statement variables. Hence, P and Q may be called the *components* of the statement variables.

4.2.5 Syntax

The study of propositional logic consists of syntax (grammar), semantics (meaning), inference rules, and derivation. In logic, syntax is the grammar, structure, or order of the elements in a language statement, and semantics is the meaning of these elements. It applies to computer languages as well as to natural languages. Propositional logic can be considered a language constructed on symbols, i.e., alphabets. The primitive symbols of propositional logic consist of propositional variables, constants, and connectives.

4.2.6 Connectives

In mathematics, the letters $x, y, z, \ldots.$ often denote variables that real numbers can replace, and these variables can be combined with the familiar operations $+, -, \times,$ and $\div$. The letters $p, q, r, \ldots.$ denote **propositional variables** in logic. Those words or symbols used to make a sentence by two sentences are called *logical connectives.* Any two statements may be combined in several ways to form new statements. To form a new statement, the words used are called connectives. The most fundamental connectives are negation, conjunction, and disjunction.

4.2.7 Logic Variables

That is, variables that statements can replace. The logical connectives used in mathematics are negation, conjunction, disjunction, conditional (implication), and biconditional (bi-implication or equivalence). There are three basic connectives. These are negation, conjunction, and disjunction and are related to English words, not, and, & or, respectively. The five logical connectives-negation, conjunction, disjunction, implication, and equivalence-are are sufficient to capture the underlying logical structure of all complex propositions. Thus, we say that they are a complete set of connectives.

4.2.8 Tautologies and Contradiction

A compound statement is called a tautology if it is valid for all truth value assignments for its components. If a compound statement is false for all such assignments, it is called a contradiction. In other words, a statement formula that is true regardless of the truth values of the statements that replace the variables in it is called a *universally valid formula or a tautology or a logical truth, or logically valid*. A statement formula that is false regardless of the truth values of the statements that replace the variables in it is called a *contradiction.* The negation of a contradiction is a tautology. Thus, we may say that a statement formula which is a "**tautology**," is identically true (T), and a "contradiction" is identically false (F). In other words, a statement true for all possible values of its propositional variables is called a tautology, and a statement that is always false is called a **contradiction** or an **absurdity**. A contradiction is also called logically false. Further, a statement that can be either true or false, depending on the truth values of its propositional variables, is called a **contingency**. In other words, a proposition that is neither a tautology nor a contradiction is called a contingency.

An effortless method to determine whether a given formula is a tautology is to construct its truth table. The number of rows in a truth table is 2^n, where n is the number of distinct variables in the formula. A simple fact about tautologies is that the conjunction of two tautologies is also a tautology. The conjunction of two contradictions is also a tautology. The conjunction of two **contingencies** always produces another contingency.

Negating a tautology is a contradiction, and that of a contradiction is a tautology.

Example 4.1

(i) The statement $(p \to q) \leftrightarrow (\sim q \to \sim p)$ and $(p \vee q) \leftrightarrow (q \vee p)$ are tautologies.
(ii) The statement $(p \wedge \sim p)$ is an absurdity.
(iii) The statement $(p \to q) \wedge (p \vee p)$, $(p \vee q) \wedge (\neg p)\&[(p \to q) \wedge q] \to p$ are contingencies.
(iv) The statement $p \wedge (\sim p), p \iff (\sim p), (p \vee q) \wedge (\sim p) \wedge (\sim q), p \Leftrightarrow (\sim p)$, and $(p \vee q) \wedge (\sim p) \wedge (\sim q)$ are contradictions.

4.2.9 Truth Table

A table giving the truth values of a compound statement in terms of its components is called a truth table. Thus, a truth table displays the relationships between the truth values of compound propositions constructed with the help of more straightforward propositions. In other words, it is a device that conveniently shows the validity of various statements considering all possible situations.

It is necessary to consider all possible truth values of p and q. These values are entered in the first two columns of the table. Thus, we have two component statements, namely p and q, so there will be 2^2(four) possible combinations of truth values. In the case of table-1, we have constructed truth table for $\neg p$. There is only one component or atomic statement, namely p, and so there are only $2^1 = 2$(two) possible truth values. In general, if there are n distinct components in a statement formula, there will be 2^n possible combination of truth values in the truth table.

Table 4.1 Truth table for negation

p	$\neg p(\sim p)$
T	F
F	T
or	
p	$\neg p(\sim p)$
1	0
0	1

4.2.10 Negation

If p is a statement, the negation of p is the statement, not p, denoted by $\neg p$ or by $\sim p$. Thus $\sim p$ is the statement "it is not the case that p." From this definition, it follows that if p is true, then $\sim p$ is false, and if p is false, then $\sim p$ is true. The truth value of $\sim p$ relative to p is given in Table 4.1.

4.2.11 Conjunction

The conjunction of two statements, p and q, is the statement $p \wedge q$, which is read as "p and q." The statement $p \wedge q$ has the truth value T whenever both p and q have truth value T; otherwise, it has the truth value F. The conjunction is defined by Table 4.2.

Table 4.2 Truth table for conjunction

p	q	$p \wedge q$
T	T	T
T	F	F
F	T	F
F	F	F
or		
p	q	$p \wedge q$
1	1	1
1	0	0
0	1	0
0	0	0

Table 4.3 Truth table for disjunction

p	q	$p \vee q$	p	q	$p \vee q$
T	T	T	1	1	1
T	F	T	1	0	1
F	T	T	0	1	1
F	F	F	0	0	0

Table 4.4 Truth table for negation, conjunction, and disjunction

p	q	$\neg p$	$p \wedge q$	$p \vee q$
T	T	F	T	T
T	F	T	F	T
F	T	F	F	T
F	F	T	F	F

4.2.12 Disjunction

The disjunction of two statements, p and q, is the statement $p \vee q$, which is read as "p or q." The statement $p \vee q$ has the truth value F only when both p and q have the truth value F; otherwise, it is true. The disjunction is defined by the Table 4.3.

Table 4.4 The truth table for all the above three connectives can be put together in the following way.

4.2.13 Conditional and Biconditional

If p and q are two statements, the statement $p \rightarrow q$ or $p \Rightarrow q$, read as "If p, then q," is a conditional statement. The statement $p \rightarrow q$ has a truth value F when q has the truth value F and p the truth value T; otherwise, it has the truth value T. **Conditional** are also known as ***implication*** propositions and are usually encountered in English literature in the form of "if,. …then" propositions. The conditional is defined by Table 4.5.

Table 4.5 Truth table for conditional

p	q	$p \rightarrow q$ $p \Rightarrow q$
T	T	T
T	F	F
F	T	T
F	F	T

Table 4.6 Truth table for biconditional

p	q	$p \leftrightarrows q$ or $p \Leftrightarrow q$
T	T	T
T	F	F
F	T	F
F	F	T

Table 4.7 Truth table for $(p \to q)(\wedge)(q \to p)$

p	q	$p \to q$	$q \to p$	$(\boldsymbol{p} \to \boldsymbol{q})(\wedge)(\boldsymbol{q} \to \boldsymbol{p})$
T	T	T	T	T
T	F	F	T	F
F	T	T	F	F
F	F	T	T	T

The statement p is called the **antecedent** or **hypothesis**, and q is the **consequent** or **conclusion** in the statements $p \to q$. This implication may be true if its antecedent is true, but its consequent is false. Again, according to the definition, there doesn't need to be any relation between p and q to form $p \to q$.

If p and q are any two statements, then the statement $p \leftrightarrows q$ or $p \Leftrightarrow q$, read as "p if and only if q" and abbreviated as "p iff q" is called a **biconditional**, also called an ***equivalence*** or ***material equivalence*** statement. The statement $p \leftrightarrows q$ has the truth value T whenever both p and q have identical truth values. The statement $p \leftrightarrows q$ is also interpreted as "p is necessary and sufficient for q". Here, we note that the truth values of $(p \to q)(\wedge)(q \to p)$ given in Table 4.7 are identical to the truth values of $p \leftrightarrows q$ given in Table 4.6.

Here, we observe that the truth values of $(p \to q)(\wedge)q \to p$ given in Table 4.7 are exactly the same to the truth values of $p \leftrightarrows q$ or $p \Leftrightarrow q$ in Table 4.6.

4.2.14 Converse, Inverse, and Contrapositive Implications

If $p \Rightarrow q$ is an implication, then the **converse** of $p \Rightarrow q$ is the implication $q \Rightarrow p$, and the **contrapositive** of $p \Rightarrow q$ is the implication $\sim q \Rightarrow \sim p$. Similarly, $\sim p \Rightarrow \sim$ q is the contrapositive of $q \Rightarrow p$. The implication $\sim p \Rightarrow \sim$ q is called the **inverse** of $p \Rightarrow q$. In other words, if $p \Rightarrow q$ is the direct statement, then (i) the statement $q \Rightarrow p$ is called its converse (ii) the statement $\sim p \Rightarrow \sim$ q is called its inverse and (iii) the statement $\sim q \Rightarrow \sim p$ is called its contrapositive. The truth table for conditional, converse, inverse, and contrapositive is as follows, Table 4.8.

Table 4.8 Truth table for conditional, converse, inverse, and contrapositive

p	q	Conditional $p \Rightarrow \mathrm{q}$	Converse $\mathrm{q} \Rightarrow \mathrm{p}$	Inverse $\sim \mathrm{p} \Rightarrow \sim \mathrm{q}$	Contrapositive $\sim \mathrm{q} \Rightarrow \sim \mathrm{p}$
T	T	T	T	T	T
T	F	F	T	T	F
F	T	T	F	F	T
F	F	T	T	T	T

From the above table, we observe that if p is true and q is false, the implication $p \Rightarrow \mathrm{q}$ and the contrapositive $\sim \mathrm{q} \Rightarrow \sim \mathrm{p}$ are false, while the converse $\mathrm{q} \Rightarrow \mathrm{p}$ and the inverse $\sim \mathrm{p} \Rightarrow \sim \mathrm{q}$ are true. For the case where p is false and q is true, the implication $p \Rightarrow \mathrm{q}$ and the contrapositive $\sim \mathrm{q} \Rightarrow \sim \mathrm{p}$ are now valid, while the converse $\mathrm{q} \Rightarrow \mathrm{p}$ and the inverse $\sim \mathrm{p} \Rightarrow \sim \mathrm{q}$ are false. When p and q are both true and false, the implication is true, as are the contrapositive, converse, and inverse.

We note that, since the truth table for $p \Rightarrow \mathrm{q}$ and $\sim \mathrm{q} \Rightarrow \sim \mathrm{p}$ are identical,

$$p \Rightarrow \mathrm{q} \equiv \sim \mathrm{q} \Rightarrow \sim \mathrm{p}$$

i.e., Contrapositive $\equiv$ Direct Statement

Similarly, the truth table for $q \Rightarrow \mathrm{p}$ and $\sim \mathrm{p} \Rightarrow \sim \mathrm{q}$ are identical and therefore

$$q \Rightarrow \mathrm{p} \equiv \sim \mathrm{p} \Rightarrow \sim \mathrm{q}$$

i.e., Converse $\equiv$ Inverse

It is not necessary that if the direct statement is true, its converse and inverse are true.

Example 4.2 Give the converse and contrapositive of the statement, "If you are a mathematician, then you are an algebraist."

Solution: Let us consider

p : You are a mathematician.

q : You are an algebraist.

The converse is

$\mathrm{q} \Rightarrow \mathrm{p}$: If you are an algebraist, then you are a mathematician.

The contrapositive is

$\sim \mathrm{q} \Rightarrow \sim \mathrm{p}$: Suppose you are not an algebraist, then you are not a mathematician.

4.2.15 Well-Formed Formula

Mathematical expressions are represented in well-defined form using parenthesis, braces, and brackets to avoid ambiguity. We use the precedence of operators to solve the expressions and follow the rules of BODMAS recursively. Similarly, in logic, a statement, which is the collection of variables or statements, is represented in well-defined form using parentheses according to the priority of the operation. These statements are known as well-formed formulas abbreviated as **wff**. We can generate well-formed formulas recursively using the following rules:

(i) An atomic statement p is a well-formed formula.
(ii) If p is a well-formed formula, $\sim$ p is a well-formed one.
(iii) If p and q are well-formed formula then, $(p \wedge q)$, $(p \vee q)$, $(p \Rightarrow q)$, and$(p \Leftrightarrow q)$ are also well-formed.
(iv) A statement consisting of variables, parentheses, and connectives is recursively a well-formed formula if and only if it can be obtained by applying the above three rules.
(v) Nothing else is a well-formed formula.

Note 1: According to rule (iv), the following are also well-formed formulas:

$$\neg(p \vee q), \neg(p \wedge q), (p \Rightarrow (p \wedge q)), (p \Rightarrow (p \vee q)), ((p \Rightarrow q) \Leftrightarrow r), \text{etc.}$$

But now, a question arises: when a formula is not well formed? The answer is

(i) When any parenthesis is missing (either at the beginning or the end of the statement).
(ii) The binary operator $(\wedge, \vee, \Rightarrow, \Leftrightarrow)$ contains single operand in place of a double. For example $(p \wedge q) \Rightarrow (\wedge p), (\neg p \wedge q, (p \vee q) \Rightarrow (q \Rightarrow r) \Rightarrow (r.$

Now we will discuss some other connectives because, by using them, some of the formulas become simpler. Other connectives which serve similar purposes will be defined in this section. Other connectives with valuable applications in computer design are called NAND and NOR.

4.2.16 NAND

The word "NAND" is a combination of "NOT" and "AND," where "NOT" stands for negation and "AND" for the conjunction. The symbol $\uparrow$ usually denotes the connective NAND. For any two statements or formulas p and q, we define the NAND of p and q by

$$p \uparrow q \Leftrightarrow \neg(p \wedge q)$$

Table 4.9 Truth table for NAND

p	q	$p \uparrow q$
T	T	F
T	F	T
F	T	T
F	F	T

The statement $p \uparrow q$ is false only when p and q both are true; otherwise, it is true. The truth table for NAND is given in Table 4.9

4.2.17 NOR (Joint Denial)

Another connective, useful in a similar context, is called NOR, a combination of "NOT" and "OR," where "OR" stands for disjunction. The symbol $\downarrow$ usually denotes the connective NOR. For any two statements or formulas p and q, we define NOR of p and q by

$$p \downarrow q \Leftrightarrow \neg(p \vee q)$$

The statement $p \downarrow q$ is valid only when p and q both are false; otherwise, it is false. The truth table for NOR is given in Table 4.10.

Here we observe that the connectives $\uparrow$ and $\downarrow$ have been defined in terms of the connectives $\wedge$, $\vee$, and $\neg$. Therefore, for any formula containing the connectives $\uparrow$ or $\downarrow$, one can obtain an equivalent formula containing the connectives $\wedge$, $\vee$, and $\neg$. We also note that $\uparrow$ and $\downarrow$ are duals of each other. Therefore, to obtain the dual of a formula containing $\uparrow$ and $\downarrow$, we usually interchange $\uparrow$ and $\downarrow$, following the other necessary interchanges.

Now we will show that the connectives $\uparrow$ and $\downarrow$ are *functionally complete*. It is enough to show that the sets of connectives $\{\wedge, \neg\}$and$\{\vee, \neg\}$ can be expressed in terms of $\uparrow$ alone or in terms of $\downarrow$ alone.

Table 4.10 Truth table for NOR

p	q	$p \downarrow q$
T	T	F
T	F	F
F	T	F
F	F	T

4.2.18 Functionally Complete Sets of Connectives

A set of connectives is considered functionally complete if every formula can be expressed in terms of an equivalent formula consisting of the connectives from this set.

We have seen that $p \Leftrightarrow q \equiv (p \rightarrow q) \wedge (q \rightarrow p)$

Example 4.3 Could you write an equivalent formula for $p \wedge (q \Leftrightarrow r) \vee (r \Leftrightarrow p)$ that does not involve the biconditional?

Solution: We have

$$p \wedge (q \Leftrightarrow r) \vee (r \Leftrightarrow p) \Leftrightarrow p \wedge ((q \Rightarrow r) \wedge (r \Rightarrow q)) \vee ((r \Rightarrow p) \wedge (p \Rightarrow r))$$

Thus, the equivalent formula is $p \wedge ((q \Rightarrow r) \wedge (r \Rightarrow q)) \vee ((r \Rightarrow p) \wedge (p \Rightarrow r))$.

Example 4.4 Write an equivalent formula for $\wedge(q \Leftrightarrow r)$, free from biconditional and conditional.

Solution: We have

$$\begin{aligned} p \wedge (q \Leftrightarrow r) &\Leftrightarrow p \wedge ((q \Rightarrow r) \wedge (r \Rightarrow q)) \\ &\Leftrightarrow p \wedge (\neg q \vee r) \wedge (\neg r \vee q) \end{aligned}$$

Thus, the equivalent formula is $p \wedge (\neg q \vee r) \wedge (\neg r \vee q)$.

We also have from De Morgan's laws that

$$p \wedge q \Leftrightarrow \neg(\neg p \vee \neg q \text{ and } p \vee q \Leftrightarrow \neg(\neg p \wedge \neg q$$

The first equivalence shows that we can obtain a formula equivalent to a formula in which conjunction is absent. A similar result holds for the elimination of disjunction.

Thus, when we replace all biconditionals, then conditionals, and all the conjunctions or disjunctions in any formula, we get an equivalent formula that consists of both the negation and disjunction only or the negation and conjunction only. By this truth, we infer that the set of connectives $\{\wedge, \neg\}$ and $\{\vee, \neg\}$ are **functionally complete** for $\uparrow$ and $\downarrow$ respectively.

Also, it is essential to mention that one can easily show that $\{\neg\}$, $\{\wedge\}$, and $\{\vee\}$ are not functionally complete, and neither is $\{\wedge, \vee\}$. Thus, from the five connectives $\wedge$, $\vee$, $\neg$, $\Rightarrow$, $\Leftrightarrow$ we have obtained *two sets of functionally complete connectives.*

4.2.19 Some Basic Properties of NAND and NOR

(i) $p \uparrow q \Leftrightarrow q \uparrow p$ and $p \downarrow q \Leftrightarrow q \downarrow p$ Commutative

(ii) $p \uparrow p \Leftrightarrow \neg p$ and $p \downarrow p \Leftrightarrow \neg p$
(iii) $p \uparrow (q \uparrow r) \Leftrightarrow (p \uparrow q) \uparrow r$ and $(p \downarrow q) \downarrow r \Leftrightarrow p \downarrow (q \downarrow r)$ Associative

Proof.

(i) $p \uparrow q \Leftrightarrow \neg(p \wedge q) \Leftrightarrow \neg(q \wedge p) \Leftrightarrow q \uparrow p$
(ii) $p \uparrow p \Leftrightarrow \neg(p \wedge p) \Leftrightarrow \neg p \vee \neg p \Leftrightarrow \neg p$
(iii)
$$\begin{aligned} & p \uparrow (q \uparrow r) \Leftrightarrow p \uparrow \neg(q \wedge r) \Leftrightarrow \neg(p \wedge \neg(q \wedge r)) \\ & \Leftrightarrow \neg p \vee \neg\neg(q \wedge r) \Leftrightarrow \neg p \vee (q \wedge r), \because \neg\neg p \equiv p \end{aligned}$$

While
$$\begin{aligned} & (p \uparrow q) \uparrow r \Leftrightarrow (\neg(p \wedge q)) \uparrow r \Leftrightarrow \neg((\neg(p \wedge q)) \wedge r) \\ & \Leftrightarrow \neg\neg(p \wedge q) \vee \neg r) \Leftrightarrow (p \wedge q) \vee \neg r \end{aligned}.$$

Thus, we observe that L.H.S. and R. H. S. are not identical. Hence the connective $\uparrow$, i.e., NAND, is not associative. Similarly, due to the property of duality, connective $\downarrow$ (NOR) is also not associative.

Note 2: Please make sure to prove the other's results directly.

4.2.20 XOR

If p and q are any two formulas or statements, then the XORing of p and q, denoted by $p \oplus q$ or by $p \overline{\vee} q$, where $\overline{\vee}$ is called exclusive OR, is true whenever either p or q is true, but not both. The exclusive OR $\oplus$ is also called **exclusive disjunction**, and its truth table is given by Table 4.11.

The following equivalences follow from its definition.

4.2.21 Some Basic Properties of Exclusive OR (XOR)

(i) $p \oplus q \equiv q \oplus p$ Commutative (Symmetric)
(ii) $(p \oplus q) \oplus r \equiv p \oplus (q \oplus r)$ Associative
(iii) $p \wedge ((q \oplus \mathrm{r})) \equiv (p \wedge q) \oplus (p \wedge r)$ Distributive
(iv) $p \oplus q \equiv (p \wedge \neg q) \vee (\neg p \wedge q)$
(v) $p \oplus q \equiv \neg(p \Leftrightarrow q)$

Table 4.11 Truth table for XOR

p	q	$p \overline{\vee} q$ $p \oplus q$
T	T	F
T	F	T
F	T	T
F	F	F

The above statements can be proved easily by using truth tables.

Next, we shall show that the single operator *NAND or NOR is functionally complete.* We have

(i) $p \uparrow p \Leftrightarrow \neg(p \wedge p) \Leftrightarrow \neg p \vee \neg p \Leftrightarrow \neg p$

(ii)
$$\begin{aligned}(p \uparrow q) \uparrow (p \uparrow q) &\Leftrightarrow \neg(p \wedge q) \uparrow \neg(p \wedge q)\\ &\Leftrightarrow \neg(\neg(p \wedge q)) \wedge (\neg(p \wedge q))\\ &\Leftrightarrow \neg\neg(p \wedge q) \wedge \neg\neg(p \wedge q)\\ &\Leftrightarrow (p \wedge q) \wedge (p \wedge q)\\ &\Leftrightarrow p \wedge q\end{aligned}$$

(iii)
$$\begin{aligned}(p \uparrow p) \uparrow (q \uparrow q) &\Leftrightarrow \neg(p \wedge p) \uparrow \neg(q \wedge q)\\ &\Leftrightarrow \neg p \uparrow \neg q\\ &\Leftrightarrow \neg(\neg p \wedge \neg q)\\ &\Leftrightarrow \neg\neg p \vee \neg\neg q\\ &\Leftrightarrow p \vee q\end{aligned}$$

The above equivalence expresses $\neg$, $\vee$, and $\wedge$ in terms of $\uparrow$ alone. One can show similarly that, the following equivalence expresses $\neg$, $\vee$, and $\wedge$ in terms of $\downarrow$ alone.

(i) $p \downarrow p \Leftrightarrow \neg(p \vee p) \Leftrightarrow \neg p \wedge \neg p \Leftrightarrow \neg p$

(ii)
$$\begin{aligned}(p \downarrow q) \downarrow (p \downarrow q) &\Leftrightarrow \neg(p \vee q) \downarrow \neg(p \vee q)\\ &\Leftrightarrow \neg(\neg(p \vee q)) \vee (\neg(p \vee q))\\ &\Leftrightarrow \neg\neg(p \vee q) \wedge \neg\neg(p \vee q)\\ &\Leftrightarrow (p \vee q) \wedge (p \vee q)\\ &\Leftrightarrow p \vee q\end{aligned}$$

(iii)
$$\begin{aligned}(p \downarrow p) \downarrow (q \downarrow q) &\Leftrightarrow \neg(p \vee p) \downarrow \neg(q \vee q)\\ &\Leftrightarrow \neg p \downarrow \neg q\\ &\Leftrightarrow \neg(\neg p \vee \neg q)\\ &\Leftrightarrow \neg\neg p \wedge \neg\neg q\\ &\Leftrightarrow p \wedge q\end{aligned}$$

Since a single operator NAND or NOR is functionally complete, we designate each set $\{\downarrow\}$ and $\{\uparrow\}$ a **minimal functionally complete set**, or in brief, a **minimal set**.

4.2.22 Logical Equivalence

Two propositions, $P(p, q, \dots)$ and $Q(p, q, \dots.)$, are said to be logically equivalent, tautologically equivalent, or simply equal if they have identical truth values or truth tables for all possible assignments provided to the variables and are denoted by

$$P(p, q, \ldots.) \equiv Q(p, q, \ldots.)$$

For example, we can see that the truth tables of $\neg(p \wedge q)$ and $\neg p \vee \neg q$ are the same. That is, both propositions are false in the first case and accurate in the other three cases. Accordingly, we can write

$$\neg(p \wedge q) \equiv \neg p \vee \neg q$$

In other words, the propositions are logically equivalent.

4.2.23 Duality Principle

Two propositions or compound statements, P and P^*, are said to be duals of each other if either of them can be derived or procured from the other by replacing $\bigvee$ by $\bigwedge$ and $\bigwedge$ by $\bigvee$. The two connectives, $\bigwedge$ and $\bigvee$, are called duals of each other. If proposition P contains some unique variables T or F then P^* its dual can be obtained by replacing T by F and F by T.

The duals of (i) $(P \wedge Q) \vee R$; (ii) $(P \vee Q) \wedge T$ and (iii) $\neg(P \wedge Q) \vee (P \wedge \neg(Q \vee \neg R))$ are given by (i) $(P \vee Q) \wedge R$; (ii) $(P \wedge Q) \vee F$ and (iii) $\neg(P \vee Q) \wedge (P \vee \neg(Q \wedge \neg R))$ respectively.

In other words, let P and Q be statements that contain no logical connectives other than $\bigwedge$ and $\bigvee$. If $P \Leftrightarrow Q$, then $P^d \Leftrightarrow Q^d$, where P^d is the dual of P.

Some important logical equivalent relations and properties are summarized, along with proof of some relations in Table 4.12.

Part (a) of all the formulas except those given in Table 4.12 (b) can be proved using the truth table, and the other will be left for the practice of readers. Relations 1–6, 13, and 14 are **the laws of logic**.

Commutative Laws: Let p and q be any two propositions, then

(a) $p \wedge q \equiv q \wedge p$
(b) $p \vee q \equiv q \vee p$ are the Commutative Laws. Now we will form the truth table for (a) Table 4.13

Thus, $p \wedge q \equiv q \wedge p$ is a tautology, as the last column consists of all the truth values "true."

Associative Laws: Let p, q, and r be any three propositions, then

(a) $p \wedge (q \wedge r) \equiv (p \wedge q) \wedge r$
(b) $p \vee (q \vee r) \equiv (p \vee q) \vee r$ are the Associative Laws.

Now we will form the truth table for (a). Table 4.14.

Thus, $p \wedge (q \wedge r) \equiv (p \wedge q) \wedge r$ is a tautology. Since the last column consists of all the truth values "true."

Table 4.12 Logical equivalent relations and properties

1	*Commutative Laws*	(a) $p \wedge q \equiv q \wedge p$	(b) $p \vee q \equiv q \vee p$	
2	*Associative Laws*	(a) $p \wedge (q \wedge r) \equiv (p \wedge q) \wedge r$	(b) $p \vee (q \vee r) \equiv (p \vee q) \vee r$	
3	*Distributive Laws*	(a) $p \wedge (q \vee r) \equiv (p \wedge q) \vee (p \wedge r)$	(b) $p \vee (q \wedge r) \equiv (p \vee q) \wedge (p \vee r)$	
4	*Idempotent Laws*	(a) $p \wedge p \equiv p$	(b) $p \vee p \equiv p$	
5	*Absorption Laws*	(a) $p \wedge (p \vee q) \equiv p$	(b) $p \vee (p \wedge q) \equiv p$	
6	*De Morgan's Laws*	(a) $\neg(p \wedge q) \equiv \neg p \vee \neg q$	(b) $\neg(p \vee q) \equiv \neg p \wedge \neg q$	
7	*Detachment*	$(p \wedge (p \to q)) \to q$		
8	*Chain Rule or Transitive Property*	$((p \to q) \wedge (q \to r)) \to (p \to r)$		
9	*Conditional Rules*	(a) $(p \to q) \equiv (\neg p) \vee q$	(b) $(p \to q) \equiv (\neg q) \to (\neg p)$	
10	*Equivalence Rules*	(a) $(p \leftrightarrow q) \equiv (p \to q) \wedge (q \to p)$	(b) $\neg(p \leftrightarrow q) \equiv (p \leftrightarrow \neg q)$	(c)$(p \leftrightarrow q) \equiv (p \wedge q) \vee (\neg p) \wedge (\neg q)$
11	*Implication Rules*	(a)$(p \to q) \to ((p \wedge r) \to (q \wedge r))$	(b)$(p \to q) \to ((p \vee r) \to (q \vee r))$	
12	*Contrapositive Inference*	$(p \to q) \wedge (\neg q) \to (\neg p)$		
13	*Involution Law* or *Law of Double Negation*	$\neg(\neg p) \equiv p$		
14	*Identities*	(i) (a) $p \wedge T \equiv p$	(b)$p \vee F \equiv p$	
	Domination Laws	*(ii)(a)*$p \wedge F \equiv F$	(b) $p \vee T \equiv T$	
	Inverse Laws	(iii) (a) $p \wedge \neg p \equiv F$	(b)$p \vee \neg p \equiv T$	

Table 4.13 $p \wedge q \equiv q \wedge p$

p	q	$p \wedge q$	$q \wedge p$	$p \wedge q \equiv q \wedge p$
T	T	T	T	T
T	F	F	F	T
F	T	F	F	T
F	F	F	F	T

Table 4.14 $p \wedge (q \wedge r) \equiv (p \wedge q) \wedge r$

p	q	r	$q \wedge r$	$p \wedge q$	$p \wedge (q \wedge r)$	$(p \wedge q) \wedge r$	$p \wedge (q \wedge r) \equiv (p \wedge q) \wedge r$
T	T	T	T	T	T	T	T
T	T	F	F	T	F	F	T
T	F	T	F	F	F	F	T
F	T	T	T	F	F	F	T
T	F	F	F	F	F	F	T
F	T	F	F	F	F	F	T
F	F	T	F	F	F	F	T
F	F	T	F	F	F	F	T
F	F	F	F	F	F	F	T

Table 4.15 $p \wedge (q \vee r) \equiv (p \wedge q) \vee (p \wedge r)$

p	q	r	$p \wedge q$	$p \wedge r$	$q \vee r$	$p \wedge (q \vee r)$	$(p \wedge q) \vee (p \wedge r)$	$p \wedge (q \vee r) \equiv$ $(p \wedge q) \vee (p \wedge r)$
T	T	T	T	T	T	T	T	T
T	T	F	T	F	T	T	T	T
T	F	T	F	T	T	T	T	T
F	T	T	F	F	T	F	F	T
T	F	F	F	F	F	F	F	T
F	T	F	F	F	T	F	F	T
F	F	T	F	F	T	F	F	T
F	F	T	F	F	F	F	F	T

Distributive Laws: (a) $p \wedge (q \vee r) \equiv (p \wedge q) \vee (p \wedge r)$ (b) $p \vee (q \wedge r) \equiv (p \vee q) \wedge (p \vee r)$

Let p, q, and r be any three propositions, then we will form the truth table for (a). Table 4.15.

Obviously, $p \wedge (q \vee r) \equiv (p \wedge q) \vee (p \wedge r)$ is a tautology as the last column consists of all the truth values "true."

Table 4.16 $p \wedge p \equiv p$

p	p	$p \wedge p$	$p \wedge p \equiv p$
T	T	T	T
F	F	F	T

Idempotent Laws: Let p be any statement, then (a) $p \wedge p \equiv p$ (b) $p \vee p \equiv p$ are called Idempotent Laws. We will form the truth table for (a). Table 4.16.

Absorption Laws: Let p and q be any two propositions, then (a) $p \wedge (p \vee q) \equiv p$ (b) $p \vee (p \wedge q) \equiv p$ are known as Absorption Laws. We will form the truth table for (a). Table 4.17.

De Morgan's Laws: Let p and q be any two propositions, then

(a) $\neg(p \wedge q) \equiv \neg p \vee \neg q$
(b) $\neg(p \vee q) \equiv \neg p \wedge \neg q$

are called De Morgan's Laws. It can be defined as "the negation of the disjunction (or conjunction) of two statements is logically equivalent to the conjunction (or disjunction) of the negations of the two statements respectively." Now we will form the truth table of it, i.e., for (a). Table 4.18.

Obviously, $\neg(p \wedge q) \equiv \neg p \vee \neg q$ is a tautology as the last column consists of all the truth values "true."

Chain Rule or Transitive Property: Let p, q, and r be any three propositions, then $((p \to q) \wedge (q \to r)) \to (p \to r)$ is called the Chain Rule or Transitive Property. Now we will form the truth table. Table 4.19.

Obviously, $((p \to q) \wedge (q \to r)) \to (p \to r)$ is a tautology as the last column consists of all the truth values "true."

Table 4.17 $p \wedge (p \vee q) \equiv p$

p	q	$p \vee q$	$p \wedge (p \vee q)$	$p \wedge (p \vee q) \equiv p$
T	T	T	T	T
T	F	T	T	T
F	T	T	F	T
F	F	F	F	T

Table 4.18 $\neg(p \wedge q) \equiv \neg p \vee \neg q$

p	q	$p \wedge q$	$\neg(p \wedge q)$	$\neg p$	$\neg q$	$\neg p \vee \neg q$	$\neg(p \wedge q) \equiv \neg p \vee \neg q$
T	T	T	F	F	F	F	T
T	F	F	T	F	T	T	T
F	T	F	T	T	F	T	T
F	F	F	T	T	T	T	T

Table 4.19 $((p \to q) \wedge (q \to r)) \to (p \to r)$

p	q	r	$p \to q$	$q \to r$	$p \to r$	$((p \to q) \wedge (q \to r)) = X$	$(X) \to (p \to r)$
T	T	T	T	T	T	T	T
T	T	F	T	F	F	F	T
T	F	T	F	T	T	F	T
F	T	T	T	T	T	T	T
T	F	F	F	T	F	F	T
F	T	F	T	F	T	F	T
F	F	T	T	T	T	T	T
F	F	T	T	T	T	T	T

Detachment Law: Let p and q be any two propositions, then $((p \to q) \wedge p) \to q$ is called the Detachment Law. Now we will form the truth table for it. Table 4.20.

Obviously, $(p \wedge (p \to q)) \to q$ is a tautology as the last column consists of all the truth values "true."

Conditional Rules: Let p, q, and r be any three propositions, then (a) $(p \to q) \equiv (\neg p) \vee q$ (b) $(p \to q) \equiv (\neg q) \to (\neg p)$ are called the Conditional Rules. Now we will form the truth table for (a). Table 4.21.

Obviously, $(p \to q) \equiv (\neg p) \vee q$ is a tautology as the last column consists of all the truth values "true."

Equivalence Rules: Let p and q be any two propositions, then (a) $(p \leftrightarrow q) \equiv (p \to q) \wedge (q \to p)$ (b) $\neg(p \leftrightarrow q) \equiv (p \leftrightarrow \neg q)$ and (c) $(p \leftrightarrow q) \equiv (p \wedge q) \vee (\neg p) \wedge (\neg q)$ are called Equivalence Rules. Now we will form the truth table for (c). Table 4.22.

Table 4.20 $(p \wedge (p \to q)) \to q$

p	q	$p \to q$	$p \wedge (p \to q)$	$(p \wedge (p \to q)) \to q$
T	T	T	T	T
T	F	F	F	T
F	T	T	F	T
F	F	T	F	T

Table 4.21 $(p \to q) \equiv (\neg p) \vee q$

p	q	$(p \to q)$	$(\neg p)$	$(\neg p) \vee q$	$(p \to q) \equiv (\neg p) \vee q$
T	T	T	F	T	T
T	F	F	F	F	T
F	T	T	T	T	T
F	F	T	T	T	T

Table 4.22 $(p \leftrightarrow q) \equiv (p \wedge q) \vee (\neg p) \wedge (\neg q)$

p	q	$p \wedge q = Y$	$(\neg p)$	$(\neg q)$	$(\neg p) \wedge (\neg q) = Z$	$(p \leftrightarrow q) = X$	$Y \vee Z$	$X \equiv Y \vee Z$
T	T	T	F	F	F	T	T	T
T	F	F	F	T	F	F	F	T
F	T	F	T	F	F	F	F	T
F	F	F	T	T	T	T	T	T

Obviously, $(p \leftrightarrow q) \equiv (p \wedge q) \vee (\neg p) \wedge (\neg q)$ is a tautology as the last column consists of all the truth values "true."

Implication Rules: Let p, q, and r be any three propositions, then

(a) $(p \to q) \to ((p \wedge r) \to (q \wedge r))$
(b) $(p \to q) \to ((p \vee r) \to (q \vee r))$

are called Implication Rules. Now we will form the truth table for (a). Table 4.23.

Evidently, the implication $(p \to q) \to ((p \wedge r) \to (q \wedge r))$ is a tautology as the last column consists of all the truth values "true."

Contrapositive Inference: Let p and q be any two propositions, then $(p \to q) \wedge (\neg q) \to (\neg p)$ is referred to as Contrapositive Inference. Now we will form the truth table for (c). Table 4.24.

Table 4.23 $(p \to q) \to ((p \wedge r) \to (q \wedge r))$

p	q	r	$(p \to q) = X$	$(p \wedge r) = Y$	$(q \wedge r) = Z$	$Y \to Z$	$X \to (Y \to Z)$
T	T	T	T	T	T	T	T
T	T	F	T	F	F	T	T
T	F	T	F	T	F	F	T
F	T	T	T	F	T	T	T
T	F	F	F	F	F	T	T
F	T	F	T	F	F	T	T
F	F	T	T	F	F	T	T
F	F	T	T	F	F	T	T

Table 4.24 $(p \to q) \wedge (\neg q) \to (\neg p)$

p	q	$(p \to q)$	$(\neg p)$	$(\neg q)$	$(p \to q) \wedge (\neg q)$	$(p \to q) \wedge (\neg q) \to (\neg p)$
T	T	T	F	F	F	T
T	F	F	F	T	F	T
F	T	T	T	F	F	T
F	F	T	T	T	T	T

Table 4.25 $\neg(\neg p) \equiv p$

p	$\neg p$	$\neg(\neg p)$	$\neg(\neg p) \equiv p$
T	F	T	T
F	T	F	T

Evidently, $(p \to q) \wedge (\neg q) \to (\neg p)$ is a tautology as the last column consists of all the truth values "true."

Involution Law: Let p be any proposition. Now we will draw truth table for the Involution Law $\neg(\neg p) \equiv p$. **Truth** Table 4.25, which is a tautology.

Identities Law: (a) $p \wedge T = T \wedge p \equiv p$ (b) $p \vee F = F \vee p \equiv p$ **Truth** Table 4.26 for (a).

(a) $p \wedge F \equiv F$ (b) $p \vee T \equiv T$ **Truth** Table 4.27 for (a).

(a) $p \wedge \neg p \equiv F$
(b) $p \vee \neg p \equiv T$ **Truth** Table 4.28 for (a) & (b).

Example 4.5 Use the truth table to characterize the following statements forms as a tautology, contradictory or contingent.

(i) $[p \to (p \to q)] \to q$
(ii) $[p \to (q \to r)] \to ((p \to q) \to (p \to r))$
(iii) $(p \to p) \to (q \wedge \neg q)$
(iv) $p \to (p \to (q \wedge \neg q))$

Table 4.26 $p \wedge T = T \wedge p \equiv p$

p	T	$p \wedge T$	$T \wedge p$
T	T	T	T
F	T	F	F

Table 4.27 $p \wedge F \equiv F$

p	F	$p \wedge F$	$F \wedge p$
T	F	F	F
F	F	F	F

Table 4.28 $p \wedge \neg p \equiv F$ and $p \vee \neg p \equiv T$

p	$\neg p$	F	$p \wedge \neg p$	$p \vee \neg p$	T
T	F	F	F	T	T
F	T	F	F	T	T

Solution: Let p, q, and r be any three propositions; then, we will form the truth table to verifying the nature of the statements described above. Truth table for (i). Table 4.29.

The truth table's last column has both values T i.e., "1" and F i.e.,"0"; hence, the given statement is contingent.

(v) **Truth** Table 4.30

Since the last column of the truth table consists of all the truth values "true," the given statement is a tautology.

(vi) **Truth** Table 4.31

The last column of the truth table contains all the truth values "False," the given statement $(p \to p) \to (q \wedge \neg q)$ is a contradiction.

Table 4.29 For contingent $[p \to (p \to q)] \to q$

p	q	$X : (p \to q)$	$Y : p \to X$	$Y \to q$
T	T	T	T	T
T	F	F	F	T
F	T	T	T	T
F	F	T	T	F

Table 4.30 For tautology $[p \to (q \to r)] \to ((p \to q) \to (p \to r))$

p	q	r	$U : p \to q$	$V : p \to r$	$X : q \to r$	$Y : p \to X$	$Z : U \to V$	$Y \to Z$
T	T	T	T	T	T	T	T	T
T	T	F	T	F	F	F	F	T
T	F	T	F	T	T	T	T	T
F	T	T	T	T	T	T	T	T
T	F	F	F	F	T	T	T	T
F	T	F	T	T	F	T	T	T
F	F	T	T	T	T	T	T	T
F	F	T	T	T	T	T	T	T

Table 4.31 For contradiction $(p \to p) \to (q \wedge \neg q)$

p	q	$\neg q$	$(p \to p)$	$(q \wedge \neg q)$	$(p \to p) \to (q \wedge \neg q)$
T	T	F	T	F	F
T	F	T	T	F	F
F	T	F	T	F	F
F	F	T	T	F	F

(iv) **Truth** Table 4.32

The last column of the truth table contains both values "T" and "F"; therefore, the provided information $p \to (p \to (q \wedge \neg q)$ is contingent.

Example 4.6 If $23x + 8 = 77$, then $x = 3$ then find the converse, inverse, and contrapositive.

Solution: Let $P(x) \equiv 23x + 8 = 77$
and $Q(x) \equiv x = 3$ then

(i) Converse: If $x = 3$ then $23x + 8 = 77$
(ii) Inverse: If $23x + 8 \neq 77$ then $x \neq 3$
(iii) Contrapositive: If $x \neq 3$ then $23x + 8 \neq 77$

Example 4.7 Give the converse and contrapositive of the statement.

If you are a mathematician, then you are a topologist.

Solution: Let us consider
$P =$ You are mathematician.
$Q =$ You are a topologist.

The converse is
$Q \to P$: If you are a topologist, then you are a mathematician.

The contrapositive is
$\neg Q \to \neg P$: If you are not a topologist, then you are not a mathematician.

Example 4.8 If I fall ill or have a severe health issue, my insurance company Max Bupa will pay for it. Write the converse, inverse, and contrapositive of the statement.

Solution: Let P : I will feel ill
Q : Sugar and B.P. will destroy my health
R : My insurance company will pay for it.
In symbolic forms, the given statement is

$$(P \vee Q) \to R$$

Table 4.32 For contingent $p \to (p \to (q \wedge \neg q)$

p	q	$\neg q$	$(q \wedge \neg q)$	$p \to (q \wedge \neg q)$	$p \to (p \to (q \wedge \neg q)$
T	T	F	F	F	F
T	F	T	F	F	F
F	T	F	F	T	T
F	F	T	F	T	T

(i) Converse is $R \Rightarrow (P \vee Q)$
(ii) Inverse is $\neg(P \vee Q) \Rightarrow \neg R$
(iii) Contrapositive is $\neg R \Rightarrow \neg(P \vee Q)$

4.3 Principle of Substitution

As we know, increasing the number of variables n to a set number of rows in the truth table simultaneously increases the ratio of $n : 2^n$. The process to prove the statement is tautologous using truth tables becomes tedious for large values of n. In such cases, another method, namely the *Principle of Substitution,* can be used to prove the tautology of the statements. According to it, "If P is a tautology and let $P_1, P_2, P_3, \ldots..$ are any propositions such that on substituting $P_1, P_2, P_3, \ldots..$ in P in place of independent input variables, we still have a tautology."

Note 3:

(i) The formula obtained after substitution is called a **substitution instance**. For example, consider $(p \rightarrow (q \rightarrow p))$, substitute $r \leftrightarrow s$ for p and $r \rightarrow s$ for q, then the *substitution instance* be given by $((r \leftrightarrow s) \rightarrow ((r \rightarrow s) \rightarrow (r \rightarrow s)))$.
(ii) Substitutions are made only for the atomic formula, not for the molecular formula.
(iii) Whenever we use the *Principle of Substitution,* the same formula is substituted for the variable throughout the statement.

Example 4.9 Show that $\neg(p \vee (\neg p \wedge q)) \equiv \neg p \wedge \neg q$, by using logical equivalence formula.

Solution: We have

$$\begin{aligned}
\neg(p \vee (\neg p \wedge q)) &\equiv \neg(p) \wedge \neg(\neg p \wedge q), && \text{by De Morgan's Law}\\
&\equiv \neg p \wedge (\neg(\neg p) \vee \neg q), && \text{by De Morgan's Law}\\
&\equiv \neg p \wedge (p \vee \neg q), && \text{by Involution Law}\\
&\equiv (\neg p \wedge p) \vee (\neg p \wedge \neg q), && \text{by Distributive Law}\\
&\equiv F \vee (\neg p \wedge \neg q), && \text{by Identity law}\\
&\equiv \neg p \wedge \neg q, && \text{by Identity law}
\end{aligned}$$

Hence, we get $\neg(p \vee (\neg p \wedge q)) \equiv \neg p \wedge \neg q$

Example 4.10 By using logical equivalence, show that $(p \wedge q) \rightarrow (p \vee q)$ is a tautology.

Solution: We have

$$\begin{aligned}(p \wedge q) \to (p \vee q) &\equiv \neg(p \wedge q) \vee (p \vee q)\\ &\equiv (\neg p \vee \neg q) \vee (p \vee q), \quad \text{by De Morgan's Law}\\ &\equiv \neg p \vee (\neg q \vee p) \vee q\\ &\equiv \neg p \vee (p \vee \neg q) \vee q, \quad \text{by Commutative law}\\ &\equiv (\neg p \vee p) \vee (\neg q \vee q), \quad \text{by Associative Law}\\ &\equiv T \vee T, \quad \text{by Identity law}\\ &\equiv T\end{aligned}$$

Example 4.11 Show the implication:

(i) $\neg(p \Leftrightarrow q) \Leftrightarrow (p \vee q) \wedge \neg(p \wedge q)$
(ii) $((p \vee \neg p) \to q) \to ((p \vee \neg p) \to r) \to (q \to r)$
(iii) $(p \to q) \to q \Rightarrow p \vee q$

Solution:

(i) We have

$$\begin{aligned}\text{L.H.S.} = \neg(p \Leftrightarrow q) &\Leftrightarrow \neg((p \wedge q) \vee ((\neg p) \wedge (\neg q))), \quad \text{by Equivalence formula (c)}\\ &\Leftrightarrow \neg(p \wedge q) \wedge \neg((\neg p) \wedge (\neg q)), \quad \text{by De Morgan's Law}\\ &\Leftrightarrow \neg(p \wedge q) \wedge \neg(\neg p) \vee \neg(\neg q), \quad \text{by De Morgan's Law}\\ &\Leftrightarrow \neg(p \wedge q) \wedge (p \vee q), \quad \text{by Involution}\\ &\Leftrightarrow (p \vee q) \wedge \neg(p \wedge q) \quad \text{by Commutative law}\\ &\Leftrightarrow \text{R.H.S.}\end{aligned}$$

(ii)

$$\begin{aligned}\text{L.H.S.} &= ((p \vee \neg p) \to q) \to ((p \vee \neg p) \to r)\\ &\Rightarrow (\neg(p \vee \neg p)) \vee q) \to (\neg(p \vee \neg p) \vee r) \quad \text{,since } p \to q = (\neg p) \vee q\\ &\Rightarrow ((\neg T) \vee q) \to ((\neg T) \vee r) \quad , p \vee \neg p \equiv T\\ &\Rightarrow (F \vee q) \to (F \vee r) \quad \text{,since } \neg T \equiv F\\ &\Rightarrow q \to r = \text{R.H.S.} \quad , F \vee q \equiv q\end{aligned}$$

(iii)

$$\begin{aligned}\text{L.H.S.} &= (p \to q) \to q\\ &\Rightarrow (\neg p \vee q) \to q\\ &\Rightarrow \neg(\neg p \vee q) \vee q\\ &\Rightarrow (\neg(\neg p) \wedge (\neg q)) \vee q\\ &\Rightarrow (p \wedge \neg q) \vee q\\ &\Rightarrow (p \vee q) \wedge (\neg q \vee q)\\ &\Rightarrow (p \vee q) \wedge T \equiv (p \vee q) = \text{R.H.S}\end{aligned}$$

Example 4.12 Simplify the following:

(i) $(p \wedge q) \vee p$

(ii) $(p \wedge q) \wedge \neg p$
(iii) $\neg(\neg p \wedge q) \wedge (\neg p \vee q) \wedge (p \vee q)$
(iv) $\neg(p \vee q) \vee (\neg p \wedge q)$

Solution:

(i)
$$\begin{aligned}(p \wedge q) \vee p &\equiv p \vee (p \wedge q), && \text{by Commutativity}\\ &\equiv (p \wedge T) \vee (p \wedge q), && \text{since } p \wedge T \equiv p\\ &\equiv (p \wedge (T \vee q) \\ &\equiv p \wedge T, && \text{since } T \vee q \equiv T\\ &\equiv p\end{aligned}$$

(ii)
$$\begin{aligned}(p \wedge q) \wedge \neg p &\equiv \neg p \wedge (p \wedge q), && \text{by Commutativity}\\ &\equiv (\neg p \wedge p) \wedge q\\ &\equiv F \wedge q \equiv F\end{aligned}$$

(iii)
$$\begin{aligned}\neg(\neg p \wedge q) \wedge (\neg p \vee q) \wedge (p \vee q) &\equiv ((\neg(\neg p) \vee \neg q) \wedge \{(\neg p \vee q) \wedge (p \vee q)\}\\ &\equiv (p \vee \neg q) \wedge \{(\neg p \wedge p) \vee q\}\\ &\equiv (p \vee \neg q) \wedge (F \vee q)\\ &\equiv (p \vee \neg q) \wedge q\\ &\equiv (p \wedge q) \vee (\neg q \wedge q)\\ &\equiv (p \wedge q) \vee F\\ &\equiv (p \wedge q)\end{aligned}$$

(iv)
$$\begin{aligned}\neg(p \vee q) \vee (\neg p \wedge q) &\equiv (\neg p \wedge \neg q) \vee (\neg p \wedge q)\\ &\equiv \neg p \wedge (\neg q \vee q)\\ &\equiv \neg p \wedge T\\ &\equiv \neg p\end{aligned}$$

We now investigate those processes accepted as valid in deriving a statement, called the **conclusion**, from other given accounts, called **premises**.

4.4 Argument, Valid Argument, and Fallacy Argument

An argument is a process by which a conclusion is obtained from a given set of premises. In other words, an argument is an assertion that a set of given propositions called premises yield another recommendation called the conclusion. If $p_1, p_2, p_3, \ldots\ldots.., p_n$ be the set of all propositions that yield the conclusion q, then it is denoted by $p_1,\ p_2,\ p_3, \ldots\ldots.p_n \vdash q$.

4.4.1 Representation of an Argument

An argument $p_1, p_2, p_3, \ldots\ldots.., p_n \vdash q$ is written as

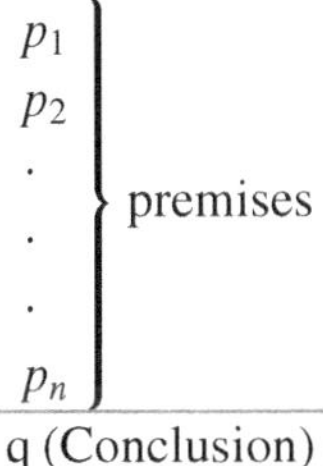

In the above representation premises are listed above the horizontal line, and the conclusion is below the horizontal line.

An argument is called a **valid argument** if the conclusion is true whenever all the premises are true. That argument is reasonable if and only if the premises imply the conclusion. An argument that yields a conclusion q from premises $p_1, p_2, p_3, \ldots\ldots.., p_n$ is valid if and only if the statement $p_1 \wedge p_2 \wedge p_3 \wedge \ldots\ldots.. \wedge p_n) \rightarrow q$ is a tautology. An argument is called a **fallacy** or an **invalid argument** if it is not a valid argument.

In general, there are three methods to check the validity of a given argument: (i) Truth table method (ii) Simplification method, and (iii) Inference Rule Method. We have discussed the first two methods. By the Inference rule method, we reduce the given argument to a series of arguments each of which is valid. This method is the simplest of the three.

The rule of detachment and the law of syllogism are the most frequently used valid arguments.

4.5 Rule of Detachment or Modus Ponens

For the first example, we consider the rule of inference called **Modus Ponens (MP)** or the Rule of Detachment. Modus Ponens comes from Latin and may be translated as "*the method of affirming.*" In symbolic form, this rule is expressed by the logical implication WFF:

$$[p \wedge (p \rightarrow q)] \rightarrow q$$

This is usually expressed in the following format:

$$\begin{array}{ll} & p \\ & p \rightarrow q \\ \hline \therefore & q \text{ (conclusion)} \end{array} \text{Premises}$$

This is known as the law of detachment, verified in Table 4.17, where the first row is the only one where both the premises p and $p \to q$, and the conclusion q are true.

This rule arises when we argue that if (i) p is true, and (ii) $p \to q$ is true (or $p \Rightarrow q$), then the conclusion q must also be true. (After all, if q were false and p were true, then we could not have $p \to q$ true.)

Modus Ponens (MP) is also written in the following form:

$$\begin{array}{l} \text{If } p \text{ then } q \\ p \\ \hline \qquad\qquad q \end{array}$$

In this format, If p, then q is called a *Rule*, p in the second line is called the *Fact,* and q in the last line is called the *Conclusion.*

Example 4.13 Do you think the following argument valid?
Drinking alcohol is healthy.
If drinking alcohol is healthy, then alcohols are prescribed by Doctors.
Hence, alcohols are prescribed by Doctors.

Solution: The argument is valid since it is of the form modus ponens. However, the conclusion is false. Here we observe that the first premise, p: drinking alcohol is healthy, is wrong. The second premise, $p \to q$, is then valid, and the conjunction of the two premises, $[p \wedge (p \to q)] \to q$, is false.

4.6 Law of Syllogism or Transitive Rule or Chain Rule

The logical implication of the WFF gives the second rule of inference:

$$((p \to q) \wedge (q \to r)) \to p \to r$$

where p, q, and r are any statements referred to as the Law of Syllogism (see Table 4.16). In tabular form, it is written as

$$\begin{array}{r} p \to q \\ q \to r \\ \hline \therefore p \to r(\text{conclusion}) \end{array} \quad \text{Premises} \quad \text{or} \quad \begin{array}{c} p \to q; q \to r \\ \hline p \to r \end{array}$$

Using the equivalence $p \to q = (\neg p) \vee q$, the above format can be further expressed as

$$\frac{\begin{array}{l}\neg p \vee q \\ \neg q \vee r\end{array}}{\neg p \vee r} \quad \text{or} \quad \frac{\neg p \vee q;\ \neg q \vee r}{\neg p \vee r}$$

The syllogism can be considered a **precursor** to the **resolution principle** in this form. Syllogism is a valid inference rule that can be proved by the truth table method, and it emerges as a tautology.

Example 4.14 For example, we may use the Law of Syllogism as follows:

(i) If the integer 35244 is divisible by 396, then the integer 35244 is divisible by **66**

$$p \rightarrow q$$

(ii) If the integer 35244 is divisible by 66, then the integer 35244 is divisible by 3

$$q \rightarrow r$$

(iii) Therefore, if the integer 35244 is divisible by 396, then the integer 35244 is divisible by 3.

$$\overline{\therefore p \rightarrow r}$$

4.7 Modus Tollens

The rule of inference is called Modus Tollens (MT); we mean the following WFF:

$$[(p \rightarrow q) \wedge (\neg q)] \rightarrow (\neg p)$$

This may also be written as

$$\frac{\begin{array}{r}p \rightarrow q \\ \neg q\end{array}}{\therefore \neg p}$$

This follows from the logical implication $[(p \rightarrow q) \wedge \neg q] \rightarrow \neg p$. Modus Tollens comes from Latin and can be translated as "*method of denying.*" This is appropriate because we deny the conclusion, q, to prove $\neg p$.

Note 4: We can also obtain this rule from the Modus Ponens by using the fact that $p \rightarrow q \Leftrightarrow \neg q \rightarrow \neg p$.

The following exemplifies the use of Modus Tollens in making a valid inference.

Example 4.15

(i) If Namrata is elected Secretary of the Delhi University female students, then Madhuri will promise that female students

$$p \to q$$

(ii) Madhuri did not assure Delhi University, female students

$$\frac{\neg q}{\neg p}$$

(iii) Therefore, Namrata was not elected Secretory of the Delhi University female student.

4.8 Some Other Basic Inference Forms Table 4.33

4.9 Resolution

In the resolution procedure of inference, there is only one deduction rule, the **resolution principle** (RP). The resolution principle can be put in only when all the premises are given as disjunctions. RP is used as the main inference engine in several automated reasoning programs. One prime instance is PROLOG, the language of artificial intelligence (AI). This principle was proposed in 1965 by **J. A. Robinson** and is the basis of many computer programs designed to automate a reasoning system.

Given primitive statements p, q, and r, the implication of $[(p \vee q) \wedge (\neg p \vee r)] \to (q \vee r)$ is a tautology. This tautology provides the inference rule known as **resolution**, where the conclusion $(q \vee r)$ is called the **resolvent**. The resolution principle has the following features:

(i) There is only one rule of inference resolution.
(ii) Resolution uses the rebuttal principle, i.e., negating the conclusion is inconsistent with the premises.
(iii) The only expressions allowed in resolution are disjunctions of WFFs or clauses. Usually, a clause is a disjunction of literals.

In applying resolution, each premises and the conclusion are written as clauses. A **clause** is a primitive statement or its negation, or it is the disjunction of terms, each of which is a primitive statement or the negation of such a statement. Hence the given rule has the clauses $(p \vee q)$ and $(\neg p \vee r)$ as premises and the clause $(q \vee r)$ as its conclusion or resolvent. If we have premise $\neg(p \wedge r)$, we replace this with the logically equivalent clause $(\neg p \vee \neg r)$, by the first of De Morgan's Laws. The two clauses $\neg p$, and $\neg q$ can replace the premise $\neg(p \vee r)$. This is due to the second De Morgan's Law and the rule of Conjunctive Simplication. For premise $p \vee (q \wedge r)$, we

Table 4.33 Some other basic inference forms

Conjunction (Conj) $\dfrac{p \quad q}{\therefore p \wedge q}$	Conjunctive Simplification or Simplification (Simp) $\dfrac{p \wedge q}{\therefore p}$
Disjunctive Amplification or Addition (Add.) $\dfrac{p}{\therefore p \vee q}$	Disjunctive Syllogism (DS) $\dfrac{p \vee q \quad \neg p}{\therefore q}$
Modus Ponens (MP) $\dfrac{p \quad p \to q}{\therefore q}$	Modus Tollens (MT) $\dfrac{p \to q \quad \neg q}{\therefore \neg p}$
Hypothetical Syllogism (HS) $\dfrac{p \to q \quad q \to r}{\therefore p \to r}$	Absorption (Abs.) $\dfrac{p \to q}{\therefore p \to (p \wedge q)}$
Constructive Dilemma (CD) $\dfrac{p \to q \quad r \to s \quad p \vee r}{\therefore q \vee s}$ $[(p \to q) \wedge (r \to s) \wedge (p \vee r)] \to (q \vee s)$ Rule of Contradiction $\dfrac{\neg p \to F}{\therefore p}$ $(\neg p \to F) \to p$	Destructive Dilemma (DD) $\dfrac{p \to q \quad r \to s \quad \neg q \vee \neg s}{\therefore \neg p \vee \neg r}$ $[(p \to q) \wedge (r \to s) \wedge (\neg q \vee \neg s)] \to (\neg p \vee \neg r)$ Rule for Proof by Cases $\dfrac{p \to r \quad q \to r}{\therefore (p \vee q) \to r}$ $[(p \to r) \wedge (q \to r)] \to [(p \vee q) \to r]$

apply the Distributive Law of $\vee$ over $\wedge$ and the rule of Conjunctive Simplication to arrive at either of the two clauses $p \vee q, p \vee r$. Finally, the premise $p \to q$ becomes the clause $\neg p \vee q$.

Example 4.16 Could you set up the validity of the following arguments using the resolution (along with rules of inference and the laws of logic)?

(i)
$$\begin{array}{c} p \vee (q \wedge r) \\ p \to s \\ \hline \therefore r \vee s \end{array}$$

(ii)
$$\begin{array}{c} p \\ p \leftrightarrow q \\ \hline \therefore q \end{array}$$

(iii)
$$\begin{array}{c} p \vee q \\ p \to r \\ r \to s \\ \hline \therefore q \vee s \end{array}$$

(iv)
$$\begin{array}{c} \neg p \vee q \vee r \\ \neg p \\ \neg r \\ \hline \therefore \neg p \end{array}$$

(v)
$$\begin{array}{c} \neg p \vee s \\ \neg t \vee (s \wedge r) \\ \neg q \vee r \\ p \vee q \vee t \\ \hline \therefore r \vee s \end{array}$$

4.10 Predicates and Quantification

Sets can be defined by a property $P(x)$ that elements of the set have in common. Thus an element of $\{x : P(x)\}$ is an object t for which the statement $P(t)$ is true. Such a statement $P(x)$ is called a predicate because, in English, the property is grammatically a predicate. $P(x)$ is also called a propositional function. After all, each choice of x, produces a proposition $P(x)$ that is either true or false.

Example 4.17 Let $A = \{x : x \text{ is an integer less than } 12\}$. Here $P(x)$ is the sentence "x is an integer less than 12." The common property is " is an integer less than 12". Since $P(1)$ is true, $1 \in A$.

4.10.1 Universal Quantification

The universal quantification of a predicate $P(x)$ is the statement, "For all values of x, $P(x)$ is true." We assume here that only values of x that make sense in $P(x)$ are considered. If we wish to restrict the value of x, we can, for example, write $\forall\, x \geq 0$, or $\forall\, n \in \mathbb{Z}$. The universal quantification of $P(x)$ is denoted by $\forall\, x\, P(x)$. The symbol $\forall$ is called the universal quantifier.

Example 4.18

(a) The sentence $P(x)\,:\,-(-x)$ is a predicate that makes sense for real numbers x. The universal quantification of $P(x), \forall\, x\ P(x)$ is true because for all real numbers $-(-x) = x$.
(b) Let $Q(x) : x + 1 < 7$. Then $\forall\, x \geq 0,\ Q(x)$ is false because $Q(9)$ is not true.

Universal quantification can also be stated in the English language as "for every x," "every x," or "for any x." A predicate may contain several variables. Universal quantification may be applied to each of the variables. For instance, a commutative property can be expressed as $\forall\, x\, \forall\, y\ \ x \circ y = y \circ x$. The order in which the universal quantifiers are considered does not change the truth value.

4.10.2 Existential Quantification

The existential quantification of a predicate $P(x)$ is the statement "there exists a value of x, $P(x)$ is true." The existential quantification of the predicate $P(x)$ is denoted by $\exists\, x\ P(x)$. The symbol $\exists$ is called the existential quantifier, which symbolizes expressions such as "there exists at least one x such that" or "there exists an x such that or" "for some x." We may include restrictions in the quantifier, such as $\exists\, x > 0$.

Example 4.19

(a) Let $Q(x) : x + 1 < 5$. The existential quantification of $Q(x)$, $\exists\, x\, Q(x)$, is a true statement because $Q(3)$ is a true statement.
(b) The statement $\exists\, z\ \ z + 3 = z$ is false. There is no value of z for which the propositional function $z + 3 = z$ produces a true statement.

In the English language, $\exists\, x$ can be read as "there is an x," "there is some x," "there exists an x," or "there is at least one x."

(c) Could you consider the following examples? There exists a man. Some men are honest. Some real numbers are integers.

$$M(x) : x \text{ is a man},\ \ C(x) : x \text{ is honest},$$

$$R(x) : x \text{ is a real number},\ \ I(x) : x \text{ is an integer}.$$

Then the above statements can be expressed as follows:
$(\exists\, x)(M\,(x))$; $(\exists\, x)(M\,(x) \wedge C(x))$; $(\exists\, x)(R(x) \wedge I\,(x))$, respectively.

4.10.3 Free and Bound Variables

We have discussed well-formed formula (wff). We shall write wff only for a formula. Now we will discuss a formula containing a part of the form $(\forall\, x)\; P(x)$ or $(\exists\, x)\; P(x)$. Here P(x) is described as **scope of the quantifier**. In this case, languages will consist of the connectives $\forall$, $\exists$ besides the other connectives.

A **string** is a finite sequence of symbols that are chosen from a set called an alphabet. A string of formulas is defined as (i) any formula is a string of formulas (ii) if α and β are a string of formulas, then α, β, and β, α are strings of formulas. (iii) Only those strings that are obtained by steps (i) and (ii) are strings of formulas, except the empty string, which is also a string of formulas.

A letter x is said to be **bound** in a string S if $(\forall\, x)$ or $(\exists\, x)$ appears in S; otherwise, x is said to be *bound free*. In other words, x is **free** in S if x does not appear in S. Every variable occurrence must be bound in a statement, and no variable should have a free event. When a free variable occurs in a formula, we have a statement function.

Example 4.20 Let us consider the following formula:

$$(\forall\, x)\; P(x, y) \tag{4.1}$$

$$(\forall x)(P(x) \Rightarrow Q(x)) \tag{4.2}$$

$$(\forall\, x)(P(x) \Rightarrow (\exists\, y)R(x, y)) \tag{4.3}$$

$$(\forall\, x)(P(x) \Rightarrow R(x)) \text{ and } (\forall\, x)(P(x) \Rightarrow Q(x)) \tag{4.4}$$

$$(\exists\, x)P(x) \wedge Q(x) \tag{4.5}$$

In (4.1), both occurrences of x are bound, while the variable y is free. In (4.2), all occurrences of x are bound. In (4.3), all occurrences of both x and y are bound. In (4.4), the scope of the first quantifier is $P(x) \Rightarrow R(x)$, and the scope of the second is $P(x) \Rightarrow Q(x)$. All occurrences of x are bound occurrences. In (4.5), the last occurrence of x in $Q(x)$ is free. Here, we observe that in the bound occurrence of a variable, the letter used to represent the variable is unimportant; in fact, any other letter can be used as a variable without affecting the formula, provided the new letter is not used elsewhere in the formula. Thus, $\forall\, x\, P(x, y)$ and $\forall\, x\, P(z, y)$ are the same. A constant cannot substitute the bound occurrence of a variable; only a free occurrence of a variable can be. For example, $\forall\, x\, P(x) \wedge Q(a)$ is a substitution instance of $(\forall\, x)P(x) \wedge Q(y)$.

4.11 Three-Valued Logic

What might now be known as the classical system of three-valued logic was proposed by **Lukasiewicz** in 1920. Further, **Lukasiewicz and Tarski** in 1930 proposed a many-valued system axiomatized by **Wajsberg** in 1932. In many ways, the classical two-valued logic can be elongated into a three-valued sense. Some three-valued reasoning, each with its inherent hypothesis, is now well-accepted. It is prevalent in these logics to denote the truth, falsity, and indeterminacy by 1, 0, and 1/2, respectively. It is also prevailing to define the negation of a proposition $\overline{x}$ of a proposition x as $1-x$. That is, $\overline{1} = 0$, $\overline{0} = 1$, and $\overline{1/2} = 1/2$. Further primitives, such as $\wedge$, $\vee$, $\Rightarrow$, and $\Leftrightarrow$, differ from one three-valued logic to another. Some of the celebrated three-valued logics, designated by the names of their architects, are defined in terms of these four primitives in the tables given as under.

Let x and y are propositions. The logical connectives, negation ($\overline{x}$), conjunction ($\wedge$), disjunction ($\vee$), implication ($\Rightarrow$), and equivalence ($\Leftrightarrow$) are defined in terms of these five primitives; defined these three-valued logics are given Tables 4.34, 4.35, and 4.36.

Truth Values for $\wedge, \vee, \Rightarrow, \Leftrightarrow$ in *Lukasiewicz and Bochvar Logics*, respectively.

Truth Values for $\wedge, \vee, \Rightarrow, \Leftrightarrow$ in *Kleene and Heyting Three-Valued Logics* respectively.

Truth Values for $\overline{x}, \wedge, \vee, \Rightarrow, \Leftrightarrow$ in *Reichenbach three-Valued Logics.*

From Tables 4.34, 4.35, and 4.36, we can easily observe that all the logic primitives for the Lukasiewicz, Bochvar, Kleene, Heyting, and Reichenbach three-valued logics fully adhere to conventional interpretations of these primitives in the classical logic for all $x, y \in \{0, 1\}$ and that they differ from each other only in their ministration of

Table 4.34 For $\overline{x}, \wedge, \vee, \Rightarrow, \Leftrightarrow$ in **Lukasiewicz's** three-valued logic **Bochvar's** three-valued logic

x	y	$x \wedge y$	$x \vee y$	$x \Rightarrow y$	$x \Leftrightarrow y$	$x \wedge y$	$x \vee y$	$x \Rightarrow y$	$x \Leftrightarrow y$
0	0	0	0	1	1	0	0	1	1
0	1/2	0	1/2	1	½	½	1/2	½	1/2
0	1	0	1	1	0	0	1	1	0
1/2	0	0	½	1/2	1/2	½	½	1/2	1/2
1/2	1/2	1/2	1/2	1	1	½	½	1/2	1/2
1/2	1	½	1	1	1/2	½	½	1/2	1/2
1	0	1	1	0	0	0	1	0	0
1	1/2	0	1	1/2	1/2	½	½	1/2	1/2
1	1	1	1	1	1	1	1	1	1

Table 4.35 For $\overline{x}$, $\wedge$, $\vee$, $\Rightarrow$, $\Leftrightarrow$ in **Kleene's** three-valued logic **Heyting's** three-valued logic

x	y	$x \wedge y$	$x \vee y$	$x \Rightarrow y$	$x \Leftrightarrow y$	$x \wedge y$	$x \vee y$	$x \Rightarrow y$	$x \Leftrightarrow y$
0	0	0	0	1	1	0	0	1	1
0	½	0	½	1	½	0	½	1	0
0	1	0	1	1	0	0	1	1	0
½	0	0	½	½	½	0	½	0	0
½	½	½	½	½	½	½	½	1	1
½	1	½	1	1	½	½	1	1	½
1	0	0	1	0	0	0	1	0	0
1	½	½	1	½	½	½	1	½	½
1	1	1	1	1	1	1	1	1	0

Table 4.36 For $\overline{x}$, $\wedge$, $\vee$, $\Rightarrow$, $\Leftrightarrow$ in **Reichenbach's** three-valued logic

x	y	$\overline{x}$	$\overline{y}$	$x \wedge y$	$x \vee y$	$x \Rightarrow y$	$x \Leftrightarrow y$
0	0	**1**	**1**	0	0	1	1
0	1/2	**1**	1/2	0	1/2	1	1/2
0	1	**1**	**0**	0	1	1	0
1/2	0	1/2	**1**	0	1/2	1/2	1/2
1/2	1/2	1/2	1/2	1/2	1/2	1	1
1/2	1	1/2	**0**	1/2	1	1	1/2
1	0	**0**	**1**	0	1	0	0
1	1/2	**0**	1/2	1/2	1	1/2	1/2
1	1	**0**	**0**	1	1	1	1

the new truth value $\frac{1}{2}$. One can substantiate that none of these three-valued logics appease the law of contradiction, i.e., $x \wedge \overline{x} = 0$, the law of excluded middle, i.e., $x \vee \overline{x} = 1$, and some other tautologies of two-valued logic. Lukasiewicz's logic L_3 satisfies all the first nine laws of classical logic. The Bochvar three-valued logic does not fulfill or meet any of the tautologies of the two-valued sense, because each of its primitives contributes the truth value $\frac{1}{2}$ on each occasion when at least one of the propositions x and y takes the value $\frac{1}{2}$.This is the reason behind the extension of the notion of the tautology to the extensive idea of a quasi-tautology.

4.11.1 Quasi-Tautology

A logic formula in a three-valued logic that does not assume the truth value falsity (0) in any way of the truth values designated to its proposition variables is called a quasi-tautology.

4.11.2 Quasi-Contradiction

A logic formula in a three-valued logic that does not assume the truth value truth (1) any way of the truth values designated to its proposition variables is called a quasi-contradiction.

Now we will erect the truth table for the compound propositions $x \wedge \overline{x}$ and $x \vee \overline{x}$ in the following Table 4.37 for $x \wedge \overline{x}$ and $x \vee \overline{x}$

Obviously, $x \wedge \overline{x}$ is the general law of quasi-contradiction, and $x \vee \overline{x}$ is the general law of quasi-tautology.

Note 5: Just as the classical set theory is related to classical logic, the 3-set theory is intimately related to Lukasiewicz's sense L_3. This has been prorated in Table 4.38.

4.12 Many-Valued Logic

When the several three-valued logics were acknowledged as significant and helpful, it allowed generalizations into n-valued logics for varying truth values ($n \geq 2$). Many n-valued logics evolved during 1930. For an arbitrary value of n, rational numbers in the unit interval [0, 1] labeled the truth values in these logics. One can obtain these values by dividing the unit interval uniformly, excluding 0 and 1. The set T_n of the truth values of an n-valued logic is defined as

$$T_n = \left\{0 = \frac{0}{n-1}, \frac{1}{n-1}, \frac{2}{n-1}, \frac{3}{n-1}, \ldots\ldots\ldots\ldots\ldots\ldots, \frac{n-2}{n-1}, \frac{n-1}{n-1} = 1\right\} \tag{4.6}$$

Table 4.37 General law of quasi-contradiction and quasi-tautology

x	$\overline{x}$	$x \wedge \overline{x}$	$x \vee \overline{x}$
0	1	0	1
$\frac{1}{2}$	$\frac{1}{2}$	$\frac{1}{2}$	$\frac{1}{2}$
1	0	0	1

Table 4.38 Relation between Lukasiewicz logic L_3 & 3-set theory $P_3(U)$

Lukasiewicz logic $\boldsymbol{L_3}$	3-set theory $\boldsymbol{P_3(U)}$
$\wedge$	$\cap$
$\vee$	$\cup$
$\neg$	$'$ or $-$
$\Rightarrow$	$\subseteq$
$\Leftrightarrow$	$=$

We elucidate these values as the *degree of truth*. Lukasiewicz propounded the n-valued logic in the forefront of 1930 as a stereotype of his three-valued logic. It employs truth values in T_n and describes the primitives by the following equations:

$$\begin{aligned} \overline{x} &= 1 - x \text{ or } T(\neg x) = 1 - T(p) \\ x \wedge y &= \min(x, y) \text{ or } T(x \wedge y) = \min\{T(x), T(y)\} \\ x \vee y &= \max(x, y) \text{ or } T(x \vee y) = \max\{T(x), T(y)\} \end{aligned} \tag{4.7}$$

$$\begin{aligned} x \Rightarrow y = \min(1, 1 + y - x) \text{ or } T(x \Rightarrow y) &= 1 - T(x) + T(y) \text{ if } T(x) > T(y) \\ &= 1 \text{ if } T(x) > T(y) \end{aligned}$$

This definition was proposed by Zadeh.

$$x \Leftrightarrow y = 1 - |x - y| \text{ or } T(x \Leftrightarrow y) = 1 - |T(x) - T(y)|$$

where x and y are fuzzy propositions and $T(x)$ and $T(y)$ are truth values that take values in $I = [0, 1]$.

Indeed, Lukasiewicz adopted only negation and implication as primitives and described other logic operations in terms of these two primitives as follows:

$$\begin{aligned} x \vee y &= (x \Rightarrow y) \Rightarrow y \\ x \wedge y &= \overline{\overline{x} \vee \overline{y}} \\ x \Leftrightarrow y &= (x \Rightarrow y) \wedge (y \Rightarrow x) \end{aligned}$$

One can substantiate that the Eq. (4.7) involves the definitions of the conventional primitives of two-valued logic for $n = 2$ and defines the primitives of Lukasiewicz's three-valued logic as proposed in Table 4.34. Symbolically L_n stands for the n-valued logic of Lukasiewicz. The truth values of L_n have been taken from T_n, and its primitives are defined by Table 4.34. The symbol L_2 stands for the classical two-valued logic, and L_∞ represents an infinite-valued logic whose truth values have been taken from the countable set T_∞ of all rational numbers in the unit interval [0, 1]. Logics L_2 and L_∞ are the two extreme cases of the sequence $(L_2, L_3, L_4, \ldots\ldots\ldots, L_\infty)$. In general, infinite-valued logic represents the logic whose truth values are represented by all the real numbers in the interval [0, 1]. This is also known as the *standard Lukasiewicz logic* L_1, the subscript 1 is an $\aleph_1$ (called aleph 1).

We know that the two-valued logic is isomorphic to the classical set theory analogously; the standard Lukasiewicz logic L_1 is isomorphic to fuzzy set theory on the standard fuzzy operators. The membership grades $A(x) \forall\, x \in X$ defined the fuzzy set A on the universe of discourse X. This can be interpreted as the truth values of the proposition "x is a member of the set A in L_1." On the contrary, the truth values $\forall\, x \in X$ of any proposition "x is P" in L_1, where P is a fuzzy predicate and can be elucidated as the membership degrees $P(x)$ by which the fuzzy set is described by the property P is defined on X.

The isomorphism follows since the logic operations of L_1, defined by Eq. (4.7) have precisely the matching mathematical form as the standard operations on fuzzy sets.

Note 6: For $n = 2$, we have $T(n) = \{0, 1, \}$. Similarly, for $n = 2$, we have $T(n) = \{0, 1/2, 1\}$ and L_n reduces to L_3. However, for $n \geq 4$, L_n is not a generalization of other three-valued logic, for instance, Bochvar's three-valued logic.

Example 4.21 Consider about the tenth-valued logic with $n = 10$ in Eq. (4.6). Then, we have

$$T_{10} = \left\{0, \frac{1}{9}, \frac{2}{9}, \frac{3}{9}, \frac{4}{9}, \frac{5}{9}, \frac{6}{9}, \frac{7}{9}, \frac{8}{9}, \frac{9}{9} = 1\right\}.$$

Now we wish to find the logical connectives.

Solution: Let us suppose that $x = \frac{4}{9}$ and $y = \frac{3}{9}$. Then $\overline{x} = 1 - \frac{4}{9} = \frac{5}{9}$

$$x \wedge y = \min(x, y) = \min\left(\frac{4}{9}, \frac{3}{9}\right) = \frac{3}{9}$$

$$x \vee y = \max(x, y) = \max\left(\frac{4}{9}, \frac{3}{9}\right) = \frac{4}{9}$$

$$x \Rightarrow y = min(1, 1 + y - x) = \min\left(1, 1 + \frac{3}{9} - \frac{4}{9}\right) = \min\left(1, \frac{8}{9}\right) = \frac{8}{9}$$

$$x \Leftrightarrow y = 1 - |x - y| = 1 - \left|\frac{4}{9} - \frac{3}{9}\right| = 1 - \frac{1}{9} = \frac{8}{9}$$

4.13 Fuzzy Logic

Lotfi Zadeh (a Professor of Electrical Engineering at the University of California, Berkeley) put the notion of fuzzy logic in his esteemed research paper: Fuzzy Logic and Approximate Reasoning, Synthese, 1975. The term fuzzy logic usually has been used in two distinct senses. So, explaining the difference between these two different usages of fuzzy logic is imperative. In a restricted sense, it refers to a logical system that induces classical two-valued logic for reasoning under uncertainty. In a sweeping sense, it refers to all of the theories and technologies that employ fuzzy sets, which are classes of unclear or blurry boundaries. The broad sense of fuzzy logic includes the restricted sense of fuzzy logic as a branch. It also includes other areas such as fuzzy pattern recognition, fuzzy mathematical programming, fuzzy control, fuzzy decision-making, fuzzy arithmetic, fuzzy neural networks, imprecise probability theory, fuzzy topology, and computing with words. In all of these areas, the notion of black-and-white is induced to a matter of degree. By accomplishing this, we perform two things:-

(i) Convenience of expressing human knowledge and comprehending unclear or cloudy concepts.
(ii) Amplify the ability to evolve a cost-effective solution to real-world problems.

4.14 Fuzzy Propositions

This section will discuss the propositional form of fuzzy logic, i.e., fuzzy propositional logic. In a broad sense, fuzzy logic holds within the logic developed so far and even more. Genuinely L_2 is a fuzzy logic in this sense. But, for the present, it is entirely appropriate to consider fuzzy logic with prejudice: fuzzy logic has truth values within [0, 1]. One of the rudimentary dissimilarities between crisp and fuzzy propositions is in the range of their truth values. Although all crisp propositions are needed to be either true or false, for the fuzzy propositions, the falsity, and truth, it is a matter of degree. It is customary to represent the degree of validity of every fuzzy proposition by a unique number in the unit interval. A **fuzzy proposition** is a statement p that acquires a fuzzy truth value $T(p)$ that ranges from 0 to 1. In other words, the truth value of the proposition p indicates the proposition's relation to the truth belonging to the unit interval.

In general, fuzzy propositions are classified into the following four categories:

(i) Unconditional and unqualified fuzzy propositions;
(ii) Unconditional and qualified fuzzy propositions;
(iii) Conditional and unqualified fuzzy propositions;
(iv) Conditional and qualified fuzzy propositions.

We will discuss each of them individually and introduce the appropriate canonical forms for them and their expositions.

4.14.1 *Unconditional and Unqualified Fuzzy Propositions*

Let p be a fuzzy proposition of such form. In canonical form, this fuzzy proposition can be expressed in the following format:

$$p : \mathcal{V} \text{ is } \mathcal{F} \tag{4.8}$$

In the above equation, $\mathcal{V}$ is the value of the variable v in the proposition p having taken from the universal set X, and $\mathcal{F}$ is a fuzzy set of X, which depicts a fuzzy predicate like costly, slow, high, young, ordinary, etc. For a specified value of $\mathcal{V}$

(assume v) with membership grade $\mathcal{F}(v) \in \mathcal{F}$, the degree of truth $T(p)$ of the proposition p is then determined by the membership grade $\mathcal{F}(v)$, precisely,

$$T(p) = \mathcal{F}(v) \tag{4.9}$$

$\forall$ specified value v of the variable $\mathcal{V}$ in the proposition p. It follows that T is a fuzzy set on [0, 1], which allocates the membership grade $\mathcal{F}(v) \forall\, v \in \mathcal{V}$.

Example 4.22 To make the notion described above clear, let us assume that the air humidity measured in Hygrometers (H_A) at some specified point on the globe be denoted by $\mathcal{V}$ and suppose that membership function represented by proposition "**dew point 75**" "unbearable" then, the corresponding fuzzy proposition p be stated as

$$p : \text{humidity}(\mathcal{V}) \text{ is unbearable } (H_A)$$

$T(p)$, the degree of truth is based on the actual value of the humidity and rests on the meaning of the predicate *unbearable* in the essence of Eq. (4.8) depicted in Fig. 4.1a and b. To exemplify it, let us assume that $v = 85$, then $\mathcal{F}(85) = 0.73 \Rightarrow T(p) = 0.73$.

In this way, we observe that the behavior pattern of the function T is to furnish us with a flyover in the company of fuzzy sets and fuzzy propositions. Even though the relationship connecting grades of membership in $\mathcal{F}$ and degrees of the truth of the linked fuzzy proposition p, as given by Eq. (4.9), is numerically frivolous for unqualified propositions. It has just theoretical importance.

In a few canonical forms of fuzzy propositions, the variable $\mathcal{V}$ values are allocated to particulars in a specified set I. In such a situation, $\mathcal{V}$ gets into a function $\mathcal{V} : I \to V$, where $\mathcal{V}(i)$ is the value of $\mathcal{V}$ for particular i in V, and therefore the canonical form of fuzzy propositions is altered to the form

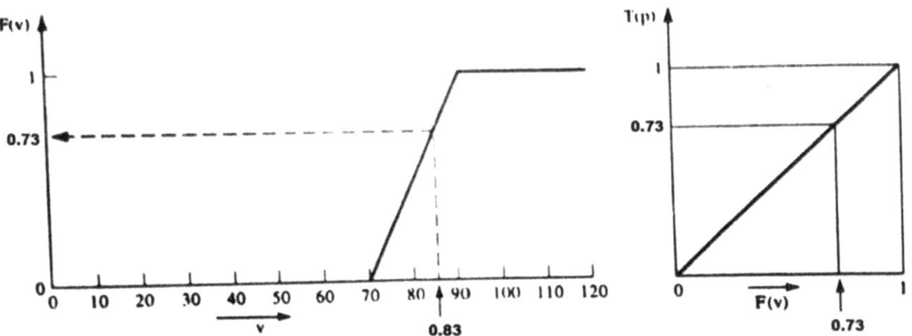

Fig. 4.1 **a** Representation of components of the fuzzy proposition p : **b** Representation of humidity ($\mathcal{V}$) is unbearable (H_A)

$$p : \mathcal{V}(i)\,\text{is}\,F, \forall\, i \in I \tag{4.10}$$

Example 4.23 To exemplify this, let us consider that I will be a collection of human beings, and each human is described by their age. Further, we assume a fuzzy set enunciating the predicate tall is specified. By designating the variable by height and the fuzzy set by tall, we can illustrate the general form given by Eq. (4.10) by the particular fuzzy proposition

$$p\text{: height}(i)\text{ is tall}$$

Then, the degree of truth $T(p)$ for this proposition can be resolved for each human $i \in I$ by the equation

$$T(p) : \text{tall}(\text{height}(i))$$

Section 3.14 of this book shows that any proposition of the configuration given by Eq. (4.8) can be explicated as a possibility distribution function r_F on V, which the equation

$$r_F(v) = F(v)\forall\, v \in V$$

can define. Indeed, one can apply this explanation to all propositions of the altered form given by (4.10) too.

4.14.2 *Unconditional and Qualified Fuzzy Propositions*

Let p be a fuzzy proposition of such form. In canonical form, this fuzzy proposition can be expressed in the following format:

$$p\text{: } \mathcal{V} \text{ is } \mathcal{F}\,\text{is}\, S \tag{4.11}$$

or in the following canonical format

$$p : \text{Pro}\{\mathcal{V} \text{ is } \mathcal{F}\}\,\text{is}\, P \tag{4.12}$$

In the above equation, $\mathcal{V}$ is the value of the variable v in the proposition p having taken from the universal set X, and $\mathcal{F}$ is a fuzzy set of X, as in Eq. (4.8) and Pro$\{\mathcal{V}$ is $\mathcal{F}\}$ stands for the probability of the fuzzy event "$\mathcal{V}$ is $\mathcal{F}$". Here S is a fuzzy truth qualifier. P is termed a fuzzy probability qualifier. In fancy, $\mathcal{V}$ may be restored with $\mathcal{V}(i)$, which has the identical connotation as in Eq. (4.10). The above two Eqs. (4.11) and (4.12) can be claimed as propositions of *truth qualified* and *probability qualified* respectively. At the same time S and P are fuzzy sets on [0, 1].

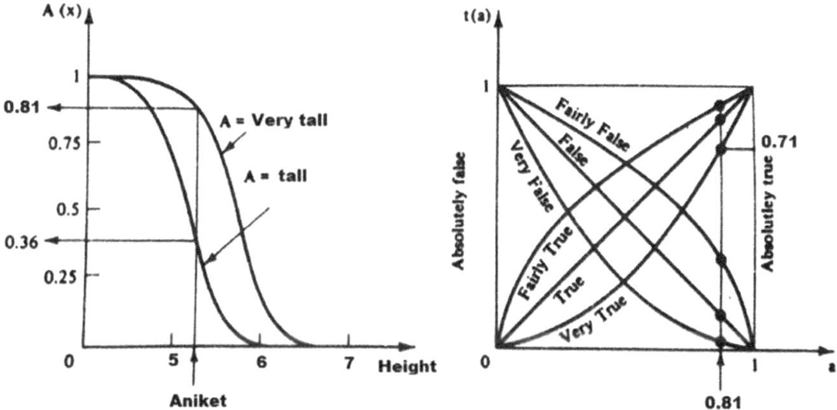

Fig. 4.2 **a** Representing fuzzy proposition tall. **b** Representing truth value "true" of the fuzzy proposition

Example 4.24 For instance, of a "truth-qualified proposition," we have the proposition "Aniket is tall is very true," where the predicate "tall" and the truth qualifier "very true" are depicted by the corresponding fuzzy sets as exemplified in Fig. 4.2. Let us presume that the height of Aniket is 6.4 feet; he is in the collection describing the predicate tall with a membership grade of 0.81. It means that our proposition belongs to the group of propositions that are very true with a membership grade of 0.71, as depicted in Fig. 4.2a and b. Suppose we customize the proposition young by altering the predicate to "very tall" or somewhat false, very accurate; we will obtain the corresponding degrees of the truth of the related proposition by the same procedure.

Primarily, the degree of truth, $T(P)$, of any truth-qualified proposition p is provided for each and every-one $v \in V$ by the equation

$$T(P) = S(F(v)) \tag{4.13}$$

Contemplating the membership function $G(v) = S(F(v)) \forall\ v \in V$ as a simple predicate, we can explicate any truth-qualified proposition of the configuration (4.11) as the unqualified proposition "$\mathcal{V}$ is G".

Here, we notice that the unqualified propositions are particular truth-qualified propositions, wherein the truth qualifier S is presumed valid. As exhibited in Fig. 4.2a and b, the membership function depicting this qualifier is the identity function. Specifically, $S(F(v)) = F(v)$ for unqualified propositions follows that, for the simple reason, we may overlook S.

Next, we will talk about probability-qualified propositions of appearance (4.12). One and all propositions of this category depict flexible limitation on feasible probability distributions on V, being provided probability distribution f on V; we encompass

$$\text{Pro}\{\mathcal{V} \text{ is } F\} = \sum_{v \in V} f(v) \cdot F(v) \tag{4.14}$$

Then, in this case, the scheme

$$T(P) = P\left(\sum_{v \in V} f(v) \cdot F(v)\right) \tag{4.15}$$

can state the degree $T(P)$ to whichever proposition p of the structure (4.12) is valid

Example 4.25 To illustrate the probability-qualified proposition, assume that the day-to-day air humidity at any place on the globe throughout the month is denoted by the variable $\mathcal{V}$ and is measured in Hygrometers (H_A). Then the probability-qualified proposition is expressed by

$$p : \text{Pro}\left\{\text{humidity is around (at a specified time and place)} 3.5\ \text{g}/m^3\right\} \text{ likely}$$

may furnish us with a powerful depiction of some climate features at a specified time and place. One may merge with similar propositions concerning extra facets like temperature, storm, cloudburst, etc. Suppose that the predicate "around 75 °F", "around 3.5 g/m^3" be depicted by the fuzzy set A on $\mathbb{R}$ described in Fig. 4.3a and the qualifier "likely" may be depicted by the fuzzy set on the closed interval [0, 1] represented in Fig. 4.3b.

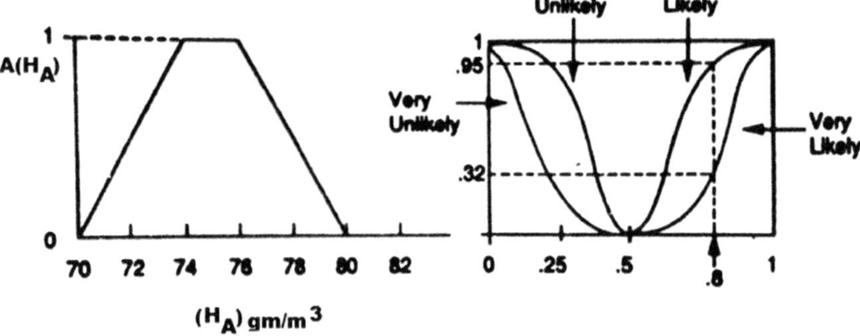

Fig. 4.3 **a** Representing probability-qualified proposition humidity. **b** Representing probability-qualified proposition qualifier likely

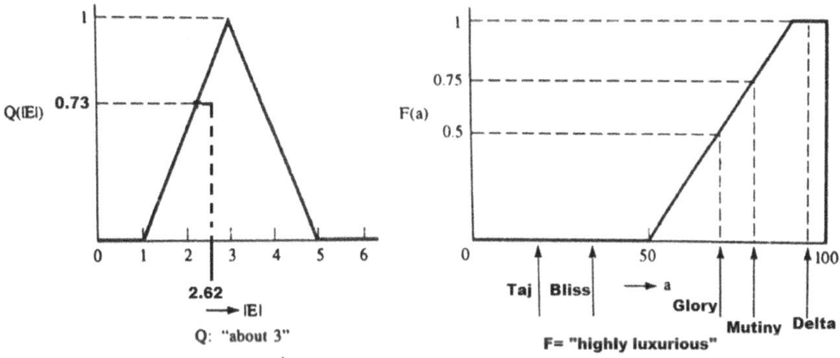

Fig. 4.4 **a** Representation of fuzzy sets in quantified fuzzy proposition (truth value of proposition). **b** Depicting the linguistic term "highly luxurious"

4.14.3 Conditional and Unqualified Fuzzy Propositions

Let p be a fuzzy proposition of such form. In canonical form, this fuzzy proposition can be demonstrated in the following format:

$$p : \text{If } \mathcal{X} \text{ is } P, \text{ then } \mathcal{Y} \text{ is } Q \tag{4.16}$$

In the above equation, $\mathcal{X}$ and $\mathcal{Y}$ are variables whose values belong to the X and Y collections, respectively, and P and Q are fuzzy sets on X and Y, respectively. One may also witness the propositions described above as the configuration

$$(\mathcal{X}, \mathcal{Y}) \text{ is } \mathbb{R} \tag{4.17}$$

Here $\mathbb{R}$ is the fuzzy set on the Cartesian product of $X \times Y$, being regulated for all $x \in X$ and $y \in Y$ by the technique

$$\mathbb{R}(\mathrm{x},\ \mathrm{y}) = \mathcal{J}\left[P(\mathrm{x}),\ Q(\mathrm{y})\right]$$

Here $\mathcal{J}$ signifies a binary composition on the closed interval [0, 1] characterizing a relevant fuzzy implication.

Now we will exemplify the relationship joining (4.16) and (4.17) for a specific fuzzy implication called the *Lukasiewicz* implication and is given by

$$\mathcal{J}(\mathrm{x},\ \mathrm{y}) = \min\left[1, 1 - \mathrm{x} + \mathrm{y}\right] \tag{4.18}$$

Example 4.26 Let us assume that $P = \frac{.2}{p_1} + \frac{.7}{p_2} + \frac{.9}{p_3}$ and $Q = \frac{.6}{q_1} + \frac{.9}{q_2}$. Then, ascertain $\mathbb{R}(\mathrm{x}\ ,\ \mathrm{y})$.

Solution: By the notion of proposition p, 'If $\mathcal{X}$ is P, then $\mathcal{Y}$ is Q', then we have

$$\mathbb{R}(\mathrm{p}_1, \mathrm{q}_1) = \min[1, 1 - .2 + .6] = \min[1, 1.4] = 1$$

$$\mathbb{R}(\mathrm{p}_1, \mathrm{q}_2) = \min[1, 1 - .2 + .9] = \min[1, 1.7] = 1$$

$$\mathbb{R}(\mathrm{p}_2, \mathrm{q}_1) = \min[1, 1 - .7 + .6] = \min[1, .9] = .9$$

$$\mathbb{R}(\mathrm{p}_2, \mathrm{q}_2) = \min[1, 1 - .7 + .9] = \min[1, 1.2] = 1$$

$$\mathbb{R}(\mathrm{p}_3, \mathrm{q}_1) = \min[1, 1 - .9 + .6] = \min[1, .7] = .7$$

$$\mathbb{R}(\mathrm{p}_3, \mathrm{q}_2) = \min[1, 1 - .9 + .9] = \min[1, 1] = 1$$

Then,

$$\mathbb{R}(\mathrm{x}\,,\ \mathrm{y}) = \frac{1}{\mathrm{p}_1,\ \mathrm{q}_1} + \frac{1}{\mathrm{p}_1,\ \mathrm{q}_2} + \frac{.9}{\mathrm{p}_2,\ \mathrm{q}_1} + \frac{1}{\mathrm{p}_2,\ \mathrm{q}_2} + \frac{.7}{\mathrm{p}_3,\ \mathrm{q}_1} + \frac{1}{\mathrm{p}_3,\ \mathrm{q}_2}$$

It implies that $T(p) = 1$ albeit $\mathcal{X} = \mathrm{p}_1$ and $\mathcal{Y} = \mathrm{q}_1$; $T(p) = 1$ when $\mathcal{X} = \mathrm{p}_1$ and $\mathcal{Y} = \mathrm{q}_2$;

$T(p) = 1$ when $\mathcal{X} = \mathrm{p}_2$ and $\mathcal{Y} = \mathrm{q}_2$; $T(p) = 1$ when $\mathcal{X} = \mathrm{p}_3$ and $\mathcal{Y} = \mathrm{q}_2$;

$T(p) = .9$ when $\mathcal{X} = \mathrm{p}_2$ and $\mathcal{Y} = \mathrm{q}_1$; $T(p) = .7$ when $\mathcal{X} = \mathrm{p}_3$ and $\mathcal{Y} = \mathrm{q}_1$.

4.14.4 Conditional and Qualified Fuzzy Propositions

Let p be a fuzzy proposition of similar form. In canonical form, such a fuzzy proposition can be depicted either in the following format:

$$p: \text{ If } \mathcal{X} \text{ is } P, \text{ then } \mathcal{Y} \text{ is } Q \text{ is } S \tag{4.19}$$

or in the canonical form

$$p: \mathrm{Pro}\{\mathcal{X} \text{ is } P : \mathcal{Y} \text{ is } Q\} \text{ is } R \tag{4.20}$$

In the above formula $\mathrm{Pro}\{\mathcal{X} \text{ is } P : \mathcal{Y} \text{ is } Q\}$ is a conditional probability.

As procedures established for the different varieties of propositions can be amalgamated to confront submissions of such form, it is optional to discuss them more.

4.15 Fuzzy Quantifiers

Fuzzy quantifiers, i.e., operators intended to provide a numerical interpretation of natural language (NL) quantifiers like "almost all," are valuable tools for image processing, mainly to express accumulative properties of fuzzy image regions. Fuzzy quantifiers like "few," "about half," "much more than," and many others abound in natural language. Humans use them to describe uncertain facts, quantitative relations, and processes. In this section, we will study the generic term "fuzzy quantifiers," which is being employed for the depiction of the collection of quantifiers in natural languages whose figurative elements are: several, most, much, not many, few, large number, approximately 12, quite a few, frequently, usually, almost all, at least about 7, about half, much more than, about 100, etc. In our approach, such quantifiers are treated as fuzzy numbers, which may be manipulated using fuzzy arithmetic and, more generally, fuzzy logic.

We have seen that the universal quantifiers for all and the existential quantifier that exists are the two quantifiers of predicate logic. The procedure of interpreting natural-language sentences in the direction of the language of quantifiers compels us to classify whether we are talking about a class's elements or only about a few specific individuals. Naturally, when we mention all members of a set, there is no difficulty in expressing the whole meaning and truth of the sentence in natural language. For instance, the sentence

"All dolphins are mammals" is undisputedly interpreted as

For all a, if a is a dolphin, then a is a mammal, or
$(\forall a)(Da \Rightarrow Ma)$

Although, we were confronted with some awkward options when interpreting a slighter specific sentence. For instance, the sentence

Some dolphins are toxic cocktails.

It is interpreted in symbolic logic in the form of an existential quantifiers as

$$(\exists a)(Da \wedge Ta)$$

From the sentence described above, we mean that at least one toxic dolphin exists and expresses a minimum explanation of the primary sentence. If the sentence provided in English is correct, then the exemplified one must be correct.

We would like to clear the primary English sentence of a specified quantity of explanation as it assigns a distinctive to a limited or small number of dolphins. To a great extent, classical logic is related much more to exhibiting the logical structure of propositions, seldom showing their essence. However, if our sentence is of the form

Almost all dolphins are cocktails.

then, we have to produce greater renunciation in terms of meaning. Then the new sentence would be interpreted as

$$(\exists a)(Da \wedge Ta)$$

as we do not promise a universal assertion.

Thus, we observe that after being confined to just two quantifiers, we have to leave a more significant amount of information regarding the matter concerned to make the logical form of our submission crystal clear. Fuzzy quantifiers are appliances for exemplifying specified statements that reduce the drop of information obligatory under the option of the quantifier.

Fuzzy propositions of any kind whatsoever are discussed in Sect. 4.13, can be quantified by a proper fuzzy quantifier. Generally, fuzzy quantifiers are fuzzy numbers that participate in several propositional forms and influence the authenticity of particular fuzzy propositions. All fuzzy quantifiers convey rough estimates of elements in a prescribed universal set that appeases the specified property. Generally, fuzzy quantifiers are classified into two categories.

(i) Absolute quantifiers
(ii) Relative quantifiers.

4.15.1 Absolute Quantifiers

Such quantifiers are defined on a set of real numbers $\mathbb{R}$ or a set of integers. They portray linguistic terms such as about a fortnight, at most about 21, much more than 67, at least about 99, somewhere around, roundabout, much greater than, and the like.

Example 4.27 For an example of fuzzy propositions with fuzzy quantifiers are
At most, six hotels are near the railway station, which is highly luxurious.
About 11 girl's students in the class are highly talented in mathematics.
It takes at least about a fortnight of days for the monsoon to come to our state.

In general, two kinds of propositions consist of absolute fuzzy quantifiers. The first one is of the form

$$p: \text{ There are } A\ i's \text{ in } I \text{ such that } \mathcal{V}(i) \text{ is } F \tag{4.21}$$

In the above formula, $\mathcal{V}$ is a variable $\forall\, i \in I$ that attains a value $\mathcal{V}(i)$, F is a fuzzy set on the values of the variable $\mathcal{V}$, and A is a fuzzy number on $\mathbb{R}$. Here, I is an index set by that well-defined estimation of variables V are esteemed.

Any proposition p of the configuration (4.21) can be transformed into another proposition of some uncomplicated arrangement as

$$p': \text{There are } A\ E's \tag{4.22}$$

In the above formula, A is the identical quantifier as in Eq. (4.21), and E is a fuzzy set on a given set I and is defined by the formation

$$E(i) = F(\mathcal{V}(i))\forall\, i \in I \tag{4.23}$$

To illustrate the above concept, consider the proposition, "There are about six motels in the vicinity of the railway station, which have highly luxurious rooms." Provided a set of motels, I, the value $\mathcal{V}(i)$ of variable $\mathcal{V}$ denotes in the proposition the degree of luxury in a Motel's room in the vicinity of a railway station i expressed by numbers in the closed interval [0, 1], F is a fuzzy set characterized by the set of values of variables $\mathcal{V}$ that represents the linguistic term high, and A is a fuzzy number demonstrating the linguistic term "about six."

For an illustration of the proposition (4.22), the proposition

p: "There are about six motels in the vicinity of the railway station,which have highly luxurious rooms" can be changed by the proposition

p': "There are about six high-luxurious room motels near the railway station."

In this proposition, E is the fuzzy set of high-luxurious room motels near the railway station.

We may contemplate proposition p' given by (4.22) as a straightforward expression of proposition p given by (4.21). Using the short form in place of the complete form is natural. We accept this utilization to make it simple. In this case, the proposition $p^{'}$ of the configuration (4.22) may be rephrased as

$$p' : \mathcal{W}\,\text{is}\,A, \tag{4.24}$$

$\mathcal{W}$ is a variable taking value in $\mathbb{R}$, which stands for the scalar cardinality (sigma count) of the fuzzy set E. That is, we have $\mathcal{W} = |E|$. Evidently

$$|E| = \sum_{i \in I} E(i) = \sum_{i \in I} F(\mathcal{V}(i))$$

Also, for every fuzzy set E, we get

$$T(p) = T\left(p'\right) = A(|E|). \tag{4.25}$$

Example 4.28 Let us talk about the proposition p, given by p : There are about three motels in I whose luxurious room $\mathcal{V}(i)$, is high.

Suppose that $I = \{\text{Taj, Bliss, Glory, Mutiny, Delta}\}$, and $\mathcal{V}$ is a variable with values in the closed interval [0, 100] that intimates the degree of luxury in the room. On comparing this proposition with its general counterpart given by (4.21), we can say in this case that, A is a fuzzy quantifier "about 3," and F is a fuzzy set on [0, 100] that represents the linguistic term "highly luxurious." We suppose that the ensuing scores are provided: $\mathcal{V}(\text{Taj}) = 11, \mathcal{V}(\text{Bliss}) = 13, \mathcal{V}(\text{Glory}) = 52, \mathcal{V}(\text{Mutiny}) = 71, \mathcal{V}(\text{Delta}) = 95$. We wish to determine the proposition's p truth value.

To determine the truth value of the proposition, firstly, we will construct the fuzzy set authenticated by Eq. (4.23)

$$E = \frac{0.2}{\text{Taj}} + \frac{0.3}{\text{Bliss}} + \frac{0.43}{\text{Glory}} + \frac{0.69}{\text{Mutiny}} + \frac{1.0}{\text{Delta}}$$

Afterward, we will compute the cardinality of the fuzzy set E:

$$|E| = \sum_{i \in I} E(i) = 0.2 + 0.3 + 0.43 + 0.69 + 1.0 = 2.62$$

Eventually, we make use of Eq. (4.25) to gather the truth value of its proposition:

$$T(p) = T\left(p'\right) = A(|E|) = A(2.62) = 0.728 = 0.73$$

Instead, we assume that the motel's scores are anonymous/unknown. So we are unable to raise the set E. In this situation, the proposition bestows us the details regarding the degrees of possibility of numerous values of the cardinality of E.

For example, as $A(3) = 1$, the possibility of $|E| = 3$ is one; since $A(5) = 0$, it is unfeasible that $|E| = 5$, and so forth.

We can also observe that the fuzzy proposition of the initial form may too emerge in the recommendations of the type

$$p : \text{There are } A\ i's\ in\ I, \text{ such that } \mathcal{V}_1\ (i) \text{ is } F_1 \text{ and } \mathcal{V}_2\ (i) \text{ is } F_2 \tag{4.26}$$

In the above equation, $\mathcal{V}_1$ and $\mathcal{V}_2$ are variables that successively capture values from the collections V_1 and V_2. The symbol I stands for the index set from which clear quantifications of variables $\mathcal{V}_1$ and $\mathcal{V}_2$ are detected, and A is a fuzzy number on $\mathbb{R}$. Also, F_1 and F_2 are the fuzzy sets on V_1 and V_2 consecutively.

Consider a quantified fuzzy proposition: "There are about six motels near the Gaya Ji railway station whose luxury rooms are high and spacious." In this manner, I here represent the index set from which motels in the stated vicinity of the railway station, motel are marked; the variables $\mathcal{V}_1$ and $\mathcal{V}_2$ portray luxury in room and area of the room, and A is a fuzzy number that apprehends the linguistic term "about 6", and F_1, F_2 are fuzzy sets that portrayed the linguistic terms "highly" and "spacious" successively.

Now, any proposition p of the configuration (4.26) can be enunciated in a simple form as

$$p' : A\ E_1's\ E_2's \tag{4.27}$$

Here A is an identical quantifier as in Eq. (4.26), and E_1 and E_2 are fuzzy sets on I, which are described by the compositions:

$$E_1(i) = F_1(\mathcal{V}_1(i)) \text{ and } E_2(i) = F_2(\mathcal{V}_2(i))\ \forall\ i \in I \tag{4.28}$$

Furthermore, (4.27) may be transliterated as

$$p' : \text{There are } A(E_1 \text{ and } E_2)'s \tag{4.29}$$

Equation (4.29) can be rephrased in the following form after comparing Eqs. (4.22) and (4.29)

$$p' : \mathcal{W} \text{ is } A \tag{4.30}$$

Here $\mathcal{W} = |E_1 \cap E_2|$ denotes the scalar cardinality of the fuzzy set $E_1 \cap E_2$ and is a variable that takes value in $\mathbb{R}$.

By standard intersection, we have

$$\mathcal{W} = |E_1 \cap E_2| = \sum_{i \in i} \min\{E_1, E_2\} = \sum_{i \in i} \min\{F_1(\mathcal{V}_1(i)), F_2(\mathcal{V}_2(i))\}, \text{ using (4.34)} \tag{4.31}$$

Moreover, for given fuzzy sets E_1 and E_2,

$$T(p) = T(p') = A(\mathcal{W}) \tag{4.32}$$

In addition, the proposition p and its equivalent proposition p' persuade a possibility distribution function, r_A, which is defined for every one $\mathcal{W} = |E_1 \cap E_2|$ by the equation

$$r_A(\mathcal{W}) = A(\mathcal{W}) \tag{4.33}$$

Now, we canvass fuzzy propositions having quantifiers of another type, usually referred to as **relative quantifiers**.

4.15.2 Relative Quantifiers

Relative quantifiers are expressed by fuzzy numbers defined on the unit interval [0, 1]. Relative quantifiers characterize quantifiers/linguistic terms such as "about 30%," "almost all," "most," "almost all," "about half," and the like.

Example 4.29

(i) Almost all students in a given class are fluent in English.
(ii) About half of the families living in the colony are fluent in English.
(iii) Most of the students in this colony are young girls.

Illustrations of fuzzy quantifiers of such type are depicted in Fig. 4.5.

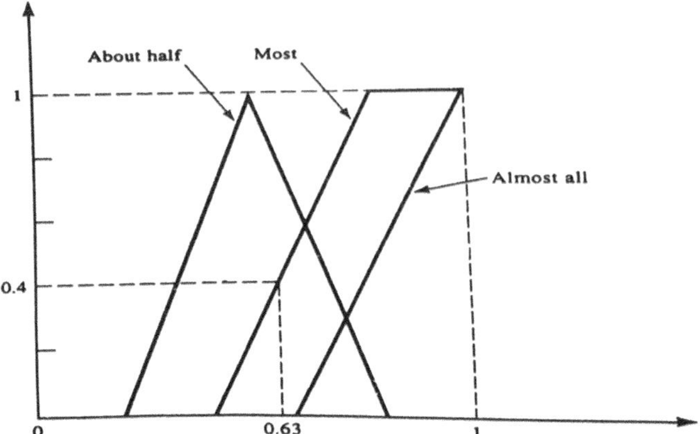

Fig. 4.5 Representation of relative quantifiers or fuzzy quantifiers of the second kind

Fuzzy propositions with relative quantifiers have a standard configuration, such as

$$p: \text{Among } i's \text{ in } I \text{ such that } \mathcal{V}_1(i) \text{ is } F_1 \text{ there are } A\ i's \text{ in } I \text{ such that } \mathcal{V}_2(i) \text{ is } F_2 \tag{4.34}$$

In the above equation, A is a fuzzy number on the unit interval [0, 1] and the other symbols have their meanings as usual.

Example 4.30 To illustrate the proposition of the form described above, consider the proposition, "Among families in a given colony whose daughters are young, almost all have a high fluency in English." We may rewrite any proposition of the configuration (4.34) in a simple way as

$$p' : A\ E_1's \text{ are } E_2's \tag{4.35}$$

In the above equation A is the identical quantifier as in (4.34), and E_1, E_2 are fuzzy sets on X and are given by

$$E_1(i) = F_1(\mathcal{V}_1(i)) \text{ and } E_2(i) = F_2(\mathcal{V}_2(i)) \forall\ i \in I \tag{4.36}$$

Equation (4.35) can be rephrased in the following form after comparing Eqs. (4.27) and (4.35) as we can observe that both these propositions have the identical configuration. On comparison of the two configurations, Eq. (4.35) can be expressed in the form

$$p' : \mathcal{W} \text{ is } A \tag{4.37}$$

Here $\mathcal{W}$ is a variable that denotes the degree of subsethood of E_2 in E_1. Therefore, we

$$
\begin{aligned}
\mathcal{W} &= \frac{|E_1 \cap E_2|}{|E_1|} \\
&= \frac{\sum\limits_{i \in i} \min\{E_1, E_2\}}{\sum\limits_{i \in i} \min F_1(\mathcal{V}_1(i))} \\
&= \frac{\sum\limits_{i \in i} \min\{F_1(\mathcal{V}_1(i)), F_2(\mathcal{V}_2(i))\}}{\sum\limits_{i \in i} \min F_1(\mathcal{V}_1(i))} \ \forall \text{ given sets } E_1 \text{ and } E_2
\end{aligned}
\tag{4.38}
$$

One can derive $T(p)$ from (4.32). The feeling about the name assigned to the quantifiers; can make some sense by comparing the definitions of the variable $\mathcal{W}$ for the ***absolute quantifiers***, given by (4.31), and the ***relative quantifiers***, provided by (4.38).

4.16 Linguistic Hedges

A hedge is a modifier of a fuzzy set. It is an operation that amends the meaning of the primary set to produce a compound fuzzy set. Hedges are like adjectives or adverbs in the English language. They modify the shape of the fuzzy set's underlying too or dilute the membership relationships, and they convert classical numbers (scalars) into fuzzy sets through approximation. They can be used in rules to modify the evaluation of both the premise and the consequent. In linguistics, a **hedge is a word or phrase used in a sentence to express ambiguity, probability, caution, or indecisiveness about the remainder of the sentence**.

In other words, we can elucidate that linguistic hedges (or plainly hedges) are exceptional linguistic terms by which other linguistic terms are being amended. Linguistic terms like slightly, remarkably, almost, mostly, approximately, low, more or less, relatively, barely, roughly, somewhat, extremely, and significantly are some commonly used hedges. That is, the singular meaning of an atomic term is amended or hedged from its original interpretation. According to **Zadeh**, (1972) manipulating the fuzzy sets as the calculus of interpretation, these linguistic hedges have the consequence of amending the membership function for a basic atomic term. Thus, we can claim that linguistic hedges can be used for amending fuzzy predicates, fuzzy truth values and fuzzy probabilities. As an illustration, let us see at the basic linguistic atom, θ and subject it to some hedges. Let us define $\theta = \int_X \gamma_\theta(x)/x$, then

Very:

$$\gamma_{\text{Very}\,\theta}(x) = \left[\gamma_\theta(x)\right]^2 \quad \text{or} \quad \text{"very"}\,\theta = \theta^2 = \int_X \frac{\left[\gamma_\theta(x)\right]^2}{x}$$

$$\text{"very, very"}\,\theta = \left(\theta^2\right)^2 = \theta^4$$

$$\text{"Plus"}\,\theta = \theta^{1.25}$$

More or less:

$$\gamma_{\text{More or Less}\,\theta}(x) = \sqrt{\gamma_\theta(x)} \quad \text{or} \quad \text{"Slight"}\,\theta = \sqrt{\theta} = \int_X \frac{\left[\gamma_\theta(x)\right]^{0.5}}{x}$$

$$\text{"Minus"}\,\theta = \theta^{0.75}$$

The expressions described above are examples of linguistic hedges and are usually referred to as ***concentrations*** (**Zadeh**, 1972) depicted in Fig. 4.6a. Concentrations influence to compression of the elements of a fuzzy set by diminishing the degree of membership of all the elements that are "partly" in the set. The less an element is in a set, the more it is reduced in membership by concentration. The above formula depicting slightly and Minus are linguistic hedges called "***dilations***" depicted in Fig. 4.6b. Dilations extend or enlarge or widen a fuzzy set by increasing the membership of elements that are *partly* in the set (**Zadeh**, 1972).

We have illustrated the hedges "very" and "More or Less" in Fig. 4.7 and performed their application to the notion of "high temperature." On observing the figure, we see that "Very" has the consequence of contracting the membership function, whereas "broadens the membership function. This is instinctively tempting because the benchmark for "Very High" must be stiffer than "High," whereas the yardstick for "must" be moderate. In general, we will have

$$\gamma_{\text{Very}\,\theta}(x) \le \gamma_\theta(x) \le \gamma_{\text{More or Less}\,\theta}(x)$$

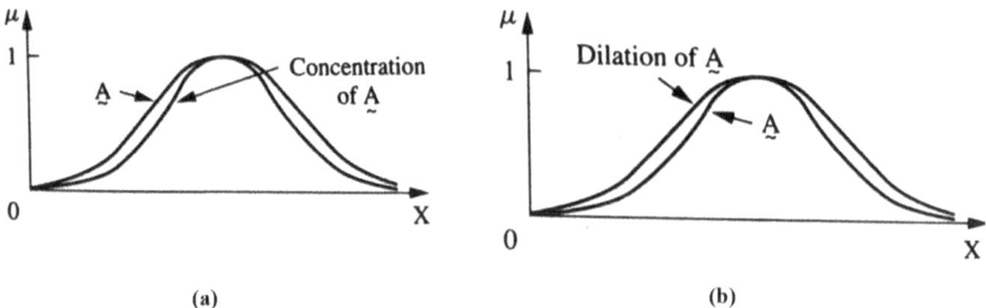

Fig. 4.6 **a** Fuzzy concentration, **b** Fuzzy dilation

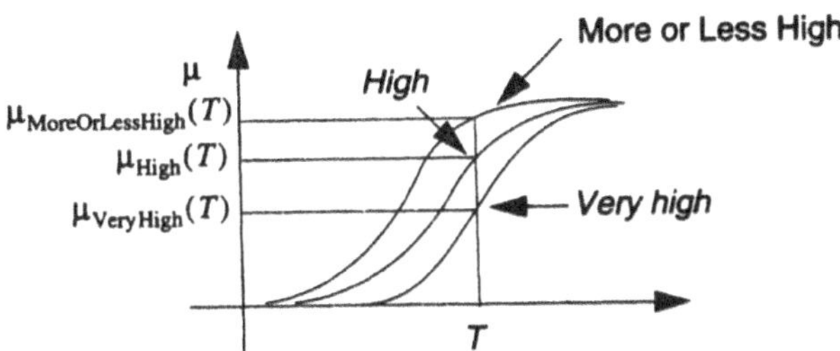

Fig. 4.7 Hedges-High, Very High, More, or Less High Temperature

In essence, a hedge can be implemented on any fuzzy set. The usual procedure can only be used when the compound term is substantial. For example, "very medium temperature" do not make much feeling.

As a consequence, the hedge "very" is mainly used with fuzzy sets on the two ends of the universe of discourse (for instance, small vs. giant, long vs. short, high vs. low).

Example 4.31 Consider, the proposition "*x* is tall", by which we mean that "*x* is tall is true", can be modified by the linguistic hedge "very" by anyway of the succeeding means

"*x* is very tall is true",

"*x* is tall is very true",

"*x* is very tall is very true".

Likewise, the proposition, "*x* is tall is likely" may be modified as *x* is tall is very likely". Normally, provided a proposition *p* such as

$$p_x\text{: '}x \text{ is } A\text{' is true}$$

And a linguistic hedge, *H* we can create a modified proposition,

$$Hp_x : \text{ '}x \text{ is } HA\text{' is true}$$

Here *HA* depicts the fuzzy predicate derived by administering the hedge *H* to the provided predicate *A*.

We may interpret, any linguistic hedge, *H*, as an *unitary operation h* on the unit interval [0, 1].

4.16.1 Modifiers

Unary operations that display linguistic hedges be usually referred to as modifiers. A modifier may enhance the ability to describe our fuzzy notions. Modifiers such as very, slightly used in the phrases as such very hot or somewhat cold amend the shape of the fuzzy set in such a way that outfit the explanation of the word used.

Example 4.32 In the absence of a straightforward membership function for a hedge, the hedge very is frequently elucidated as a unary operation $h(x) = x^2 \; \forall \; x \in [0, 1]$. Consequently, the hedge "very" has the net effect of squaring the truth value of the predicate it amends. The hedge fairly is frequently expounded as $h(x) = \sqrt{x} \; \forall \; x \in$ [0,1]. Thus, very, fairly are modifiers.

When a fuzzy predicate A is given w.r.t. the universal set X and a modifier h displaying a linguistic hedge H, the modified fuzzy predicate HA is resolved for all $x \in X$ by the equation

$$HA(x) = h(A(x))$$

It precisely hints that one can learn the characteristics and consequences of fuzzy hedges by examining the truth-changing properties of modifiers.

4.16.2 Strong Modifiers

Let h is a modifier. Hence, it is an increasing bijection (one–one onto). If a modifier h is such that $h(a) < a$ for all a in the unit interval is referred to as a strong modifier. That is, a modifier that strengthens a fuzzy predicate to which it has been enforced and, consequently, it decreases the truth value of the connected proposition is called as strong modifier. For example, extremely or very strengthens the fuzzy predicate to which they are applied.

Example 4.33 The proposition "Aniket is very tall' is very true" has a lower truth value than the proposition "Aniket is very tall" is true".

4.16.3 Weak Modifiers

A modifier h be dubbed as a weak modifier, if $h(a) > a$ for all a in the unit interval. On the contrary, a weak modifier weakens the predicate, and consequently, the proposition's truth value increases.

Example 4.34 Let us consider the following three fuzzy propositions: (all of whichever are true)

$$p_1 : \text{Brajesh is tall}$$

$$p_2 : \text{Brajesh is very tall}$$

$$p_3 : \text{Brajesh is fairly tall}$$

Moreover, assume that linguistic hedges fairly and very be displayed by the strong modifier a^2 and the weak modifier $\sqrt{a}$, respectively. We now assume that Brajesh is 6.3*ft* and that in accordance with the endorsed fuzzy set, TALL portraying the fuzzy predicate tall, *TALL*(6.3) $= (0.82)^2 = 0.67$, and *FAIRLY TALL*(6.3) $= \sqrt{0.82} = 0.91$. Hence, $T(p_1) = 0.82$, $T(p_2) = 0.67$, $T(p_3) = 0.91$. These values conform to our instinct. It means that the strong modifier reduces the truth value of the associated proposition, and the weak modifier increases the truth value of the related proposition. In other words, given the same data, the stronger statement is less accurate, and the weaker statement is more authentic.

4.16.4 *Identity Modifier*

The modifier for which $h(a) = a$ for all a in the unit interval is called the **identity (vacuous) modifier**.

Note 7: The linguistic term *not very* may be considered as the *negation* of the hedge very. Some people assert it as a new hedge that is some *weaker* than the hedge very.

Remarks 4.2 Every modifier h appeases the under-noted conditions:

(i) $h(0) = 0$ and $h(1) = 1$
(ii) The modifier h is a continuous function.
(iii) If h is weak, then h^{-1} is strong and conversely.
(iv) If h and g are two modifiers, their compositions $h \circ g$ and $g \circ f$ are also modifiers. Furthermore, if h and g are weak modifiers, then their composition is again weak and conversely.

4.16.5 *Class of Modifiers*

An advantageous class of functions that persuade the above-listed conditions is the class given by

$$h_\alpha(a) = a^\alpha \ \forall\, \alpha \in R^+$$

is a parameter by which particular modifiers in this class are discriminated and $a \in [0,1]$. Moreover, if

(i) $\alpha < 1$, h_α is a weak modifier.
(ii) $\alpha > 1$, h_α is a strong modifier.
(iii) h_1 is the identity modifier.

4.17 Inference from Conditional Fuzzy Propositions

All inference rules in classical logic are positioned on numerous tautologies. The inference rules can be generalized to accelerate the approximate reasoning within walls of fuzzy logic. Approximate reasoning is an essential application area of fuzzy set theory. It is necessary for modeling human common-sense reasoning. It is also decisive for expert systems in dealing with reasoning under an ambiguous or fuzzy environment. In this segment, we depict the observation of three classical inference rules, *modus ponens, modus tollens, and hypothetical syllogism.* These observations are constructed on the professed compositional rule of inference.

Contemplate variables $\mathcal{X}$ and $\mathcal{Y}$ that extract values from sets X and Y accordingly. Presume that the variables are interconnected by a functional relation $y = f(x)$, $\forall\, x \in X$ and $y \in Y$. Then, provided $\mathcal{X} = x$, we could deduce that $\mathcal{Y} = f(x)$ as depicted in Fig. 4.8a. Likewise, experiencing that the value of $\mathcal{X}$ is in a specified set A, we can deduce that the value of $\mathcal{Y}$ is in the set $B = \{y \in Y : y = f(x), x \in A\}$ as depicted in Fig. 4.8b.

Let the variables be connected by an arbitrary relation on the Cartesian product of X and Y, i.e., $X \times Y$. It doesn't have to be a function. Given $\mathcal{X} = u$ and a relation

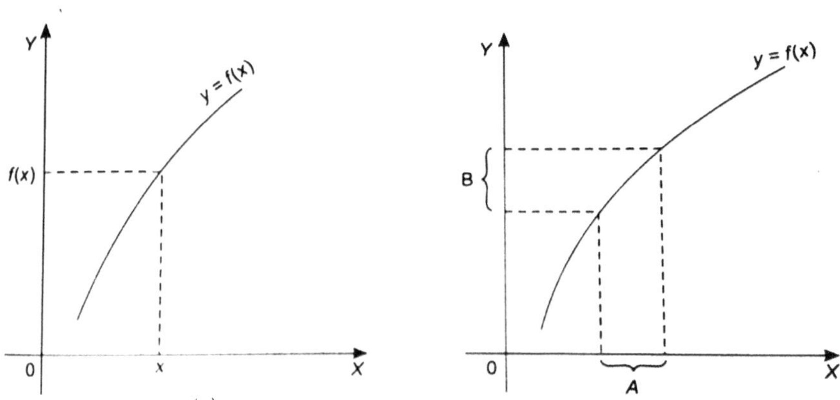

Fig. 4.8 Functional relation between two variables: **a** $x \to y$, where $y - f(x)$; **b** $A \to B$, where $B = \{y \in Y : y = f(x), x \in A\}$

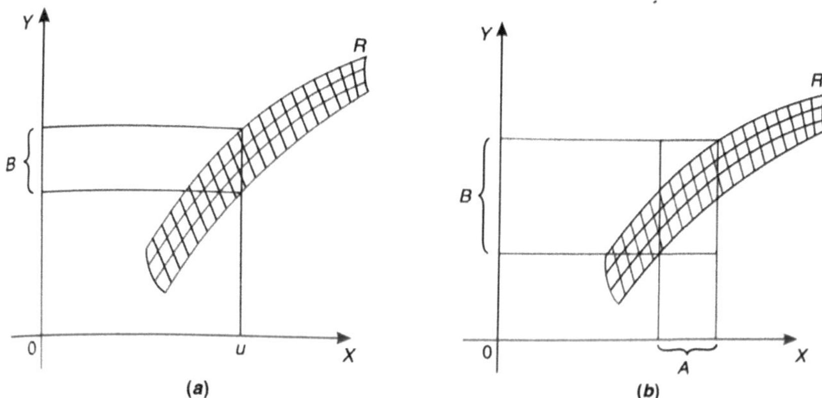

Fig. 4.9 **a** and **b** Inference expressed by Eq. (4.51)

R, we can deduce that $\mathcal{Y} \in B$, where $B = \{y \in Y :< x, y >\in R\}$, as depicted in Fig. 4.8a. Likewise, if we have the value of $\mathcal{X} \in A$, we can derive that $\mathcal{Y} \in B$, where $B = \{y \in Y :< x, y >\in R\}, x \in A$ as depicted in Fig. 4.8b. Here we perceive that this inference can be demonstrated alike effectively in the style of characteristic functions ψ_A, ψ_B, and ψ_C of sets A, B, and R respectively, by the equation

$$\psi_B(y) = \overset{\text{Sup}}{x \in X}\left[\psi_A(x), \psi_R(x, y)\right] \forall\, y \in Y \tag{4.39}$$

Suppose R is a fuzzy relation on $X \times Y$, and A' and B' are fuzzy sets on X and Y, respectively. Furthermore, if we have R and A', then we can derive B' by the equation

$$B'(y) = \overset{\text{Sup}}{x \in X} \min\left[A'(x), R(x, y)\right] \forall\, y \in Y \tag{4.40}$$

This equation has been obtained by displacing the characteristic function in Eq. (4.39) with the corresponding membership functions. In matrix notation, Eq. (4.40) can be rewritten as

$$B' = A' \circ R$$

and is usually referred to as the ***compositional rule of inference*** (Fig. 4.10).

4.17.1 Generalized Modes Ponens (GMP)

In our humdrum affairs of life, we usually make use of common-sense reasoning as elucidated by the following inference:

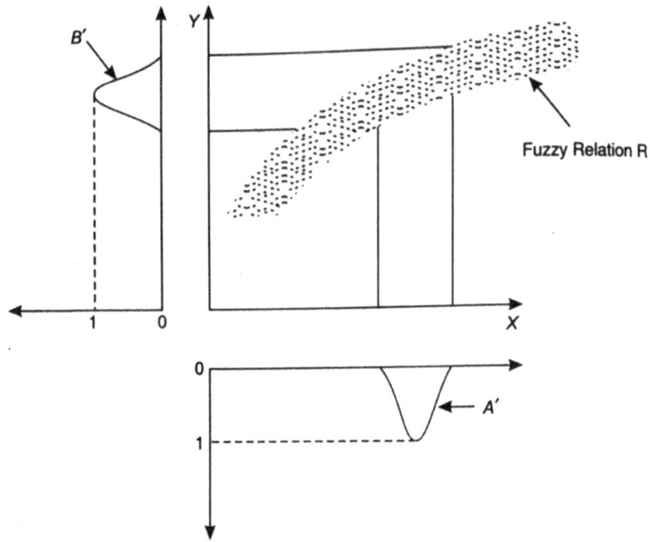

Fig. 4.10 **a**, **b**, and **c** Displaying composition rule of Inference expressed by Eq. (4.40)

$$\begin{array}{ll} \text{Rule:} & \text{If the length of the vehicle is large, then it is expensive} \\ \text{Fact:} & \text{Car } x \text{ is fairly large} \\ \hline \text{Conclusion:} & \text{Car } x \text{ is fairly expensive} \end{array} \quad (4.41)$$

In terms of classical two-valued logic, such types of inference cannot be composed. The first reason is that the concepts such as fairly large, large, and expensive are fuzzy notions and, as such, cannot be composed of two-valued propositions. The second cause is that the crisp modus ponens needs the premise to match the antecedent of the if–then rule. In inference (4.40), this is not the case; hence, the crisp modus ponens is irrelevant.

The inference (4.40) is an example of the *generalized modus ponens* in approximate reasoning. Conventionally, we can display it by the scheme:

$$\begin{array}{ll} \text{Rule :} & \text{If } \mathcal{X} \text{ is } A, \text{ then } \mathcal{Y} \text{ is } B \\ \text{Fact :} & \mathcal{X} \text{ is } A' \\ \hline \text{Conclusion :} & \mathcal{Y} \text{ is } B' \end{array} \quad (4.42)$$

where $\mathcal{X}$ and $\mathcal{Y}$ are variables that take values in the universal sets X and Y, respectively; A and A' are fuzzy sets on X; and B and B' are fuzzy sets on Y. The conclusion is calculated for all $y \in Y$ by the formula

$$B'(y) = \overset{\text{Sup}}{x \in X} \min\left[A'(x), I(A(x), B(y))\right] \quad (4.43)$$

Equation (4.43) is usually referred to as the *compositional rule of inference,* and I represent an approximate fuzzy implication.

Alternatively, consider a relation R that is embedded in a conditional fuzzy proposition p of the form

$$p: \text{If } \mathcal{X} \text{ is } A, \text{ then } \mathcal{Y} \text{ is } B$$

is determined $\forall\, x \in X$ and $y \in Y$ by the formula

$$R(x, y) = \mathcal{J}\big[A(x), B(y)\big] \tag{4.44}$$

here $\mathcal{J}$ represents a fuzzy implication.

By using a relation R derived from the given proposition p with the help of Eq. (4.44) along with an additional proposition q of the configuration

$$q: \mathcal{X} \text{ is } A',$$

By the composition rule of inference, we may deduce that $\mathcal{Y}$ is B', this course of action is called *generalized modus ponens.* It is important to mention here that (4.44) reduces to the *crisp modus ponens* if the sets are crisp and $A' = A$ and $B' = B$.

Example 4.35 If $<$ mango is yellow $>$ then $<$ it must be ripe $>$

This mango is yellow

$\therefore$ Mango is very ripe

Example 4.36 If $<$ a man drinks alcohol very heavily $>$ then

$<$ there is a possibility of his getting liver disease/cancer $>$

Bradley Cooper drinks more or less heavily

$\therefore$ *It is more or less possible that* Bradley Cooper gets liver cancer

The aforesaid decoration of inference/analysis is called "generalized" as it enlarges the Modus Ponens law of crisp logic so that if p, p', q, and q' are crisp propositions and if $p = p'$ then we will get $q = q'$.

Thus, in a *generalized modus ponens,* the first statement is called a *rule*, the second statement is called *premise* or *fact*, and the last statement is a *conclusion.*

Example 4.37 Suppose sets of values of variables $\mathcal{X}$ and $\mathcal{Y}$ be given by $= \{x_1, x_2, x_3\}$, and $Y = \{y_1, y_2, y_3\}$, respectively. Let us consider a proposition "If $\mathcal{X}$ is A, then $\mathcal{Y}$ is B", where, we have $A = \frac{0.6}{x_1} + \frac{0.9}{x_2} + \frac{0.7}{x_3}$ and $B = \frac{1.0}{y_1} + \frac{0.5}{y_2} + \frac{0.8}{y_3}$. Then, provided a truth communicated by the proposition "$\mathcal{X}$ is A", where $A' =$

$\frac{0.7}{x_1} + \frac{1.0}{x_2} + \frac{0.8}{x_3}$. Now we wish to utilize the *generalized modus ponens* to obtain a conclusion in the conformation "$\mathcal{Y}$ is B".

By employing Lukasiewicz's implication

$$\mathcal{J}(x, y) = \min(1, 1 - x + y)$$

We will have,

$$\mathcal{J}(x, y) = \min(1, 1 - x + y) = \min(1, 1 - 0.6 + 1) = \min(1, 1.4) = 1$$

and $R = \frac{1}{x_1, y_1} + \frac{0.9}{x_1, y_2} + \frac{1}{x_1, y_3} + \frac{1}{x_2, y_1} + \frac{0.6}{x_2, y_2} + \frac{0.9}{x_2, y_3} + \frac{1}{x_3, y_1} + \frac{0.8}{x_3, y_2} + \frac{1.0}{x_3, y_3}$

Now, by composition rule of inference (4.40), we will have

$$\begin{aligned} B'(y_1) &= \underset{x \in X}{\text{Sup}} \min\left[A'(x), R(x, y_1)\right] \forall\, y \in Y \\ &= \max[\min(0.7, 1.0, \min(1.0, 1.0), \min(0.8, 1.0))] \\ &= \max[0.7, 1.0, 0.8] = 1.0 \end{aligned}$$

$$\begin{aligned} B'(y_2) &= \underset{x \in X}{\text{Sup}} \min\left[A'(x), R(x, y_2)\right] \forall\, y \in Y \\ &= \max[\min(0.7, 0.9, \min(1.0, 0.6), \min(0.8, 0.8))] \\ &= \max[0.7, 0.6, 0.8] = 0.8 \end{aligned}$$

$$\begin{aligned} B'(y_3) &= \underset{x \in X}{\text{Sup}} \min\left[A'(x), R(x, y_3)\right] \forall\, y \in Y \\ &= \max[\min(0.7, 1.0, \min(1.0, 0.9), \min(0.8, 1.0))] \\ &= \max[0.7, 0.9, 0.8] = 0.9 \end{aligned}$$

Thus, we may conclude that $\mathcal{Y}$ is B', where $B' = \frac{1.0}{y_1} + \frac{0.8}{y_2} + \frac{0.9}{y_3}$.

4.17.2 Generalized Modus Tollens (GMT)

An additional inference rule in the field of fuzzy logic is called *generalized modus tollens*. In the fuzzy implication

$$\mathcal{X} \text{ is } A \rightarrow \mathcal{Y} \text{ is } B$$

If it is known that the *consequent* is false, then it can be derived by *modus tollens* that the *antecedent* is false. Given a fuzzy set B', which is not similar to B, such that

$$\mathcal{Y} \text{ is } B'$$

is false, then the intention is to derive a fuzzy set A' such that

$$\mathcal{X} \text{ is } A'$$

is false. Usually, we say that the fuzzy set A' is derived by engaging *generalized modus tollens.*

Alternatively, the configuration for *generalized modus tollens/fuzzy modus tollens is* almost similar to that of generalized modus ponens (GMP). It is given by

$$\begin{array}{ll} \text{Rule :} & \text{If } \langle \mathcal{X} \text{ is} A\rangle \text{ then } \langle \mathcal{Y} \text{ is } B\rangle \\ \text{Fact :} & \langle \mathcal{Y} \text{ is } B\rangle \\ \hline \text{Conclusion :} & \therefore \ \langle \mathcal{X} \text{ is } A'\rangle \\ \hline \end{array}$$

In this case, the compositional rule of inference has the configuration

$$A'(x) = \underset{y \in Y}{\text{Sup}} \min\left[B'(y), R(x, y)\right] \tag{4.45}$$

In the above equation, the relation R is resolved by Eq. (4.44). Here, as in GMP, A, B, A' and B' are fuzzy predicates such that A and A' are the values of one linguistic variables, and B and B' are the values of another linguistic variable. When the sets are classical and $A' = \overline{A}$, $B' = \overline{B}$, we will get the *classical modus tollens.*

Example 4.38 Let us assume that $X, Y, \mathcal{J}, A$ and B are identical, as in example (4.37). Consequently, R will also be the same as in example (4.37). Furthermore, we assume that a fact is demonstrated by the proposition "$\mathcal{Y}$ is B" is given, where $B' = \frac{0.9}{y_1} + \frac{0.6}{y_2} + \frac{0.8}{y_3}$. Now we wish to utilize the *generalized modus tollens* to obtain a conclusion in the conformation "$\mathcal{X}$ is A'".

By employing Lukasiewicz's implication

$$\mathcal{J}(x, y) = \min(1, 1 - x + y)$$

We will have,

$$\mathcal{J}(x, y) = \min(1, 1 - x + y) = \min(1, 1 - 0.6 + 1) = \min(1, 1.4) = 1$$

and $R = \frac{1.0}{x_1, y_1} + \frac{0.9}{x_1, y_2} + \frac{1.0}{x_1, y_3} + \frac{1.0}{x_2, y_1} + \frac{0.6}{x_2, y_2} + \frac{0.9}{x_2, y_3} + \frac{1.0}{x_3, y_1} + \frac{0.8}{x_3, y_2} + \frac{1.0}{x_3, y_3}$

Now, by composition rule of inference (4.33), we will have

$$\begin{aligned} A'(x_1) &= \underset{y \in Y}{\text{Sup}} \min\left[B'(y), R(x_1, y)\right] \\ &= \max[\min(0.9, 1.0), \min(0.6, 0.9), \min(0.8, 1.0))] \end{aligned}$$

$$= \max[0.9, 0.6, 0.8] = 0.9$$

$$A'(x_2) = \underset{y \in Y}{\text{Sup}} \min\left[B'(y), R(x_2, y)\right] \forall\, y \in Y$$
$$= \max[\min(0.9, 1.0), \min(0.6, 0.6), \min(0.8, 0.9))]$$
$$= \max[0.9, 0.6, 0.8] = 0.9$$

$$A'(x_3) = \underset{y \in Y}{\text{Sup}} \min\left[B'(y), R(x_3, y)\right] \forall\, y \in Y$$
$$= \max[\min(0.9, 1.0), \min(0.6, 0.8), \min(0.8, 1.0))]$$
$$= \max[0.9, 0.8, 0.8] = 0.9$$

Thus, we may conclude that "$\mathcal{X}$ is A", where $A' = \frac{0.9}{x_1} + \frac{0.9}{x_2} + \frac{0.9}{x_3}$.

4.17.3 Generalized Hypothetical Syllogism

Eventually, allow us to deliberate an abstraction of *hypothetical syllogism* constructed on two conditional fuzzy propositions. We can demonstrate the generalized hypothetical syllogism by the subsequent arrangement/schema:

$$\begin{array}{rl} \text{Rule1}: & \text{If } \langle \mathcal{X} \text{ is } A\rangle \text{ then } \langle \mathcal{Y} \text{ is } B\rangle \\ \text{Rule 2}: & \text{If } \langle \mathcal{Y} \text{ is } B\rangle \text{ then } \langle \mathcal{Z} \text{ is } C\rangle \\ \hline \text{Conclusion}: & \text{If } \langle \mathcal{X} \text{ is } A\rangle \text{ then } \langle \mathcal{Z} \text{ is } C\rangle \end{array} \tag{4.46}$$

In the above Eq. (4.46), $\mathcal{X}$, $\mathcal{Y}$, and $\mathcal{Z}$ are variables having values in sets X, Y, and Z, respectively, and A, B, and C are fuzzy sets on sets X, Y, and Z, respectively. Thus, we observe that the format of fuzzy logic is again similar to that of classical syllogism.

For every conditional fuzzy proposition depicted in Eq. (4.46), a fuzzy relation is resolved by (4.32). These fuzzy relations are determined for all $x \in X$, $y \in Y$, and $z \in Z$ by the equations

$$R_1(x, y) = \mathcal{J}\left[A(x), B(y)\right],$$

$$R_2(y, z) = \mathcal{J}\left[B(y), C(z)\right],$$

$$R_3(x, z) = \mathcal{J}[A(x), C(z)],$$

R_1, R_2, and R_3 are derived from the above equations; we claim that the generalized hypothetical syllogism holds good if we have

$$R_3(x, z) = \underset{y \in Y}{\text{Sup}} \min\left[R_1(x, y), R_2(y, z)\right] \tag{4.47}$$

This is, again, the compositional rule of inference. In matrix representation, the above equation may be rewritten in the configuration

$$R_3 = R_1 \circ R_2$$

Example 4.39 suppose that $\mathcal{X}, \mathcal{Y}$, and $\mathcal{Z}$ are sets of variables and are given by $X = \{x_1, x_2, x_3\}$, $Y = \{y_1, y_2, y_3\}$, and $Z = \{z_1, z_2, z_3\}$ respectively. We have $A = \frac{0.6}{x_1} + \frac{0.9}{x_2} + \frac{0.7}{x_3}$, $B = \frac{1.0}{y_1} + \frac{0.5}{y_2} + \frac{0.8}{y_3}$, and $C = \frac{0.3}{z_1} + \frac{0.8}{z_2} + \frac{0.7}{z_3}$ and $\mathcal{J} = \begin{cases} 1 \text{ if } a \leq b \\ b \text{ if } a > b \end{cases}$. Then, we wish to determine $R_3(x, z)$.

Solution: We have $\mathcal{J} = \begin{cases} 1 \text{ if } a \leq b \\ b \text{ if } a > b \end{cases}$

$$R_1(x, y) = \begin{bmatrix} 1 & 0.5 & 1 \\ 1 & 0.5 & 0.8 \\ 1 & 0.5 & 1 \end{bmatrix} \quad \text{and} \quad R_2(y, z) = \begin{bmatrix} 0.3 & 0.8 & 0.7 \\ 0.3 & 1 & 1 \\ 0.3 & 1 & 0.7 \end{bmatrix}$$

Then, we get

$$\begin{aligned} R_3(x, z) &= R_1(x, y) \circ R_2(y, z) \\ &= \begin{bmatrix} 1 & 0.5 & 1 \\ 1 & 0.5 & 0.8 \\ 1 & 0.5 & 1 \end{bmatrix} \circ \begin{bmatrix} 0.3 & 0.8 & 0.7 \\ 0.3 & 1 & 1 \\ 0.3 & 1 & 0.7 \end{bmatrix} \\ &= \begin{bmatrix} 0.3 & 1.0 & 0.7 \\ 0.3 & 0.8 & 0.7 \\ 0.3 & 1.0 & 0.7 \end{bmatrix} \end{aligned}$$

As we have,

$$\max\{\min(1.0, 0.3), \min(0.5, 0.3), \min(1.0, 0.3)\} = \max\{0.3, 0.3, 0.3\} = 0.3$$

$$\max\{\min(1.0, 0.8), \min(0.5, 1.0), \min(1.0, 1.0)\} = \max\{0.8, 0.5, 1.0\} = 1.0$$

$$\max\{\min(1.0, 0.7), \min(0.5, 1.0), \min(1.0, 0.7)\} = \max\{0.7, 0.5, 0.7\} = 0.7$$

$$\max\{\min(1.0, 0.3), \min(0.5, 0.3), \min(0.8, 0.3)\} = \max\{0.3, 0.3, 0.3\} = 0.3$$

$$\max\{\min(1.0, 0.8), \min(0.5, 1.0), \min(0.8, 1.0)\} = \max\{0.8, 0.5, 0.8\} = 0.8$$

$$\max\{\min(1.0, 0.7), \min(0.5, 1.0), \min(0.8, 0.7)\} = \max\{0.7, 0.5, 0.7\} = 0.7$$

$$\max\{\min(1.0, 0.3), \min(0.5, 0.3), \min(1.0, 0.3)\} = \max\{0.3, 0.3, 0.3\} = 0.3$$

$$\max\{\min(1.0, 0.8), \min(0.5, 1.0), \min(1.0, 1.0)\} = \max\{0.8, 0.5, 1.0\} = 1.0$$

$$\max\{\min(1.0, 0.7), \min(0.5, 1.0), \min(1.0, 0.7)\} = \max\{0.7, 0.5, 0.7\} = 0.7$$

Therefore,

$$R_3(x, z) = \frac{0.3}{x_1, z_1} + \frac{1.0}{x_1, z_2} + \frac{0.7}{x_1, z_3} + \frac{0.3}{x_2, z_1} + \frac{0.8}{x_2, z_2} + \frac{0.7}{x_2, z_3} + \frac{0.3}{x_3, z_1} + \frac{1.0}{x_3, z_2} + \frac{0.7}{x_3, z_3}$$

Thus, the generalized hypothetical syllogism holds since $R_1 \circ R_2 = R_3$.

4.18 Inference from Conditional and Qualified Propositions

In this section, our concern is the rule of inference which encompasses conditional fuzzy propositions and truth quantifiers. Assume that we have a conditional (dependent) and qualified fuzzy proposition p of the configuration

$$p : \text{If } \mathcal{X} \text{ is } A, \text{ then } \mathcal{Y} \text{ is } B \text{ is } S. \tag{4.48}$$

In the above, equation S stands for fuzzy truth quantifier, and a piece of evidence is in the form "$\mathcal{X}$ is A."We wish to derive an inference in the format "$\mathcal{Y}$ is B."

One procedure/technique evolved for this objective is referred to as a *method of truth-value restrictions* founded on manipulating linguistic truth values. This particular procedure implicates four successive steps.

Step I. Firstly, we derive the relative fuzzy truth value of A' w. r. t. A, which is usually signified by $RT(A'/A)$ and is a fuzzy set on the unit interval [0, 1] expounded by

$$RT\big(A'/A\big)(a) = \operatorname*{Sup}_{x : A(x) = a} A'(x) \forall\, a \in [0, 1] \tag{4.49}$$

The relative fuzzy truth value $RT(A'/A)$ exhibits the degree to which the fuzzy proposition (4.48) is true when a certitude "$\mathcal{X}$ is A" is prescribed.

Step II. Secondly, we sort out an appropriate fuzzy implication $\mathcal{J}$ by that of the fuzzy proposition "If $\mathcal{X}$ is A, then $\mathcal{Y}$ is B is S" is elucidated. This is quite similar to picking fuzzy implication as

$$\mathcal{J}(a, b) = \min(1, 1 - a + b)$$

The justification for this is to exhibit a conditional but unqualified fuzzy proposition as a fuzzy relation.

Step III. The third step is required to enumerate the relative truth value $RT(B'/B)$ by the modus operandi

$$RT\left(B'/B\right)(b) = \underset{a \in [0,1]}{\text{Sup}} \min\left[RT\left(A'/A\right)(a), S(\mathcal{J}(a, b))\right] \forall\, b \in [0, 1] \quad (4.50)$$

where S is the fuzzy qualifier in (4.48). The sole responsibility of qualifier S is to customize the truth value of $\mathcal{J}(a, b)$. Here we observe that when S stands for authentic, i.e., $S(a) = a \;\forall\, a \in [0, 1]$, then

$$S(\mathcal{J}(a, b)) = \mathcal{J}(a, b)$$

Thus, the relative truth value $RT(B'/B)$ demonstrates the degree to which the extract/outcome of the fuzzy proposition (4.48) is accurate.

Step IV. Finally, we enumerate the set B' entangled in the inference "$\mathcal{Y}$ is B" by the equation

$$B'(y) = RT\left(B'/B\right)(B(y)) \forall\, y \in Y \quad (4.51)$$

Example 4.40 Let us assume that we have prescribed a fuzzy conditional and qualified proposition,

$$p : \text{If}\, \mathcal{X} \text{ is } A \text{ then } \mathcal{Y} \text{ is}\, B \text{ is very true},$$

whence $A = \frac{0.6}{x_1} + \frac{1.0}{x_2} + \frac{0.9}{x_3}$, $B = \frac{0.6}{y_1} + \frac{1.0}{y_2}$, and S represents "very true"; consider that $S(a) = a^2, \forall\, a \in [0, 1]$. A certitude "$\mathcal{X}$ is A" is prescribed, where $A' = \frac{0.5}{x_1} + \frac{0.9}{x_2} + \frac{1.0}{x_3}$ is specified. Then, we wish to find B and will draw the inference that "$\mathcal{Y}$ is $\dot{B}$."

Solution:

Step I. We now enumerate the relative fuzzy truth value RT(A'/A) using Eq. (4.49).

$$RT\left(A'/A\right)(0.6) = A'(x_1) = 0.5$$

$$RT\big(A'/A\big)(1.0) = A'(x_2) = 0.9$$

$$RT\big(A'/A\big)(0.9) = A'(x_3) = 1.0$$

$$RT\big(A'/A\big)(a) = 0, \forall a \in [0, 1] - \{0.6, 0.9.1.0\}$$

Step II. We now choose the Lukasiewicz fuzzy implication $\mathcal{J}$ expounded by

$$\mathcal{J}(a, b) = \min(1, 1 - a + b)$$

We will get a fuzzy proposition in the form of a fuzzy relation:

$$R = \frac{1.0}{x_1, y_1} + \frac{1.0}{x_1, y_2} + \frac{0.6}{x_2, y_1} + \frac{1.0}{x_2, y_2} + \frac{0.7}{x_3, y_1} + \frac{1.0}{x_3, y_2}$$

Step III. Then we will compute $RT(B'/B)(b)$ using the Eq. (4.50).

$$RT\big(B'/B\big)(b) = \underset{a \in [0,1]}{\text{Sup}} \min\big[RT\big(A'/A\big)(a), S(\mathcal{J}(a, b))\big] \forall\, b \in [0, 1]$$

$$\begin{aligned} RT\big(B'/B\big)(b) &= \max\begin{bmatrix} \min\{0.5, S(\mathcal{J}(0.5, b))\}, \min\{0.9, S(\mathcal{J}(0.9, b))\}, \\ \min\{1.0, S(\mathcal{J}(1.0, b))\} \end{bmatrix} \\ &= \max\begin{bmatrix} \min\{0.5, \mathcal{J}(0.5, b)\}, \min\{0.9, \mathcal{J}(0.9, b)\}, \\ \min\{1.0, \mathcal{J}(1.0, b)\} \end{bmatrix} \\ &= \max[\min\{0.5, 1\}, \min\{0.9, 0.2\}, \min\{1.0, 0\}] \\ &= \max[0.5, 0.2, 0] = 0.5 \text{ and so on.} \end{aligned}$$

$$\therefore RT(B'/B)(b) = \begin{cases} (0.5 + b)^2 \text{ for } b \in [0, .207] \\ 0.5 \text{ for } b \in [.207, .608] \\ (0.1 + b)^2 \text{ for } b \in [.608, .848] \\ 0.9 \text{ for } b \in [.848, .949] \\ (0 + b)^2 \text{ for } b \in [.949, .999] \\ 1 \text{ for } b \in [.999, 1] \end{cases}$$

Step IV. Finally, we will enumerate B' by $B'(y) = RT(B'/B)(B(y)) \;\forall\, y \in Y$

$$B'(y_1) = RT\big(B'/B\big)(B(y_1)) = RT\big(B'/B\big)(0.6) = 0.5$$

$$B'(y_2) = RT\big(B'/B\big)(B(y_2)) = RT\big(B'/B\big)(1.0) = 1.0$$

Hence, we have $B' = \frac{0.5}{y_1} + \frac{1.0}{y_2}$ and we make the inference that "$\mathcal{Y}$ is B."

Theorem 4.1 Assume that a fuzzy proposition p: "If $\mathcal{X}$ is A, then $\mathcal{Y}$ is B in S" is given, whither S is the identity function (i.e., $S(a) = a \Rightarrow S$ is true). Furthermore, a fact is provided in the format "$\mathcal{X}$ is A," where

$$\underset{x : A(x) = a}{\text{Sup}} A'(x) = A'(x_0) \forall\, a \in [0, 1] \tag{4.52}$$

with some x_0 such that $A(x_0) = a$, then the inference "$\mathcal{Y}$ is B" procured by the procedure of truth-value restrictions is equal to the one procured by the generalized modus ponens, provided that the same fuzzy implication be used in both inference methods.

Proof. According to our assumption, S stands for authentic, i.e., $S(a) = a \;\forall\, a \in [0, 1]$ and B' is defined by the relation

$B'(y) = RT\big(B'/B\big)(B(y)) \forall\, y \in Y$, using Eq. (4.51)

We also have

$RT\big(B'/B\big)(b) = \underset{a \in [0, 1]}{\text{Sup}} \min\big[RT\big(A'/A\big)(a), S(\mathcal{J}(a, b))\big] \forall\, b \in [0, 1]$, using Eq. (4.50)

$$\Rightarrow RT\big(B'/B\big)(B(y)) = \underset{a \in [0, 1]}{\text{Sup}} \min\big[RT\big(A'/A\big)(a), S(\mathcal{J}(a, B(y)))\big], \text{ replacing } b \text{ by } B(y)$$

$$\Rightarrow B'(y) = \underset{a \in [0, 1]}{\text{Sup}} \min\big[RT\big(A'/A\big)(a),\; \mathcal{J}(a, B(y))\big], \text{ using the value of } B'(y)\ \&\ S(a) = a \tag{4.53}$$

$$\Rightarrow B'(y) = \underset{x \in X}{\text{Sup}} \min\big[A'(x), S(\mathcal{J}(a, B(y)))\big], \text{ using Eq. (4.49)}$$

$$\Rightarrow B'(y) = \underset{x \in X}{\text{Sup}} \min\big[A'(x), \mathcal{J}(a, B(y))\big],$$

$$\Rightarrow B'(y) = \underset{x \in X}{\text{Sup}} \min\big[A'(x), \mathcal{J}(A(x), B(y))\big],\; \forall\, y \in Y \tag{4.54}$$

We would like to show that Eqs. (4.53) and (4.54) described the identical membership function B' to establish the result. Let us assume that B_1' and B_2' represent the functions specified by Eqs. (4.53) and (4.54), respectively. Since

$$A'(x) \le \underset{x' : A(x') = A(x)}{\text{Sup}} A'(x') = RT\big(A'/A\big)(A(x)) \forall\, x \in X$$

We have

$$\min\left[A'(x), \mathcal{J}(A(x), B(y))\right] \leq \min\left[RT\left(A'/A\right)(A(x)), \mathcal{J}(A(x), B(y))\right] \forall\, y \in Y \tag{4.55}$$

Therefore,

$$\begin{aligned} B_2'(y) &= \underset{x \in X}{\text{Sup}} \min\left[A'(x), \mathcal{J}(A(x), B(y))\right] \\ &\leq \underset{x \in X}{\text{Sup}} \min\left[RT\left(A'/A\right)(A(x)), \mathcal{J}(A(x), B(y))\right], \text{using (4.61)} \\ &\leq \underset{a \in [0,1]}{\text{Sup}} \min\left[RT\left(A'/A\right)(a), \mathcal{J}(a, B(y))\right] \\ &= B_1'(y), \forall\, y \in Y, \text{using (4.59)} \end{aligned}$$

Alternatively, under condition (4.52), we have

$$\begin{aligned} \min\left[RT\left(A'/A\right)(a), \mathcal{J}(a, B(y))\right] &= \min\left[\underset{x : A(x) = a}{\text{Sup}} A'(x), \mathcal{J}(a, B(y))\right] \\ &= \min\left[A'(x_0), \mathcal{J}(A(x_0), B(y))\right] \\ &\leq \underset{x \in X}{\text{Sup}} \min\left[A'(x), \mathcal{J}(A(x), B(y))\right] \\ &= B_2'(y), \forall\, y \in Y \end{aligned}$$

Thus, finally, we have

$$B_1'(y) = \underset{a \in [0,1]}{\text{Sup}} \min\left[RT\left(A'/A\right)(a), \mathcal{J}(a, B(y))\right] \leq B_2'(y) \forall\, y \in Y$$

As a consequence, we have

$$B_1'(y) = B_2'(y) \forall\, y \in Y \Rightarrow B_1' = B_2'$$

4.19 Inference from Quantified Propositions

We, however, know that entirely quantified propositions can be put into the format

$$p : \mathcal{W} \text{ is } A$$

In the above equation, $\mathcal{W}$ is a variable whose values are inferred either as $|E|$ (where E is defined by Eq. (4.17), when A is an absolute quantifier, or as $Prop(E_2/E_1) = |E_1 \cap E_2|/|E_1|$ (where E_1, E_2 are defined by Eq. (4.24), in this case, A is a relative quantifier. Here, we also note that $\text{Prop}(E_2/E_1) = S(E_1, E_2)$.

Generally, we can state the issue of inference from quantified fuzzy propositions in the following way:

When n quantified propositions of the format

$$p_i : \mathcal{W}_i \text{ is } A_i \forall\, i \in \mathbb{N}_i \tag{4.56}$$

are provided, where Q_i is either an absolute or a relative quantifier, and $\mathcal{W}_i$ is a variable that is compatible with the quantifier Q_i, $\forall\, i \in \mathbb{N}_i$. The principle that addresses such query is usually called *quantifier extension principle.*

4.19.1 Quantifier Extension Principle

Let us consider that the anticipated inference is expressed as a quantified proposition of the configuration as

$$p : \mathcal{W} \text{ is } Q \tag{4.57}$$

This principle can be stated in the following way: If we have a function of the formulation $f : \mathbb{R}^n \to \mathbb{R}$ such that $\mathcal{W} = f(\mathcal{W}_1, \mathcal{W}_2, \mathcal{W}_3, \ldots\ldots, \mathcal{W}_n)$ and $Q = f(Q_1, Q_2, Q_3, \ldots\ldots, Q_n)$, wherever the extension principle describes the interpretation of $f(Q_1, Q_2, Q_3, \ldots\ldots, Q_n)$, then we can form an opinion that p pursue from $p_1, p_2, p_3, \ldots\ldots.p_n$. This principle can also be formulated alternatively in the following way: If we have two functions $f : \mathbb{R}^n \to \mathbb{R}$ and $g : \mathbb{R}^n \to \mathbb{R}$ in such a way that

$$f(\mathcal{W}_1, \mathcal{W}_2, \mathcal{W}_3, \ldots\ldots, \mathcal{W}_n) \leq \mathcal{W} \leq g(\mathcal{W}_1, \mathcal{W}_2, \mathcal{W}_3, \ldots\ldots, \mathcal{W}_n)$$

Then one can infer that p pursuing from $p_1, p_2, p_3, \ldots\ldots.p_n$ and Q involves in the proposition p is a quantifier represented by

$$Q = [\geq f(Q_1, Q_2, Q_3, \ldots\ldots, Q_n)] \cap [\leq g(Q_1, Q_2, Q_3, \ldots\ldots, Q_n)]$$

And this signify "at least $f(Q_1, Q_2, Q_3, \ldots\ldots, Q_n)$ and at most $g(Q_1, Q_2, Q_3, \ldots\ldots, Q_n)$." In this case, the extension principle can derive the fuzzy sets $f(Q_1, Q_2, Q_3, \ldots, Q_n)$ and $g(Q_1, Q_2, Q_3, \ldots, Q_n)$ again.

Example 4.41 Now we will discuss an example of the quantifier extension principle. For this, let us assume that the subsequent quantified fuzzy proposition is given:

$$p_1 : \text{There are about 20 students in this classroom.}$$

$$p_2 : \text{About half of them in the classroom are girls}$$

If we wish to draw inference in terms of the proposition

$$p : \text{There are } Q \text{ girls in the classroom,}$$

we will need to find out Q. To achieve this with the help of the quantifier extension principle and to ease our conversation, we assume that the quantifiers "about 20" and "about half" are represented by Q_1 and Q_2. Further, we assume that the set of students and the set of girls in the classroom be represented by symbols E and F, respectively. We are now in a position to express the given propositions in terms of these representations in the formulation as:

$$p_1 : W_1 \text{ is } Q_1$$

$$p_2 : W_2 \text{ is } Q_2$$

$$p : W \text{ is } Q$$

here W_1, W_2, and W are variables whose values are $|E|$, $\frac{|E \cap F|}{|E|}$, and$|F|$, respectively.

Since there exists a function $f : \mathbb{R}^2 \rightarrow \mathbb{R}$ such that $W = f(W_1, W_2)$ for the variables in the problem. If it is the product function, $f(xy) = xy$:

$$\begin{aligned} f(W_1, W_2) &= W_1 W_2 \\ &= |E| \frac{|E \cap F|}{|E|} = |E \cap F| = |F| = W \end{aligned}$$

Thus, we conclude from the quantifier extension principle that if $Q = Q_1.Q_2$ is the quantifier in proposition p, where $Q_1.Q_2$ is the arithmetic product of two fuzzy numbers Q_1 and Q_2 utilized in the given proposition, then the correct inference drawn from p_1 and p_2 is p.

Remark 4.3 We can use the quantifier extension principle to procure several inference rules for quantified fuzzy propositions. Now we will elaborate on two of them.

4.19.2 Intersection/Product Syllogism

The inference rule intersection/product syllogism can be expressed in the following formulation schema:

$$\begin{array}{l} p_1\colon Q_1\ E's \text{ are } F's \\ p_2\colon Q_2 \text{ (E and F)'s are G's} \\ \hline p\colon Q_1 \cdot Q_2 \text{ E's are (F and G)'s} \end{array} \quad (4.58)$$

Here E, F, and G are fuzzy sets on a universe of discourse X, Q_1, and Q_2 are relative quantifiers, i.e., fuzzy numbers on the closed interval [0, 1], and $Q_1 \cdot Q_2$ stands for the arithmetic product of the quantifiers.

Let us assume that p, p_1 and p_2 are propositions, W_1, W_2 and W are the variables with values $W_1 = \text{Prop}(F/E)$, $W_2 = \text{Prop}(G/E \cap F)$, and $W = \text{Prop}(F \cap G/E)$. We now claim that inference schema (4.58) is authentic. To do this, we shall show that by the quantifier extension principle, $W = W_1.W_2$.

We have

$$\begin{aligned} W_1 \cdot W_2 &= \text{Prop}(F/E) \cdot \text{Prop}(G/E \cap F) \\ &= \frac{|E \cap F|}{|E|} \cdot \frac{|E \cap F \cap G|}{|E \cap F|} \\ &= \frac{|E \cap F \cap G|}{|E|} \\ &= \text{Prop}\left(\frac{F \cap G}{E}\right) = W \end{aligned}$$

It follows that the inference schema (4.58) is authentic.

Example 4.42 Now, suppose that the two quantified fuzzy propositions p_1 and p_2 are given such that

p_1: Most persons are tall.

p_2: About half tall persons are females.

We may conclude by intersection/product syllogism (4.58), that the proposition

p: Q of persons are tall and females,

Here Q is a quantifier deduced by considering the arithmetic product of fuzzy numbers that display the quantifiers *most* and about half.

4.19.3 Consequent Conjunction Syllogism

The consequent conjunction syllogism can be demonstrated by the subsequent schema:

$$\begin{array}{c} p_1 : Q_1\ E's \text{ are } F's \\ p_2 : Q_2\ E' \text{ s are G's} \\ \hline p : Q \text{ E' s are (F and G)'s} \end{array} \tag{4.59}$$

Here E, F, and G are fuzzy sets on the universe of discourse X and Q_1, Q_2 are relative quantifiers. Further, the relative quantifier Q is given by

$$Q = \left[\geq \text{MAX}(0, Q_1 + Q_2 - 1)\right] \cap \left[\leq \text{MIN}(Q_1, Q_2)\right]$$

That is, Q is at least MAX$(0, Q_1 + Q_2 - 1)$ and at most MIN(Q_1, Q_2). Here, MIN and MAX are the extensions of min and max operations on real numbers to fuzzy numbers.

Example 4.43 Let p be a fuzzy conditional and qualified proposition defined by

$$p\text{: "If } \alpha \text{ is } A \text{ then } \beta \text{ is } B \text{ is very true"}$$

where $A = \frac{1.0}{\alpha_1} + \frac{0.5}{\alpha_2} + \frac{0.7}{\alpha_3}$, $B = \frac{0.6}{\beta_1} + \frac{1.0}{\beta_2}$ and S stands for very true. Let $S(a) = a^2, \forall\, a \in [0, 1]$. Given a fact "$\alpha$ is A," where $A' = \frac{0.9}{\alpha_1} + \frac{0.6}{\alpha_2} + \frac{0.7}{\alpha_3}$ is given. Then, we conclude that "β is B'," and calculate B'.

Solution: To calculate B', we will follow the following four steps:

We will follow step by step in this direction.

Step I. We now enumerate the relative fuzzy truth value $RT(A'/A)$, using Eq. (4.49).

$$RT\left(A'/A\right)(0.6) = A'(\alpha_1) = 0.9$$

$$RT\left(A'/A\right)(1.0) = A'(\alpha_2) = 0.6$$

$$RT\left(A'/A\right)(0.9) = A'(\alpha_3) = 0.7$$

$$\therefore \quad RT\left(A'/A\right)(a) = 0, \forall\, a \in [0, 1] - \{0.5, 0.7.1.0\}$$

Step II. We now choose the Lukasiewicz fuzzy implication $\mathcal{J}$ expounded by

$$\mathcal{J}(a, b) = \min(1, 1 - a + b)$$

We will get a fuzzy proposition in the form of a fuzzy relation:

$$R = \frac{0.6}{\alpha_1, \beta_1} + \frac{1.0}{\alpha_1, \beta_2} + \frac{1.0}{\alpha_2, \beta_1} + \frac{1.0}{\alpha_2, \beta_2} + \frac{0.9}{\alpha_3, \beta_1} + \frac{1.0}{\alpha_3, \beta_2}$$

Step III. Then we will compute $RT(B'/B)(b)$ using the Eq. (4.50), Fig. 4.11

$$RT\left(B'/B\right)(b) = \underset{a \in [0,1]}{\text{Sup}} \min\left[RT\left(A'/A\right)(a), S(\mathcal{J}(a,b))\right] \forall\, b \in [0,1]$$

$$\begin{aligned}
RT\left(B'/B\right)(b) &= \max\begin{bmatrix} \min\{0.9, S(\mathcal{J}(0.9,b))\}, \min\{0.6, S(\mathcal{J}(0.6,b))\}, \\ \min\{0.7, S(\mathcal{J}(0.7,b))\} \end{bmatrix} \\
&= \max\begin{bmatrix} \min\{0.9, \mathcal{J}(0.9,b)\}, \min\{0.6, \mathcal{J}(0.6,b)\}, \\ \min\{0.7, \mathcal{J}(0.7,b)\} \end{bmatrix} \\
&= \max[\min\{0.9, 0.7\}, \min\{0.6, 1.0\}, \min\{0.7, 0.3\}] \\
&= \max[0.7, 0.6, 0.3] = 0.7 \text{ and so on.}
\end{aligned}$$

$$\therefore \quad RT(B'/B)(b) = \begin{cases} (0.4+b)^2 & \text{for } b \in [0, .375] \\ 0.6 & \text{for } b \in [.375, .475] \\ (0.3+b)^2 & \text{for } b \in [.475, .537] \\ 0.7 & \text{for } b \in [.537, .737] \\ (0.1+b)^2 & \text{for } b \in [.737, .849] \\ 0.9 & \text{for } b \in [.849, 1] \end{cases}$$

Step IV. Finally, we will enumerate B' by $B'(y) = RT(B'/B)(B(y)) \forall\, y \in Y$

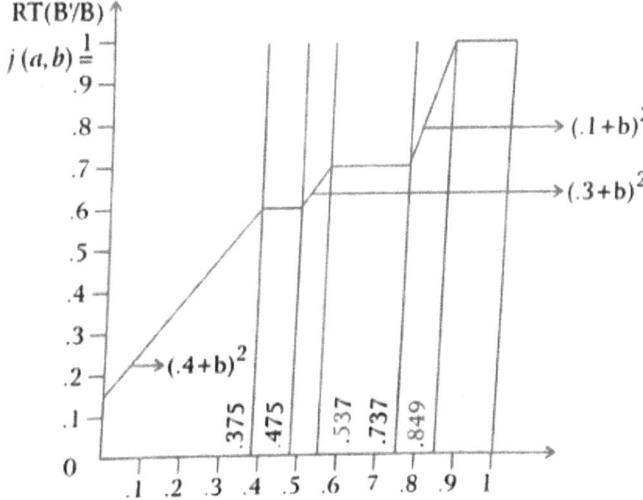

Fig. 4.11 Representation of the function $RT(B'/B)$

$$B'(y_1) = RT\big(B'/B\big)(B(y_1)) = RT\big(B'/B\big)(0.6) = 0.7$$

$$B'(y_2) = RT\big(B'/B\big)(B(y_2)) = RT\big(B'/B\big)(1.0) = 0.9$$

Hence, we have $B' = \frac{0.7}{\beta_1} + \frac{0.9}{\beta_2}$ and we make the inference that "$\mathcal{Y}$ is B."

Exercise-4

1. Explain what do you mean by fuzzy if–then rules?
2. Explain the different types of fuzzy rules?
3. Explain in your own words the various implications of rules?
4. Explain the basic notions of fuzzy logic with the help of suitable examples?
5. Discuss the different objectives of fuzzy logic.
6. What do you mean by linguistic values and linguistic variables?
7. Explain the basic concept of truth values in fuzzy logic.
8. What are the differences in the truth values in classical and fuzzy logic?
9. Explain the difference between smooth boundaries and sharp boundaries?
10. Show that a compound statement is a tautology.
11. Explain approximate reasoning and cite a good example supporting your answer?
12. What do you mean by aggregation of fuzzy rules?
13. Show that De Morgan's laws are dual.
14. Prove that the compound proposition $((p \to q) \wedge (r \to \overline{s}) \wedge (q \to r)) \to (p \to \overline{s})$ is a tautology.
15. Prove the following statements by contradiction.

    (i) $((p \to q) \wedge p) \to q$
    (ii) $((p \to \overline{q}) \wedge (q \vee \overline{r}) \wedge (r \to \overline{q}) \wedge (p \vee r)) \to r$ is not a tautology (i.e., a fallacy) by developing a counter example.

16. Discuss some myths about fuzzy logic.
17. Explain unconditional and unqualified fuzzy propositions in your own words.
18. Discuss the conditional and unqualified fuzzy propositions in your own words.
19. Describe unconditional and unqualified fuzzy propositions with the help of an apt example in your own words.
20. Could you furnish an illustration from a lifestyle of each category of fuzzy proposition discussed in this chapter and express the proposition in canonical form?
21. Presume four kinds of fuzzy predicates appropriate to a person's age, height, weight, and education level and give them membership functions.
22. Could you explain the concept of fuzzy quantifiers and cite examples in favor of your answer?
23. Illustrate linguistic hedges and modifiers with the help of suitable examples. Further distinguish between weak modifiers, strong modifiers, and identity modifiers. Discuss properties of modifiers.

24. Could you discuss general modus ponens and modes of Tollens and generalized hypothetical syllogism in your words and give good examples?
25. In the fuzzy implication $x \text{ is } A \rightarrow y \text{ is } B$. Let $A = \{(1{,}0.6), (2{,}0.8), (3{,}1), (4{,}0.8), (5{,}0.6), (6{,}0)\}$ and $B = \{(1{,}0.4), (2{,}0.6), (3{,}0.8), (4{,}1), (5{,}0.8), (6{,}0.6), (7{,}0.4)\}$. For the definition of R given by

$$R(i,j) = \begin{cases} 1 \text{ if } A(x_i) \leq B(y_j) \\ 0 \text{ if } A(x_i) > B(y_j) \end{cases}$$

for $1 \leq i \leq n$ and $1 \leq j \leq m$. Find out whether modus ponens holds. If not, give the reason why modus ponens does not holds.

26. Give an example of a hedge "very." Explain by the example of building a relation.
27. Provide an example of $A' = A$. Explain by the example of building a relation.
28. Discuss an example of a hedge "somewhat" and explain it by building a relation.
29. Define two fuzzy sets A and B such that

$A = {}^{0.0}/_{(-2)} + {}^{0.5}/_{(-1)} + {}^{1.0}/_{0} + {}^{0.5}/_{1} + {}^{0.0}/_{2}$ and

$$B = {}^{0.0}/_{0} + {}^{0.5}/_{1} + {}^{1.0}/_{2} + {}^{0.5}/_{3} + {}^{0.0}/_{4}$$

Construct the relation "If A, then B" using the following relations:

$$R(x, y) = \min[A(x), B(y)]$$

$$R(x, y) = [A(x), B(y)]$$

30. With respect to train speed, the following linguistic variables are given

$$A = \text{Express} = \frac{0.0}{0} + \frac{0.1}{10} + \frac{0.2}{20} + \frac{0.3}{30} + \frac{0.4}{40} + \frac{0.5}{50} + \frac{0.6}{60} + \frac{0.7}{70} + \frac{0.8}{80} + \frac{0.9}{90} + \frac{1.0}{100}$$

$$B = \text{Passenger} = \frac{1.0}{0} + \frac{0.9}{10} + \frac{0.8}{20} + \frac{0.7}{30} + \frac{0.6}{40} + \frac{0.5}{50} + \frac{0.4}{60} + \frac{0.3}{70} + \frac{0.2}{80} + \frac{0.1}{90} + \frac{0.0}{100}$$

Using these terms, find the membership function for the following linguistic terms. (i) Very fast (ii) very-very fast (iii) fairly fast $(-(\text{fast})^{2/3})$ (iv) not very slow and not very fast (v) slow or not very slow.

31. Find the truth value for the expression $(p \Rightarrow q) \Rightarrow (\overline{r}) \Rightarrow p$.
32. Let $X = [1, 2, 3, 4]$ and $Y = [1, 2, 3, 4, 5, 6]$ be two universes. Let A, B, and C be any three fuzzy sets defined by $A = \frac{0.6}{2} + \frac{1}{3} + \frac{0.2}{4}$, $B = \frac{0.4}{2} + \frac{1}{3} + \frac{0.8}{4} + \frac{0.3}{5}$ and $C = \frac{0.3}{1} + \frac{0.5}{2} + \frac{0.6}{3} + \frac{0.6}{4} + \frac{0.5}{5} + \frac{0.3}{6}$. Apply the modus ponens rule and find the relation (i) If x is A, then y is B(ii) If x is A, then y is B else y is C.

33. Let If $X = \{1,2,3,4,5,6\}$ be the universe and small integers be defined as $A = \{\frac{1}{1}, \frac{2}{5}, \frac{3}{4}, \frac{4}{2}\}$. The fuzzy relation "almost equal" be defined by the matrix relation as

$$R = \begin{array}{c} \\ 1 \\ 2 \\ 3 \\ 4 \end{array} \begin{array}{c} \begin{array}{cccc} 1 & 2 & 3 & 4 \end{array} \\ \begin{bmatrix} 1 & .8 & 0 & 0 \\ .8 & 1 & .8 & 0 \\ 0 & .8 & 1 & .8 \\ 0 & 0 & .8 & 1 \end{bmatrix} \end{array}$$

What is the membership function of the fuzzy set $B =$ "rather small integer", if it is interpreted as the composition $A \circ R$.

Objective-Type Questions and Answers

1. Which one is a tautology?

   (a) $p \vee \neg(p \wedge q)$
   (b) $(p \to p) \to (q \wedge \sim q)$
   (c) $p \wedge \sim q$
   (d) $((\neg q \wedge p) \wedge q)$

2. Which one is a well-formed formula?

   (a) $\neg(p \wedge q)$
   (b) $(p \to q) \to (\wedge q)$
   (c) $(p \vee q)$
   (d) none of these

3. Which one is a well-formed formula?

   (a) $(p \wedge q) \to q$
   (b) $(p \to q$
   (c) $\neg p \wedge q$
   (d) $\neg(p \wedge q)$

4. Which one is a dual of $(p \vee q) \wedge r$.

   (a) $(p \vee q) \vee r$
   (b) $(p \wedge q) \wedge r$
   (c) $(p \wedge q) \vee r$
   (d) none of these

5. The number of rows in a truth table is 2^n, where n is the number of

   (a) Columns
   (b) rows
   (c) number of distinct variables in the formula
   (d) none of these

6. Which one of the sets is functionally complete?

   (a) $\{\vee, \neg\}$
   (b) $\{\wedge\}$
   (c) $\{\vee\}$
   (d) none of these

7. Which one of the sets is functionally complete?

   (a) $\{\wedge\}$
   (b) $\{\vee\}$
   (c) $\{\neg\}$
   (d) $\{\wedge, \neg\}$

8. Choose the sets which need to be functionally completed.

   (a) $\{\wedge\}$
   (b) $\{\neg\}$
   (c) $\{\vee\}$
   (d) $\{\vee, \neg\}$

9. Which one is the law of Modus Ponens?

   (a) $[(F \to G) \wedge (NG)] \to (NF)$
   (b) $[(F \to G) \wedge (G \to H)] \to (F \to H)$
   (c) $[F \wedge (F \to G)] \to G$
   (d) none of these

10. Which one is a contradiction?

    (a) $(p \vee q) \wedge (\sim p) \wedge (\sim q)$
    (b) $(\sim p) \vee q$
    (c) $\sim \{p \wedge (\sim p)\}$
    (d) none of these

11. Which one is a contradiction?

    (a) $p \Rightarrow (\sim p) \Leftrightarrow (\sim p)$
    (b) $p \Leftrightarrow (\sim p)$
    (c) $(p \Leftrightarrow q) \Leftrightarrow (p \Rightarrow q) \wedge (q \Rightarrow p)$
    (d) none of these

12. Are the following propositions?

    (a) $x + 2 \leq 5$
    (b) 6 is an even number
    (c) $21 < 7$
    (d) $2 + 3 = 5$

13. $\sim (p \vee q)$ is equivalent to

    (a) $(p \Rightarrow q)$

(b) $\sim p \wedge q$
(c) $p \wedge (\sim q)$
(d) $\sim p \wedge \sim q$

14. The functionally complete set for $\uparrow$ is

(a) $\{\neg, \vee\}$
(b) $\{\neg, \leftarrow\}$
(c) $\{\leftrightarrow, \vee\}$
(d) $\{\neg, \wedge\}$

15. The functionally complete set for $\downarrow$ is

(a) $\{\neg, \uparrow\}$
(b) $\{\neg, \vee\}$
(c) $\{\leftrightarrow, \wedge\}$
(d) $\{\neg, \leftrightarrow\}$

16. Which one is a functionally complete set?

(a) $\{\vee\}$
(b) $\{\neg\}$
(c) $\{\neg, \rightarrow\}$
(d) $\{\wedge\}$

17. $(p \downarrow p) \downarrow (q \downarrow q)$ is equal to

(a) $p \vee q$
(b) $p \wedge q$
(c) $\sim p \wedge q$
(d) $p \wedge \sim q$

18. In $(\exists x)P(x) \wedge Q(y)$

(a) x and y are bound variables
(b) x is bound and y is a free variable
(c) x and y both are free variables
(d) x is the free variable and y is the bound variable

19. Existential quantifiers can be expressed in any of the following ways:

(a) $(\exists x \in A)P(x)$
(b) $(\exists x)P(x)$
(c) $\exists x \in A$ such that $P(x)$
(d) all of the three

20. Which one of the following is not a linguistic hedge?

(a) Very
(b) less
(c) fairly
(d) talented

21. Which one of the following linguistic terms is not meaningful?

    (a) Very rectangular
    (b) extremely
    (c) more
    (d) none of these

22. Any linguistic hedges, H, may be interpreted as a

    (a) Binary operation
    (b) unary operation
    (c) ternary operation
    (d) none of these

23. Linguistic hedges fairly be represented by

    (a) a^2
    (b) $\sqrt{a}$
    (c) $a^{3/2}$
    (d) none of these

24. Which one statement is not true for every modifier h

    (a) h is a continuous function
    (b) $h(0) = 0$ and $h(1) = 1$
    (c) Composition of any two modifiers is again a modifier
    (d) h is an even function

25. Out of the following, which one is not a fuzzy quantifier?

    (a) About 10
    (b) much more than 100
    (c) good
    (d) at least about 5

26. A desirable requirement for linguistic variables is that they must be

    (a) Quantifiable
    (b) not quantifiable
    (c) not quantitative
    (d) measurable

27. Out of the following given words which one is not a linguistic variable?

    (a) Pressure
    (b) mood
    (c) height
    (d) temperature

28. Out of the following given words, which one is not a linguistic value?

    (a) Fast

(b) warm
(c) medium
(d) speed

29. A generalized modus ponens consists of the following in order:

(a) Conclusion
(b) premise
(c) threshold value
(d) rule

30. A number expresses the degree of truth of each fuzzy proposition in the interval

(a)]0, 1]
(b)]0, 1[
(c) [0, 1[
(d) [0, 1]

31. Out of the following given fuzzy propositions, which one is not a kind of fuzzy proposition?

(a) Unconditional and unqualified
(b) Unconditional and qualified
(c) Fuzzy quantifiers
(d) conditional and unqualified

32. Fuzzy quantifiers of the first kind are defined on

(a) [0, 1]
(b) $\mathbb{N}$
(c) $\mathbb{R}$
(d) (0, 1)

33. Fuzzy quantifiers of the second kind are defined on

(a) $\mathbb{R}$
(b) (0, 1)
(c) $\mathbb{N}$
(d) [0, 1]

34. Fuzzy quantifiers of the first kind characterize linguistic terms such as

(a) Almost all
(b) about 99
(c) much more than 1000
(d) at least about 7

35. Fuzzy quantifiers of the second kind characterize linguistic terms such as

(a) Almost all
(b) at most about
(c) about half

(d) most

36. Propositions p of the canonical form p: If x is A, then y is B is S or p: Pro{x is A: y is B} is P represents or characterize

    (a) Conditional unqualified
    (b) Conditional and qualified
    (c) Unconditional and unqualified
    (d) Unconditional and qualified

37. Propositions p of the canonical form p: $\mathcal{V}$ is F is S or the canonical form p: Pro{$\mathcal{V}$ is F} is P is of the type

    (a) Conditional and qualified
    (b) Unconditional and qualified
    (c) Conditional unqualified
    (d) Unconditional and unqualified

38. The linguistic hedges are not applicable to

    (a) Classical predicates
    (b) truth values
    (c) probabilities
    (d) relative quantifiers

39. Let h be an increasing bijection. Then, h is called the strong modifier if

    (a) $h(a) > a, \forall\, a \in [0, 1]$
    (b) $h(a) < a, \forall\, a \in [0, 1]$
    (c) $h(a) = a, \forall\, a \in [0, 1]$
    (d) none of these

40. Let h be an increasing bijection. Then, h is called the weak modifier if

    (a) $h(a) < a, \forall\, a \in [0, 1]$
    (b) $h(a) > a, \forall\, a \in [0, 1]$
    (c) $h(a) = a, \forall\, a \in [0, 1]$
    (d) none of these

Answers

1. (a); 2. (c); 3. (d); 4. (c); 5. (c); 6. (a); 7. (d); 8. ((a), (b) & (c)); 9. (c); 10. (a); 11. (b); 12. (a); 13. (d); 14. (d); 15. (b); 16. (c); 17. (b); 18. (b); 19. (d); 20. (d); 21. (a); 22. (a); 23. (b); 24. (d); 25. (c); 26. (a & d); 27. (b); 28. (d); 29. (d, b, & a); 30. (d); 31. (c); 32. (c); 33. (d); 34. (b, c, & d); 35. (a, c, & d); 36. (b); 37. (b); 38. (a, b, & c); 39. (b); 40. (b).

True/False Type Statements and Answers

1. Propositional logic deals with statements that can be only either True or False but not both.

2. By Modus Tollens, we mean the following WFF: $[F \wedge (F \to G)] \to G$.
3. By Syllogism, we mean the following WFF: $[(F \to G) \wedge (G \to H)] \to (F \to H)$.
4. When the variable represented by a logic formula is always true regardless of the truth values assigned to the variables participating in the formula, it is called a tautology.
5. Dualization $\overline{a+b} = \overline{a} \cdot \overline{b}$ is not a tautology.
6. Various forms of tautologies used for making deductive inferences are called as inference rules.
7. The inference rule given by $(x \wedge (x \Rightarrow y)) \Rightarrow y$ is called Modus Ponens.
8. A predicate becomes a proposition that is either true or false when a particular subject form X is not substituted for x.
9. Every tautology remains a tautology when any of its variables is replaced with any arbitrary logic formula. This powerful rule of inference is referred to as the rule of substitution.
10. Only two kinds of quantification are predominantly used for predicates: existential and universal.
11. A well-formed formula (WFF) is a proposition, but when a proposition is substituted in place of a propositional variable, a proposition is obtained.
12. A formula with a sum of elementary products equivalent to the given formula is termed as a disjunctive normal form.
13. A formula with a product of elementary sums and equivalent to the given formula is labeled as a conjunctive normal form.
14. Fuzzy inference is also referred to as approximate reasoning.
15. In fuzzy logic Unknown is interpreted as True or false (T or F).
16. In fuzzy logic, Undecided is signified by $\ominus$.
17. Zadeh proposed truth tables for determining truth values for operation by using four-valued logic, including the truth values, True, False, Undecided, and Unknown.
18. The fundamental difference between classical and fuzzy propositions is in the range of their truth values.
19. Fuzzy propositions are segregated into four types.
20. A fuzzy proposition is a statement that does not acquire a fuzzy truth value.
21. Fuzzy quantifiers are fuzzy numbers that take part in the fuzzy proposition.
22. Fuzzy quantifiers are of more than two kinds.
23. Absolute quantifiers are elucidated on $\mathbb{R}$ and characterize linguistic terms such as almost and about.
24. Relative quantifiers are defined on [0, 1] and characterize linguistic terms such as most, almost all, and about half.
25. The linguistic hedge H is a unary operation h on the interval [0, 1].
26. The unary operation that characterizes linguistic hedges is referred to as a modifier.
27. The linguistic hedges are also applicable to classical predicates, truth values, or probabilities.
28. The hedge very is interpreted as the unary operation $h(a) = a^2$.

29. The hedge fairly is interpreted as the unary operation $h(a) = \left(\sqrt{a}\right)^2$.
30. The modifier for which $h(a) = a$ is usually known as an identity modifier.
31. The truth or falsity of fuzzy propositions is a matter of degree.
32. This wristwatch is expensive, or it is really hot are fuzzy predicates or linguistic values.
33. Height is a linguistic variable. Its values are: tall, medium, short, and dwarfish.
34. A (atomic) fuzzy proposition is of the form x is P' where 'P' is a fuzzy predicate. Such propositions can be combined by $\wedge$, $\vee$, and $\neg$.
35. Approximate reasoning is not essential for modeling human common-sense reasoning.
36. A weak modifier weakens the fuzzy predicate and, hence proposition's truth value increases.
37. A strong modifier strengthens the fuzzy predicate to which it is applied and, consequently, it increases the truth value of the associated proposition.
38. Each fuzzy quantifier expresses an approximate number of elements or an approximate proportion of elements in a given universal set that are claimed to satisfy a given property.
39. The modifier h satisfies the properties $h(0) = 0$ and $h(1) = 1$ and that it is continuous.
40. If h and g are two modifiers, then their composite $h \circ g$ and $g \circ h$ are not again modifiers.

Answers

1. (T); 2. (F); 3. (T); 4. (T); 5. (F); 6. (T); 7. (T); 8. (F); 9. (T); 10. (T); 11. (F); 12. (T); 13. (T); 14. (T); 15. (F); 16. (T); 17. (T); 18. (T); 19. (T); 20. (F); 21. (T); 22. (F); 23. (F); 24. (T); 25. (T); 26. (T); 27. (F); 28. (T); 29. (f). 30. (T); 31. (T); 32. (T); 33. (T); 34. (T); 35. (F); 36. (T); 37. (F); 38. (T); 39. (T); 40. (F).

Chapter 5
Fuzzy Topological Spaces

This chapter reviews the concepts and notions of fuzzy topological spaces and some of their properties, also available in their classical counterpart.

5.1 Introduction

Zadeh introduced the fundamental concept of fuzzy sets, which formed the backbone of fuzzy mathematics. The theory of fuzzy topological spaces is a branch of such mathematics. General topology can be regarded as a particular case of fuzzy topology where all membership functions in context take values 0 and 1 only. Therefore, we expect weaker results in the case of fuzzy topology. The notion of a fuzzy topological space was introduced by **C. L. Chang** in 1968 and is regarded as a generalization of the concept of ordinary topological space. A one–one correspondence exists between the family of all subsets of a set X and the set of all characteristic functions C which have X as its domain. Therefore, a topology on X may be regarded as a family of characteristic functions with usual set operations $\subset$, $\cup$, $\cap$, and complementation replaced by function operations $\leq$, $\vee$, $\wedge$, and $1 - f_C(x)$. Since a more general function characterizes fuzzy sets than the characteristic function, a very natural generalization of topological space is possible, which we call a fuzzy topological space.

5.2 Fuzzy Topology and Its Fundamental Properties

5.2.1 Fuzzy Topology

A fuzzy topology is a family $\mathcal{T}$ of fuzzy sets on X satisfying the following conditions:

M. K. Singh, *Applied Fuzzy Mathematics*, Forum for Interdisciplinary Mathematics,
https://doi.org/10.1007/978-981-97-3257-9_5

(i) $\phi \in \mathcal{T}, X \in \mathcal{T}$ *or* $0, 1 \in \mathcal{T}$
(ii) If $A \in \mathcal{T}$ and $B \in \mathcal{T}$, then $A \cap B \in \mathcal{T}$
(iii) If $A_i \in \mathcal{T}$ for every $i \in I$ where I is an arbitrary set, then

$$U\{A_i : i \in I\} \in \mathcal{T}$$

Then $\mathcal{T}$ is called a fuzzy topology on X and the pair $\{X, \mathcal{T}\}$ is said to be a fuzzy topological space or fts for short. The member of $\mathcal{T}$ are called $\mathcal{T}$-open fuzzy sets (or simply open fuzzy sets). A fuzzy set is called topologically closed or $\mathcal{T}$—closed if and only if its complement is topologically open or $\mathcal{T}$-open or open fuzzy sets. As ordinary topologies, the fuzzy topology $\mathcal{T}$ is termed "discrete" if it contains all of the fuzzy sets on X and is termed "indiscrete" if it has only ϕ and X. For instance, {X, P (X)} is a fuzzy topological space, namely the discrete fuzzy topology of X.

Example 5.1 Let $X = \{p, q, r\}$ be a set, and let $A = \{(p, 0), (q, 0.4), (r, 1)\}$ be a fuzzy set in X. Let $\mathcal{T} = \{0, A, 1\}$. Then $(X, \mathcal{T})$ is a fuzzy topological space, which is not a topological space.

5.2.2 Comparison of Fuzzy Topologies

Let U and $\mathcal{T}$ be two fuzzy topologies on X. If the inclusion relation $U \subset \mathcal{T}$ holds, we say that U is coarser, weaker or smaller than $\mathcal{T}$ or that $\mathcal{T}$ is finer, stronger or larger than U if and only if $U \subset \mathcal{T}$ or $U \leq \mathcal{T}$. On X, the indiscrete fuzzy topology is the coarser, and the discrete fuzzy topology is the finest.

Theorem 5.1 If $\{\mathcal{T}_i$: $i \in I\}$ where I is an arbitrary set, be a collection of fuzzy topologies on X. Then the intersection of $\mathcal{T}_i$, i.e.; $\bigcap_{i \in I} \mathcal{T}_i$ is also a fuzzy topology on X.

Proof

Let $\{\mathcal{T}_i\ i \in I\}$ be a collection of fuzzy topologies on X. Then we claim that $\bigcap_{i \in I} \mathcal{T}_i$ is also a fuzzy topology on X.

Since $\mathcal{T}_i$ is a fuzzy topology for all $i \in I$, it follows that $\phi, X \in \mathcal{T}_i\ \forall\, i \in I$.

Hence $\phi \in \mathcal{T}_i\ \forall\, i \in I \Rightarrow \phi \in \bigcap_{i \in I} \mathcal{T}_i$

and $X \in \mathcal{T}_i\ \forall\, i \in I \Rightarrow X \in \bigcap_{i \in I} \mathcal{T}_i$

Hence, condition (í) is satisfied.

Let $A, B \in \bigcap_{i \in I} \mathcal{T}_i$. Then $A, B \in \mathcal{T}_i\ \forall\, i \in I$

Since T_i is a fuzzy topology on X for all i ∈ I, it follows that

$$A \cap B \in \mathcal{T}_i \; \forall \; i \in I.$$

Hence, $A \cap B \in \bigcap_{i \in I} \mathcal{T}_i, \forall i \in I.$. Therefore, condition (íí) is satisfied.

Finally, let $A_\alpha \in \bigcap_{i \in I} \mathcal{T}_i$ for all α ∈ I, where I is an arbitrary set.

Then $A_\alpha \in \mathcal{T}_i \; \forall \; i \in I$ and $\forall \; \alpha \in I$. Since each $\mathcal{T}_i$ is a fuzzy topology on X., it follows that.

$$\bigcup_{\alpha \in I} A_\alpha \in \mathcal{T}_i \; \forall \; i \in I$$

Hence, $\bigcup_{\alpha \in I} A_\alpha \in \bigcap_{i \in I} \mathcal{T}_i \forall \; i \in I.$

Thus $\bigcap_{i \in I} \mathcal{T}_i$ is a fuzzy topology on X.

Theorem 5.2 If X is fixed and θ is the collection of all fuzzy topologies on X, then $(\theta, \supset)$ is a complete lattice.

Proof

Evidently, $(\theta, \supset)$ is a partially ordered set (poset). Since θ has the most significant element, the fuzzy topology consisting of all fuzzy sets on X; every non-empty subset of θ has a least upper bound. Hence, $(\theta, \supset)$ is a complete lattice.

5.2.3 Cover and Subcover of a Fuzzy Set

Let A and $\{A_i : i \in I\}$ be fuzzy sets in X. Then, $\{A_i : i \in I\}$ is called a cover of A if and only if $\cup\{A_i : i \in I\} \supset A$. We say that the cover of the fuzzy set is open if and only if each member of A_i is an open fuzzy set. If there exists a subset I_1 *of I such that* $\{A_i : i \in I_1\} \supset A$ then $\{A_i : i \in I_1\}$ is called a Subcover.

5.2.4 Fuzzy Point

A fuzzy set in X is called a fuzzy point if and only if it takes the value 0 for all $y \in X$ except one, say $x \in X$. If its value at x is λ where $(0 < \lambda \leq 1)$, we denote this fuzzy point by x_λ, where the point x is called its support.

5.2.5 Fuzzy Point Contained in a Fuzzy Set

The fuzzy point x_λ is said to be contained in a fuzzy set A or to belong to A iff λ < A(x) and is denoted by $x_\lambda \in A$. Every fuzzy set A can be expressed as the union of all the fuzzy points that belong to A.

5.2.6 Neighborhood of a Fuzzy Point

A fuzzy set A in (X,$\mathcal{T}$) is called a neighborhood of fuzzy point x_λ iff there exists $V \in \mathcal{T}$ such that $x_\lambda \in V \subseteq A$. A neighborhood A is said to be open iff A is open. The family consisting of all the neighborhoods of x_λ is called the system of neighborhoods of x_λ and is denoted by $N(x_\lambda)$.

Remark 5.1 A remarkable deviation from general topology arises in the property of neighborhood systems $N(x_\lambda)$ in fuzzy topological space. In fuzzy topological space (X,$\mathcal{T}$), the neighborhood system $N(x_\lambda)$ may not be closed under arbitrary union, which the following example shows.

Example 5.2 Let X be a non-empty set and x ∈ X. Let $\mathcal{T} = \{x_\lambda : \lambda < 0.5\} \cup \{0, 1\}$. Then (X,$\mathcal{T}$) is a fuzzy topological space. Let us consider the neighborhood system of the fuzzy point $x_{0.3} \cdot x_\lambda \in N(x_{0.3}), \forall\ \lambda\ with\ 0.3 \leq \lambda$.

5.2.7 Quasi-Coincident Fuzzy Point

A fuzzy point x_λ is said to be quasi-coincident with a fuzzy set A iff $\lambda > A'(x), or\ \lambda + A(x) > 1$, and is denoted by $x_\lambda \uparrow A$.

5.2.8 Intersecting Fuzzy Sets

Two fuzzy sets A and B in X are said to be intersecting iff there exists a point x ∈ X such that $(A \cap B)(x) > 0$. For such a case, we say that A and B intersect at x.

5.2.9 Quasi-Coincident Fuzzy Sets

Fuzzy set A is said to be quasi-coincident with B iff there exists x ∈ X such that $A(x) > B'(x), or\ A(x) + B(x) > 1$ and is denoted by $A \uparrow B$. If this is true, we also say that A and B are quasi-coincident (with each other) at x. It is clear that if A and B are quasi-coincident at x, both A(x) and B(x) are not zero; hence, A and B intersect at x.

5.2.10 *Q-Neighborhood of a Fuzzy Point*

A fuzzy set A in $(X,\mathcal{T})$ is called a Q-neighborhood of x_λ iff there exists a $B \in \mathcal{T}$ *such that* $x_\lambda \uparrow B \subseteq B$. The family consisting of all the Q-neighborhoods of x_λ is called the system of Q-neighborhood of x_λ.

5.2.11 *Neighborhood of a Point x (∈ X)*

Let $(X,\mathcal{T})$ be a fuzzy topological space and let $x \in X$. A fuzzy set N in fuzzy topological space $(X,\mathcal{T})$ is said to be a neighborhood of x iff there exists a fuzzy open set $G \in \mathcal{T}$ such that $G \leq N$ and $f_{\mathbf{N}}(x) = f_{\mathbf{G}}(x) > 0$.

The neighborhood of a point x is denoted by $N_{\mathbf{x}}$ and is said to be open iff $N_{\mathbf{x}} \in \mathcal{T}$.

Theorem 5.3 If N_x and P_x are neighborhoods of a point $x \in X$, then so is $N_x \wedge P_x$.

Proof

Let G and H be any two elements of $\mathcal{T}$ satisfying

$$G \leq N_x,\ H \leq P_x\ ,\ {}^{\mathbf{f}}N_x^{(x)} = f_G(x) > 0 \text{ and } {}^{\mathbf{f}}P_x(x) = f_H(x) > 0$$

Then $G \wedge H \leq N_x \wedge P_x$ and

$${}^{f}(N_x \wedge P_x)(x) = (f_{\mathbf{G}} \wedge f_{\mathbf{H}})(x) > 0$$

Hence the result.

5.2.12 *Neighborhood of a Fuzzy Set*

A fuzzy set C in a fuzzy topological space $(X,\mathcal{T})$ is a neighborhood (abbreviated as nbhd) of a fuzzy set A iff there exists an open fuzzy set B such that $A \subset B \subset C$ or $A \leq B \leq C$.

The neighborhood system of a fuzzy set A is the family of all neighborhoods of the fuzzy set.

Theorem 5.4 Let (X, $\mathcal{T}$) be a fuzzy topological space and let A, B be two fuzzy sets on X, then A is a neighborhood of B iff given x ∈ X satisfying f_B (**x**) > 0 then there exists $N_x \in \mathcal{T}$ such that $f_B(x) \le f_{N_x(x)}$ and $N_x \le A$.

Proof (⇒)
Suppose A is a neighborhood of a fuzzy set A, and G ∈ T satisfies B ≤ G ≤ A. For given x ∈ X for which $f_B(x) > 0$, we choose $N_x = G$

$$(\Leftarrow)\ \text{If}\ G = \vee\{\text{open}\ N_x : 0 < f_B(x) \le {}^{f}N_x(x)\ \text{and}\ N_x \le A\}$$

Then B ≤ G ≤ A and G ∈ T.

Theorem 5.5 Let (X,T) be a fuzzy topological space, then A ∈ T if and only if whenever B is a fuzzy set on X satisfying B ≤ A, then A is a neighborhood of B.

Proof
(⇒) Suppose that A ∈ T and B ≤ A. Then A is a neighborhood of B.
(⇐) Since A ⊂ A, therefore there exists an open fuzzy set say 0, such that

$$A \subset \mathbf{0} \subset A.$$

Hence A = 0, and therefore A is open.

Theorem 5.6 A fuzzy set A in X is open in the fuzzy topological space (X,T) iff for every x ∈ X satisfying $f_A(x) > 0$, there is $N_X \le A$ such that $f_{N_X}(x) = f_A(x)$.

Proof
(⇒) Let us suppose that A is an open fuzzy set in fuzzy topological space (X,T) and for x ∈ X, $f_A(x) > 0$. Now we take $N_x = A$, then.
$N_x \le A$ and $f_{N_X}(x) = f_A(x)$.
(⇐) Let G = ∨{open $N_X \le A$: f_A (x) > 0 and f_{A_X} (x) = $f_A(x)$}.
Then G ∈ T and G = A.

Theorem 5.7 Let (X, T) be a fuzzy topological space.

(i) Then ∅, X are closed fuzzy sets.
(ii) If A, B are closed fuzzy sets, then A ∨ B is a closed fuzzy set.
(iii) If C_i, i ∈ I, where I is an arbitrary set, are closed fuzzy sets, then $\underset{i \in I}{\wedge} C_i$ is a closed fuzzy set.

Proof

The condition (i) and (ii) are obviously true. It only remains to show that (iii) is also True. If S is a non-empty set of real numbers, then we have

$$\inf\{x : x \in S\} = -\mathrm{Sup}\{-x : x \in S\}$$

and $\inf\{1 + x\colon x \in S\} = 1 + \inf\{x\colon x \in S\}$

If $D_i = 1 - C_i$.

then $\wedge\{C_i\colon i \in I\} = \wedge\{1 - D_i\colon i \in I\}$

$$= 1 + \wedge\{-D_i : i \in I\}$$
$$= 1 - \vee\{D_i : i \in I\}$$

where each $D_i\colon i \in I$ is an open fuzzy set showing that $\wedge\{C_i\colon i \in I\}$ is a closed fuzzy set.

5.3 Interior of a Fuzzy Set

Let (X,T) be a fuzzy topological space and let A and B be fuzzy sets in (X,T) such that $A \geq B$. Then B is called an interior fuzzy set of A if and only if A is nbhd of B. The supremum (least upper bound) of all interior fuzzy sets of A is called the interior of A. In other words, the union of all interior fuzzy sets of A is called the interior of A and is symbolically denoted byA°.

Theorem 5.8 Let A be a fuzzy set in a fuzzy topological space (X,T). Then A° is open and is the largest open fuzzy set contained in A. The fuzzy set A is open iff $A = A^\circ$.

Proof (⇒)

Let $A^\circ = \sup B$, where B is an interior fuzzy set of A.

Hence, $B \leq A$. Therefore, an open fuzzy set G exists such that $B \leq G \leq A$, where $G \in T$.

$$\therefore \quad \mathrm{Sup}B \leq \mathrm{Sup}G \leq A$$

Hence, there exists an open set $0 = \sup G$ such that

$$A \leq 0 \leq A \tag{5.1}$$

But $0 \leq A^\circ$ for 0 is an interior fuzzy set of A.

$\therefore A^\circ = 0$ (open set) And $0 = \mathrm{Sup}G$, where G is open and $G \leq A$. Therefore, A° is the largest open set contained in A.

$$\text{Now if A is open, } A \leq A^{\circ} \tag{5.2}$$

Eqs. (5.1) and (5.2) $\Rightarrow A = A^{\circ}$

($\Leftarrow$) Conversely, if $A^{\circ} = A$ then, since A° is open, A is also an open fuzzy set.

5.4 Closure of a Fuzzy Set

Let (X, T) be a fuzzy topological space and A be a fuzzy set in (X, T). Then $\wedge$\{closed fuzzy set B in X: B $\geq$ A\} is called the closure of the fuzzy set A and is denoted by $\bar{A}$.

$$\text{i.e., } \quad \overline{A} = \inf\{B : B \geq A, 1 - B \in T\}$$

Theorem 5.9 Let A be a fuzzy set in a fuzzy topological space (X, T). Then $\bar{A}$ is closed and is the least closed fuzzy set, greater than or equal to A. Also the fuzzy set A is closed iff $A = \bar{A}$.

Proof

We have, by definition, of closure of a fuzzy set

$$\overline{A} = \wedge\{\text{closed fuzzy set B in X} : B \geq A\}$$

Hence, $\bar{A}$ is a closed fuzzy set, and $\bar{A} \geq A$.

Since $\bar{A}$ is the infimum of such sets, hence $\bar{A}$ is the least closed fuzzy set containing A, i.e.,

$$A \geq \overline{A}.$$

Conversely, if $\bar{A} = A$ is closed, since $\bar{A}$ is a closed fuzzy set.

Again if A is a closed fuzzy set and

$$\overline{A} = \wedge\{\text{closed fuzzy set B}, \ B \geq A\}$$

Now, since A is closed and $A \geq A$

$$\therefore \overline{A} \leq A$$

We also have $\bar{A} \geq A$

$$\therefore A = \overline{A}$$

Theorem 5.10 Let A and B be fuzzy sets in a fuzzy topological space (X, T). If A $\leq$ B, then $\overline{A} \leq \overline{B}$ and $A^\circ \leq B^\circ$. Also $\overline{\overline{A}} = \overline{A}$ *and* $(A^\circ)^\circ = A^\circ$.

Proof

We have from the definition

$$\overline{\mathrm{A}} = \wedge\{\text{closed fuzzy set C} : \mathrm{C} \geq \mathrm{A}\} \tag{5.3}$$

$$\mathrm{B} = \wedge\ \{\text{closed fuzzy set D} : \mathrm{D} \geq \mathrm{B}\} \tag{5.4}$$

But B $\geq$ A.
Hence, every D satisfying (5.4) and satisfies (5.3).
Therefore, by taking infimum of sets in (5.3) and (5.4), we have

$$\overline{A} \leq \overline{B}$$

Again, from the definition of the interior of a fuzzy set, we have

$$A^\circ = \sup\ \{\mathrm{C} : \mathrm{C} \leq \mathrm{A} \text{ and A is a neighborhood of C}\} \tag{5.5}$$

$$B^\circ = \mathrm{Sup}\ \{\mathrm{D} : \mathrm{D} \leq \mathrm{B} \text{ and B is a neighborhood of D}\} \tag{5.6}$$

But, we have A $\leq$ B.
Therefore, every C satisfying (5.5) meets (5.6).
Hence, on taking the supremum of sets in (5.5) and (5.6), we get

$$A^\circ \leq B^\circ$$

Evidently, we have $\overline{\overline{A}} = \overline{A}$ *and* $(A^\circ)^\circ = A^\circ$, from the above Theorem 5.10.

Theorem 5.11 Let (X, T) be a fuzzy topological space and A, B be any fuzzy sets in (X, T). Then.

(i) $\overline{A} \vee \overline{B} = \overline{A \vee B}$ *and* $\overline{A} \wedge \overline{B} = \overline{A \wedge B}$
(ii) $A^\circ \wedge B^\circ = (A \wedge B)^\circ$ *and* $A^\circ \vee B^\circ \leq (A \vee B)^\circ$
(iii) $(1 - A)^\circ = 1 - \overline{A}$
(iv) $\overline{1 - A} = 1 - A^\circ$

Proof

(i) Since $\overline{A} \vee \overline{B}$ is closed and $\overline{A} \vee \overline{B} \geq A \vee B$ and from the definition of the closure of $A \vee B$, we have.

$$\overline{A} \vee \overline{B} \geq \overline{A \vee B} \tag{5.7}$$

On the other hand, $A \vee B \geq A \Rightarrow \overline{A \vee B} \geq \overline{A}$
and $A \vee B \geq B \Rightarrow \overline{A \vee B} \geq \overline{B}$
On combining these two results, we have

$$\overline{A \vee B} \geq \overline{A} \vee \overline{B} \tag{5.8}$$

By virtue of Eqs. (5.7) and (5.8), we get that $\overline{A} \vee \overline{B} = \overline{A \vee B}$

Since $\overline{A}$ *and* $\overline{B}$ are closed fuzzy sets and $\overline{A} \wedge \overline{B} \geq A \wedge B$, hence from the definition of closure of the fuzzy set $A \wedge B$, we have

$$\overline{A \wedge B} \leq \overline{A} \wedge \overline{B}$$

(ii) Since $A^\circ \wedge B^\circ$ is an open set and $A^\circ \wedge B^\circ \leq A \wedge B$

We have

$$A^\circ \wedge B^\circ \leq (A \wedge B)^\circ \tag{5.9}$$

Again $A \wedge B \leq A \Rightarrow (A \wedge B)^\circ \leq A^\circ$
and $A \wedge B \leq B \Rightarrow (A \wedge B)^\circ \leq B^\circ$
Hence,

$$(A \wedge B)^\circ \leq A^\circ \wedge B^\circ \tag{5.10}$$

Therefore, from Eqs. (5.9) and (5.10), we have

$$(A \wedge B)^\circ = A^\circ \wedge B^\circ$$

We also have $A^\circ \leq A$ *and* $B^\circ \leq B$
$\therefore A^\circ \vee B^\circ \leq A \vee B$ *and* $A^\circ \vee B^\circ$ is an open set

$$\therefore \mathrm{A}^\circ \vee \mathrm{B}^\circ \leq (\mathrm{A} \vee \mathrm{B})^\circ$$

(iii) We have

$$\begin{aligned}
1 - \overline{A} &= 1 - \wedge\{D : 1 - D \in \mathcal{T} \text{ and } D \geq A\} \\
&= \vee\{1 - D : 1 - D \in \mathcal{T} \text{ and } D \geq A\} \\
&= \vee\{C : C \in \mathcal{T} \text{ and } C \leq 1 - A \text{ where } C = 1 - D\} \\
&= (1 - A)^\circ
\end{aligned}$$

(iv) Similarly, we have

$$\begin{aligned}1 - A^{\circ} &= 1 - \vee\{B : B \leq G \leq A \, for\, some\, G \in \mathcal{T}\}\\ &= \wedge\{1 - B : B \leq G \leq A \, for\, some\, G \in \mathcal{T}\}\\ &= \wedge\{1 - B : 1 - B \geq 1 - G \geq 1 - A \, for\, some\, G \in \mathcal{T}\}\\ &= \overline{(1 - A)}\end{aligned}$$

5.5 Fuzzy Limit Point

Let A be a fuzzy set in a fuzzy topological space $(X,\mathcal{T})$. A point $x \in X$ is said to be a fuzzy limit point of a iff whenever $f_A(x) = 1$, then for each N_X, there exists $y \in X - \{x\}$ such that ${}^{f}N_x(y) \wedge f_A(y) \neq 0$ or whenever $f_A(x) \neq 1$, then $f_{\overline{A}}(x) > 0$ and for each open N_x satisfying $\mathbf{1} - {}^{f}N_x(x) = f_A(x)$, $\exists\, y \in X - \{x\}$ such that ${}^{f}N_x(y) \wedge f_A(y) \neq 0$.

5.6 Derived Fuzzy Set

Let A be a fuzzy set in a fuzzy topological space $(X, \mathcal{T})$. Then the derived fuzzy set of A is denoted by A' and is defined as

$$f_{A'}(x) = \begin{cases} f_{\overline{A}}(x) \; if \; x \; is \; a \; fuzzy \; limit \; point \; of \; A \\ 0 \; otherwise \end{cases}$$

Theorem 5.12 Let A be a fuzzy set in a fuzzy topological space $(X,\mathcal{T})$. Then A is closed iff $A' \leq \mathbf{A}$. Furthermore, $A'' \leq \overline{A} = A \vee A'$

Proof

($\Rightarrow$) Let A be a closed fuzzy set in a fuzzy topological space $(X, \mathcal{T})$. Hence $A = \bar{A}$.
Since $A' \leq \overline{A}$, which implies that $A' \leq \overline{A} \leq A$.

($\Leftarrow$) Conversely, we suppose that $A' \leq A$, then we have to show that A is closed, i.e., $A = \bar{A}$.

Now, if $f_{A'}(x) = f_{\overline{A}}(x)$, then as $A' \leq A \leq \overline{A}$

Therefore, we have $f_A(x) = f_{\bar{A}}(x)$

Again, if $f_{A'}(x) \neq f_{\overline{A}}(x)$, then x is not a fuzzy limit point of A and $f_{\overline{A}}(x) > 0$.

when $f_A(x) = 1$, then $f_{\overline{A}}(x) = f_A(x)$. Thus, we suppose that x is not a fuzzy limit point of A, $f_{\overline{A}}(x) > 0$ and $f_A(x) \neq 1$. Then by the definition of a limit point, there is an open N_x such that $1 - {}^{f}N_x(x) = f_A^{(x)}$, and if $y \in X - \{x\}$, then ${}^{f}N_x(y) \wedge f_A(y) = 0$.

Hence, if $y \neq x$, then $1 - {}^{f}N_x(y) \geq f_A(y)$. Since $1 - N_x$ is closed, therefore $\bar{A} \leq 1 - N_x$. Hence

$$f_{\overline{A}}(x) \leq 1 - {}^{f}N_x(x) = f_A(x)$$

$$\therefore \overline{A} = A$$

Hence A is a closed fuzzy set in (X, $\mathcal{T}$).

Next, we shall show that $A'' \leq \overline{A}$.

By definition, we have $A'' \leq \overline{A}$ and $A' \leq \overline{A}$.

Hence, it follows that $\overline{A}' \leq \bar{A}$.

Therefore, $A'' \leq \overline{A}$

To see that $\overline{A} = A \vee A'$, we have if $f_{A'}(x) = f_A(x)$, *then* $f_{\overline{A}}(x) = f_{(A \vee A')}(x)$.

On the other hand if $f_{A'}(x) \neq f_A(x)$, *then* $f_{A'}(x) = 0$ *and hence* $f_{\overline{A}}(x) = f_A(x)$.

Theorem 5.13 Let (X,$\mathcal{T}$) be a fuzzy topological space and let A ⊂ X. Then the family $\mathcal{T}_A$ = {G/A: G ∈ $\mathcal{T}$} is a fuzzy topology on A, where G/A is the restriction of G to A.

Proof

We wish to show that $\mathcal{T}_A$ = {G/A: G ∈ $\mathcal{T}$} satisfies all conditions for a fuzzy topology on A since Ø, X ∈ $\mathcal{T}$. Therefore $\varnothing/A \in \mathcal{T}_A$ and X/A ∈ $\mathcal{T}_A$.

Hence condition (i) is satisfied.

Again if G_1/A and G_2/A ∈ $\mathcal{T}_A$ where $G_1, G_2 \in \mathcal{T}$ implies $G_1 \cap G_2 \in \mathcal{T}$. Therefore,

$$(G_1 \cap G_2)/A \in \mathcal{T}_A$$

i.e., $(G_1/A \cap G_2/A) \in \mathcal{T}_A$

Hence (ii) condition is satisfied.

Finally, if G_i/A ∈ $\mathcal{T}_A$ where $G_i \in \mathcal{T}$, then $\cup G_i \in \mathcal{T}$

$\therefore \cup G_i/A \in \mathcal{T}_A$. Hence $\mathcal{T}_A$ be a fuzzy topology on A and is called relative fuzzy topology on A induced by fuzzy topology $\mathcal{T}$ on X.

5.7 Basis for a Fuzzy Topology

Let $B \subset \mathcal{T}$, where $\mathcal{T}$ is a fuzzy topology on X. Then B will be called as a basis for T if and only if each element of $\mathcal{T}$ is a supremum of members of B.

5.8 Subbasis for a Fuzzy Topology

Let (X, T) be a fuzzy topological space and let$S \subset \mathcal{T}$. Then we say that S is a subbasis of $\mathcal{T}$ if and only if the family of all finite infimums of elements of S is a basis for$\mathcal{T}$.

Theorem 5.14 Let $\mathcal{T}$ be a fuzzy topology on X and let $B \subset \mathcal{T}$. Then the following two properties of B are equivalent:

(i) B is a basis for $\mathcal{T}$.
(ii) For each G $\in \mathcal{T}$, for each x $\in$ X such that f_G (x) > 0 and for each real number $\in > 0$, there is A $\in$ B such that $A \leq G$ and $f_G(x) - f_A(x) < \in$.

Proof
Firstly, we shall show that (i) $\Rightarrow$ (ii).
Let G $\in \mathcal{T}$, $f_G(x) > 0$ and B is a basis for $\mathcal{T}$, it follows that

$$f_G(x) = \vee\{f_A(x) : A \leq G \text{ and } A \in B\}.$$

Thus, there is A $\in$ B satisfying $A \leq G$ and $f_G(x) - f_A(x) < \in$.
(ii) $\Rightarrow$(i)
Let G $\in \mathcal{T}$ and let $f_G(x) > 0$, then there is $A_{x,n} \in$ B such that $A_{x,n} \leq G$ and

$$f_G(x) - f_{A_{x,n}}(x) < 1/n.$$

Hence, $G = \vee_{x,n}\{A_{x,n}: f_G(x) > 0 \text{ and } n = 1,2,3,\ldots\}$.

Corollary 5.1 If $B \subset \mathcal{T}$ is a basis for a fuzzy topology $\mathcal{T}$ on X and A is a fuzzy set in X, then A is open if and only if for each x $\in$ X such that $f_A(x) > 0$ and each $\in > 0$, there is C $\in$ B such that

$$C \leq A \text{ and } f_A(x) - f_C(x) < \in .$$

Theorem 5.15 Let B_1, B_2 are the basis for fuzzy topologies S_1 and S_2 on X respectively, then $S_1 \subset S_2$ if and only if for each $A_1 \in B_1$, each x $\in$ X for which $f_{A_1}(x) > 0$ and each $\in > 0$ there is $A_2 \in B_2$ such that $A_2 \leq A_1$ and $f_{A_1}(x) - f_{A_2}(x) < \in$.

Proof
Let us assume that $S_1 \subset S_2$. Let $A_1 \in B_1$ and x $\in$ X for which $f_{A_1}(x) > 0$, then $A_1 \in S_2$. Now B_2 is a basis for S_2 there is $A_2 \in B_2$ such that

$$f_{A_1}(x) - f_{A_2}(x) < \in \tag{5.11}$$

Again, let B_1 and B_2 be the basis for fuzzy topologies S_1 and S_2, respectively. Let $A_1 \in B_1$ and suppose that for each x $\in$ X for which $f_{A_1}(x) > 0$ and each $\in > 0$ there is $A_2 \in B_2$ (and hence $A_2 \in S_2$) such that $A_2 \leq A_1$ and $f_{A_1}(x) - f_{A_2}(x) < \in$. Then A_1 is an open set for the fuzzy topology S_2.

$$\therefore B_1 \subset S_2 \text{ and } S_1 \subset S_2.$$

5.9 Fuzzy Set Induced by Mapping

Let g be a mapping from X to Y. Let B be a fuzzy set in Y with membership function $f_B(y)$. Then the inverse of B, denoted as $g^{-1}(B)$, is a fuzzy set in X whose membership function is defined by

$$f_{g^{-1}(B)}(x) = f_B(g(x))\text{for all } x \in X$$

Conversely, if g is a mapping from X to Y and A is a fuzzy set in X with membership function $f_A(x)$. The image of A under g, written as g (A), is a fuzzy set in Y whose membership function is given by

$$\begin{aligned} f_{g(A)}(y) &= \underset{x \in g^{-1}[y]}{Sup} \{f_A(x)\} \text{ if } g^{-1}[y] \text{ is not empty} \\ &= 0 \quad otherwise \end{aligned}$$

for all y in Y, where $g^{-1}[y] = \{x : g(x) = y\}$

Theorem 5.16 Let g be a function from X to Y. Then the following holds:

(a) $g^{-1}[B'] = \{g^{-1}[B]\}'$ for any fuzzy set B in Y where B' stands for the complement of B.
(b) $g[A'] \supset \{g[A]\}'$ for any fuzzy set A in X, where A' is the complement of A.
(c) $B_1 \subset B_2 \Rightarrow g^{-1}(B_1) \subset g^{-1}(B_2)$ where B_1 and B_2 are fuzzy sets in Y.
(d) $A_1 \subset A_2 \Rightarrow g(A_1) \subset g(A_2)$ where A_1 and A_2 are fuzzy sets in X.
(e) $B \supset g[g^{-1}[B]]$ for any fuzzy set B in Y.
(f) $A \subset g^{-1}[g[A]]$ for any fuzzy set A in X.
(g) Let g be a function from X to Y and h be a function from Y to Z. Then $(h \circ g)^{-1}(C) = g^{-1}[h^{-1}(C)]$ for any fuzzy set C in Z, where $h \circ g$ is the composition of h and g.

Proof

(a) Let g be a function from X to Y and suppose that B is a fuzzy set in Y. Obviously, B', i.e., the complement of B, is also a fuzzy set in Y.

Hence, $f_{g^{-1}(B')}(x) = f_{B'}(g(x))$ for all $x \in X$

$$\begin{aligned} &= 1 - f_B(g(x)), \forall x \in X \\ &= 1 - f_{g^{-1}(B)}(x), \forall x \in X \\ &= f_{(g^{-1}(B))'}(x), \forall x \in X \end{aligned}$$

$$\therefore \quad g^{-1}(B') = (g^{-1}(B))'$$

(b) Let A be any fuzzy set in X. Then A', the complement of A, is also a fuzzy set in X. Also, suppose that g is a function from X to Y. Suppose that $\mathbf{y} \in$ Y and $g^{-1}(y)$ is non-empty. Hence, membership function of $g(A')$ is given by

$$\begin{aligned} f_{g(A')}(y) &= \underset{z \in g^{-1}(y)}{Sup} \{f_{A'}(z)\} \\ &= \underset{z \in g^{-1}(y)}{Sup} \{1 - f_A(z)\} \\ &= 1 - \underset{z \in g^{-1}(y)}{inf} \{f_A(z)\} \end{aligned} \tag{5.12}$$

Again g(A) is a fuzzy set in Y. Therefore, $\{g(A)\}'$ is also a fuzzy set in Y.

$$\begin{aligned} \therefore f_{g(A)'}(y) &= 1 - f_{g(A)}(y) \\ &= 1 - \underset{z \in f^{-1}(y)}{Sup} \{f_A(z)\} \end{aligned} \tag{5.13}$$

From Eqs. (5.12) and (5.13), we have

$$\begin{aligned} f_{g(A')}(y) &\geq f_{\{g(A)\}'}(y) \\ &\Rightarrow g(A') \supset \{g(A)\}' \end{aligned}$$

(c) Let us assume that $\mathbf{B_1}$ and $\mathbf{B_2}$ are two fuzzy sets in Y, and $\mathbf{B_1} \subset \mathbf{B_2}$ and g are functions from X to Y.

Then $f_{g^{-1}(B_1)}(x) = f_{B_1}\{g(x)\}, \forall\, x \in X$
and $f_{g^{-1}(B_2)}(x) = f_{B_2}\{g(x)\}, \forall\, x \in X$
But $B_1 \subset B_2 \Rightarrow f_{B_1}\{g(x)\} \leq f_{B_2}\{g(x)\}, \forall\, x \in X$

$$\begin{aligned} \therefore\ f_{g^{-1}(B_1)}(x) &\leq f_{g^{-1}(B_2)}(x), \forall\, x \in X \\ &\Rightarrow g^{-1}(B_1) \subset g^{-1}(B_2) \end{aligned}$$

(d) Let g be a function from X to Y, and let $\mathbf{A_1}$, $\mathbf{A_2}$ are two fuzzy sets in X such that

$$\mathbf{A_1} \subset \mathbf{A_2}.$$

Further, suppose that y $\in$ Y, then

$$f_{g(A_1)}(y) = \underset{z \in g^{-1}(y)}{Sup} \{f_{A_1}(z)\},\ if\ g^{-1}(y) \neq \emptyset$$

and $f_{g(A_2)}(y) = \underset{z \in g^{-1}(y)}{Sup} \{f_{A_2}(z)\},\ if\ g^{-1}(y) \neq \varnothing$

Since $A_1 \subset A_2 \Rightarrow f_{A_1}(z) \leq f_{A_2}(z)$

$$\begin{aligned} \therefore \ & f_{g(A_1)}(y) \leq f_{g(A_2)}(y), \ \forall y \in Y \\ & \Rightarrow g(A_1) \subset g(A_2) \end{aligned}$$

(e) Let B be a fuzzy set in Y, then $g^{-1}(B)$ is a fuzzy set in X. Let $g^{-1}(y) \in X$ for all y ∈ Y.

$$\begin{aligned} \therefore f_{g\{g^{-1}(B)\}(y)} &= \underset{z \in g^{-1}(y)}{Sup} \left\{ f_{g^{-1}(B)}(z) \right\}, \ if \ g^{-1}(y) \neq \emptyset \\ &= \underset{z \in g^{-1}(y)}{Sup} \{ f_B(g(z)) \} \\ &= f_B(y) \\ \therefore g\{g^{-1}(B)\} &= B \end{aligned}$$

If $g^{-1}(y) = \phi \ then \ f_{g(g^{-1}(B))}(y) = 0. \ But \ f_B(y) may \ not \ be \ zero.$
Hence $g\{g^{-1}(B)\} \leq B$

$$\Rightarrow g\{g^{-1}(B)\} \subset B$$

(f) Let A be a fuzzy set in X, then g(A) will be a fuzzy set in Y. Hence

$$\begin{aligned} f_{g^{-1}[g(A)]}(x) &= f_{g(A)}\{g(x)\} for \ all \ x \in X \\ &= \underset{z \in g^{-1}(g(x))}{Sup} \{ f_A(z) \}, \forall z \in X \\ &\geq \{ f_A(x) \}, \ \forall x \in X \\ &\text{i.e., } A \subset g^{-1}(A) \end{aligned}$$

If g is injective that is if g is one–one, then

$$\begin{aligned} & z \in g^{-1}(g(x)) \Rightarrow z = x \\ & \therefore f_{g^{-1}(g(A))}(x) = f_A(x) \end{aligned}$$

If for each y ∈ Y, $g^{-1}(y) \neq \phi$ and if f_A is constant on$g^{-1}(y)$, then $f_A(z)$ is constant for each$z \in g^{-1}(g(x))$.

$$\therefore g^{-1}(g(A)) = A$$

(g) Let g be a function from X to Y and h be a function from Y to Z, then (hog) is a composite function from X to Z. Let C be any fuzzy set in Z.

Then,

$$
\begin{aligned}
f_{(h\circ g)^{-1}C}(x) &= f_C\{(h\circ g)(x)\},\ \forall x\in X\\
&= f_C\big[h[g(x)]\big],\ \forall x\in X\\
&= f_{g^{-1}[C]}\{g(x)\},\ \forall x\in X\\
&= f_{(g^{-1}\circ h^{-1})[C]}(x),\ \forall x\in X\\
&= f_{g^{-1}[h^{-1}[C]]}(x),\ \forall x\in X\\
\therefore (h\circ g)^{-1}[C] &= (g^{-1}\circ h^{-1})[C]\\
&= g^{-1}\big[h^{-1}[C]\big]
\end{aligned}
$$

Theorem 5.17 Let g be a function from X to Y. If A and A_i, i ∈ I are fuzzy sets in X, and if B and B_j, j ∈ J are fuzzy sets in Y, then the following relations hold:

(i) $g(\wedge A_i) \le \wedge g(A_i)$
(ii) $g^{-1}(\wedge B_j) = \wedge g^{-1}(B_j)$
(iii) $g(\vee A_i) = \vee g(A_i)$
(iv) $g^{-1}(\vee B_j) = \vee g^{-1}(B_j)$
(v) $g\{g^{-1}(B)\wedge A\} = B\wedge g(A)$

Proof

(i) Suppose thatA_i, $i\in I$, be a family of fuzzy sets in X. Therefore, $g(A_i)$, i ∈ I are fuzzy sets in Y.

Now $f_g(\wedge A_i)(y_i) = \vee\left[\overset{\wedge_i}{\underset{x}{}}\{f_{A_i}(x) : g(x) = y\}\right]$

$$
\begin{aligned}
&\le \underset{i}{\wedge}\,[\,\underset{x}{\vee}\{f_{A_i}(x) : g(x) = y\}\\
&= \wedge f_{g(A_i)}(y),\ \forall y\in Y\\
&\therefore g(\wedge A_i) \quad \le \quad \wedge g(A_i)
\end{aligned}
$$

(ii) Since $B_j, j\in J$ are fuzzy sets in Y; therefore, $g^{-1}(B_j)$ are fuzzy sets in X.

Now $f_{g^{-1}(\wedge B_j)}(x) = f_{\wedge B_j}(g(x))$

$$= \underset{j}{\wedge} f_{B_j}(g(x))$$

Hence $f_{g^{-1}(\wedge B_j)}(x) = \underset{j}{\wedge} f_{g^{-1}(B_j)}(x),\ \forall x\in X$

$$\therefore g^{-1}(\wedge B_{j)} = \wedge g^{-1}(B_j)$$

(iii) Since A_i, $i \in I$ are fuzzy sets in X; therefore $\underset{i}{\vee} A_i$ is also a fuzzy set in X, and $g(\vee A_i)$ is a fuzzy set in Y. Hence its membership function is given by

$$\begin{aligned}
f_{g(\vee A_i)}(y) &= \underset{x}{Sup} \{f_{(\vee A_i)}(x) : g(x) = y\} \\
&= \underset{x}{Sup} \left\{ \underset{I}{Sup} f_{A_i}(x) : g(x) = y \right\} \\
&= \underset{I}{Sup} \left\{ \underset{x}{Sup} \{f_{A_i}(x) : g(x) = y\} \right\} \\
&= \vee f_{g(A_i)}(x) \\
\therefore \quad g(\vee A_i) &= \vee g(A_i)
\end{aligned}$$

(iv) Let$B_j, j \in J$, be a fuzzy sets in Y, then $\vee B_j$ is also a fuzzy set in Y. Hence $g^{-1}(\vee B_j)$ be the fuzzy set in X, and its membership function is given by

$$\begin{aligned}
f_{g^{-1}(\vee B_j)}(x) &= f_{\vee B_j}(g(x)), \forall x \in X \\
&= \underset{J}{Sup} \{f_{B_j}(g(x))\}, \forall x \in X \\
&= \underset{J}{Sup} f_{g^{-1}(B_j)}(x), \forall x \in X \\
&= \vee f_{g^{-1}(B_j)}(x), \forall x \in X \\
\text{i.e., } g^{-1}(\vee B_j) &= \vee g^{-1}(B_j)
\end{aligned}$$

(v) Let A be a fuzzy set in X and B be a fuzzy set in Y, then $g^{-1}(B)$ is a fuzzy set in X and hence $(g^{-1}(B) \wedge A)$ be a fuzzy set in X. Therefore, $g(g^{-1}(B) \wedge A)$ be a fuzzy set in Y, and its membership function be given by

$$\begin{aligned}
f_{g(g^{-1}(B) \wedge A)}(y) &= Sup\left\{f_{(g^{-1}(B) \wedge A)}(x) : g(x) = y\right\} \\
&= \vee \{f_B(g(x) \wedge f_A(x)) : f(x) = y\} \\
&= f_B(y) \wedge [\vee f_A(x) : g(x) = y] \\
&= f_B(y) \wedge f_{g(A)}(y) \\
&= f_{[B \wedge g(A)]}(y) \\
\therefore \quad g\{g^{-1}(B) \wedge A\} &= B \wedge g(A)
\end{aligned}$$

5.10 Fuzzy Continuous Functions

Any function g from a fuzzy topological space (X,$\mathcal{T}$) to a fuzzy topological space (Y,U) is said to be fuzzy continuous (or F-continuous) if and only if the inverse of each U—open fuzzy set is $\mathcal{T}$-open fuzzy set.

Theorem 5.18 Let (X,$\mathcal{T}$) and (Y, U) are fuzzy topological spaces, and g be a function from X to Y i.e.,$g : X \rightarrow Y$; then the following conditions are equivalent:

(a) The function g is F—continuous.
(b) The inverse image of every closed fuzzy set is closed.
(c) The inverse image of every element of a subbasis for U is in$\mathcal{T}$.
(d) For every x ∈ X and every nbhd N of $g(x)$, $g^{-1}(N)$ is a nbhd of x.
(e) For every x ∈ X and every nbhd N of g(x), there is a nbhd M of x such that

$$g(M) \leq N \text{ and } f_M(x) = f_{g^{-1}(N)}(x)$$

(f) For every fuzzy set A in X, $g(\overline{A}) \leq \overline{g(A)}$
(g) For every fuzzy set B in Y, $\overline{\{g^{-1}(B)\}} \leq g^{-1}(\overline{B})$

Proof (a) ⇒ (b)

Suppose that g is a function from a fuzzy topological space (X,T) to a fuzzy topological space (Y,U). Further we assume that B is a fuzzy set in Y. Then we have

$$g^{-1}(B') = \left[g^{-1}(B)\right]'$$

Now we assume that B is closed. Hence B' will be an open fuzzy set in Y, and if g is F-continuous, then $g^{-1}(B')$ is open in X, i.e., $\left[g^{-1}(B)\right]'$ open in X which implies that $g^{-1}(B)$ is a closed fuzzy set in X, i.e., the inverse of U-closed fuzzy set is $\mathcal{T}$-closed fuzzy set in X.

(b) ⇒ (a)

Let us suppose that the inverse of each U-closed fuzzy set is $\mathcal{T}$-closed in X, and let G be an open fuzzy set in Y.

Hence G' be a closed fuzzy set in Y ⇒ $g^{-1}(G')$ is a closed fuzzy set in X (by hypothesis)

⇒$\left[g^{-1}(G)\right]'$ is a closed fuzzy set in X

⇒$g^{-1}(G)$ is an open fuzzy set in X.

Thus, G is an open fuzzy set in Y ⇒ $g^{-1}(B)$ is an open fuzzy set in X.

Hence. g is F-Continuous.

(a) ⇒ (c)

It is obviously true. Since every element of a subbasis for U is an open fuzzy set and g is F-Continuous, therefore the inverse image of every element of a subbasis is also an open fuzzy set in $\mathcal{T}$.

(c)⇒ (a)

Suppose that s_1, s_2, s_3,…, s_n are elements of a subbasis S_1 for U and let S_2 be the basis of finite infimums of elements of S_1. Then, $g^{-1}(\wedge S_x) = \wedge g^{-1}(S_x)$ is in $\mathcal{T}$. Therefore, every member of S_2 has the property that its inverse image is in $\mathcal{T}$. Now let $S \in U$. Then

$S = \vee\{B_i : i \in I_s, B_i \in S_2\}$. It follows that

$g^{-1}(S) = \vee\{g^{-1}(B_i)\}$ is in $\mathcal{T}$.

(a)⇒ (d)

Let N is a neighborhood of g(x). Then by the definition, there exists on open set G ∈ U such that G ≤ N and $f_N\{g(x)\} = f_G(g(x)) > 0$. Since g is F-continuous and G ≤ N, $g^{-1}(G) \in \mathcal{T}$ and $g^{-1}(G) \leq g^{-1}(N)$ and $f_{g^{-1}(N)}(x) = f_{g^{-1}(G)}(x) > 0$

Therefore, $g^{-1}(N)$ is a neighborhood of x.

(d) ⇒ (e)

Let x be any element of X and N be any neighborhood of g(x). Let us suppose that $g^{-1}(N)$ be any neighborhood of x. Let $M = g^{-1}(N)$, then M is a neighborhood of x and g(M) = $g(g^{-1}(N)) \leq N$, by **Theorem 5.16** and also $f_{g^{-1}(N)}(x) = f_M(x)$.

(e) ⇒ (a)

Let G ∈ U and x ∈ X such that $f_{g^{-1}(G)}(x) > 0$. Then $f_G(g(x)) > 0$ and G is a neighborhood of g(x). Therefore, by **Theorem 5.6**, $g^{-1}(G) \in \mathcal{T}$ and hence g is F-continuous.

(b) ⇒ (f)

Firstly, we assume that the inverse image of every closed fuzzy set is closed. Let g(A) be any fuzzy set in Y where A is any fuzzy set in X. Now, by taking the closure of g(A); we have, by definition

$$\overline{g(A)} = \wedge\{C : C \geq g(A)\, and\; 1 - C \in U\}$$

Hence $g^{-1}(\overline{g(A)}) = \wedge\{g^{-1}(C) : C \geq g(A)\; and\; 1 - C \in U\}$

Since $g^{-1}(C)$ is a closed fuzzy set and $g^{-1}(C) \geq A$, it follows that $g^{-1}(C) \geq \overline{A}$, and hence

$$\wedge\{g^{-1}(C)\} \geq \overline{A}.$$

Thus, $g^{-1}(\overline{g(A)}) \geq \overline{A}$

$$\therefore\; \overline{g(A)} \;\geq\; g(\overline{A})$$

i.e., $g(\overline{A}) \leq \overline{g(A)}$.

(f) ⇒ (g)

Let B be any fuzzy set in Y; then $g^{-1}(B)$ is a fuzzy set in X. For any fuzzy set A in X, we have

$$g(\overline{A}) \leq \overline{g(A)}$$

$\therefore g\{\overline{g^{-1}(B)}\} \leq \overline{g(g^{-1}(B))} \leq \overline{B}$, by Theorem 5.16 (e) and (f)
Hence, $\overline{g^{-1}(B)} \leq g^{-1}(\overline{B})$

(g) ⇒ (b)

Let C be a closed fuzzy set in Y. Then, $\overline{g^{-1}(C)} \leq g^{-1}(\overline{C}) = g^{-1}(C)$, since C is a closed fuzzy set. Thus $g^{-1}(C)$ is a closed fuzzy set in X. Hence the inverse image of every closed fuzzy set is closed.

Theorem 5.19 If X and Y are fuzzy topological spaces and g is a function from X to Y, then the under-noted conditions are related as follows: (i) imply (ii) and (ii) and (iii) are equivalent.

(i) The function g is F-continuous
(ii) For each fuzzy set A in X, the inverse of every neighborhood of g(A) is a neighborhood of A.
(iii) For each fuzzy set A in X and each neighborhood V of g(A), there is a neighborhood W of A such that $g(W) \subset V$

Proof

(i) ⇒ (ii) If g is F-continuous and A is a fuzzy set in X and if V is a neighborhood of g(A), then by definition, V contains an open fuzzy set W such that $g(A) \subset W \subset V$, $g^{-1}(g(A)) \subset g^{-1}(W) \subset g^{-1}(V)$. But $A \subset g^{-1}(g(A))$, by **Theorem 5.16(f)**. Hence $A \subset g^{-1}(W) \subset g^{-1}(V)$ and fuzzy continuity of g implies that $g^{-1}(W)$ is an open fuzzy set in X. Consequently, $g^{-1}(V)$ is a neighborhood of A.
(ii) ⇒ (iii) Let us assume that for each fuzzy set, A in X, and any neighborhood V of g(A), $g^{-1}(V)$ is a neighborhood of A. Let $W = g^{-1}(V)$, then W is a neighborhood of A and

$$g(W) = g(g^{-1}(V)) \subset V \Rightarrow g(W) \subset V.$$

(iii) ⇒ (ii) Let V is a neighborhood of g(A). Then there is a neighborhood W of A such that $g(W) \subset V$. Hence,

$g^{-1}(g(W)) \subset g^{-1}(V)$but $W \subset g^{-1}(g(W))$. Hence $W \subset g^{-1}(V)$, i.e., $g^{-1}(V)$ is a neighborhood of A.

Composition Theorem

Theorem 5.20 If g is an F-Continuous function from a fuzzy topological space X to a fuzzy topological space Y and h is an F-Continuous function from fuzzy topological space Y to fuzzy topological space Z, then (hog) is an F-Continuous function from the fuzzy topological space X to fuzzy topological space Z.

Proof
Let us assume that g is a function from a fuzzy topological space X to a fuzzy topological space Y and h be a function from a fuzzy topological space Y to a fuzzy topological space Z. Then the composite function (hog) be a function from the fuzzy topological space X to the fuzzy topological space Z. Hence for any fuzzy set C in Z, we have

$$(h \circ g)^{-1}[C] = g^{-1}\left[h^{-1}(C)\right]$$

Let C be an open fuzzy set in Z and h be an F-Continuous function from the fuzzy topological space Y to the fuzzy topological space Z. Then $h^{-1}(C)$ be an open fuzzy set in Y. Again, let g be an F-continues function from a fuzzy topological space X to a fuzzy topological space Y. This implies that $g^{-1}\left[h^{-1}(C)\right]$ is an open fuzzy set in X, i.e., $(h \circ g)^{-1}[C]$ is an open fuzzy set in X for any open fuzzy set C in Z suggests that (hog) is an F-continuous function.

5.11 Fuzzy Homeomorphism

Let (X,$\mathcal{T}$) and (Y,U) be two fuzzy topological spaces and let g be a mapping of X onto Y. Then g is said to be a fuzzy homeomorphism iff

(i) g is bijective, that is, g is one–one and onto
(ii) g is F-continuous
(iii) g^{-1} is F-continuous

In other words, a fuzzy homeomorphism g is a fuzzy continuous one-to-one mapping of a fuzzy topological space (X,$\mathcal{T}$) onto a fuzzy topological space (Y,U) such that the inverse of the mapping, i.e., g^{-1}, is also fuzzy continuous i.e., F-continuous.

Suppose there exists a fuzzy homeomorphism of one fuzzy space onto another fuzzy space. In that case, the two fuzzy spaces are said to be fuzzy homeomorphic or F-homeomorphic, and we say that each is a fuzzy homeomorph of the other. We say that two fuzzy topological spaces are topologically F-equivalent if and only if they are fuzzy homeomorphic.

Theorem 5.21 The necessary and sufficient condition for a one–one mapping $g : X \longrightarrow Y$ from a fuzzy topological space X onto a fuzzy topological space Y to be fuzzy homeomorphic is that

$$g\left(\overline{A}\right) = \overline{g(A)}$$

, where A is any fuzzy set in X.

Proof

The condition is necessary:

Let us suppose that (X,$\mathcal{T}$) and (Y,U) be two fuzzy topological spaces and let the one–one onto mapping $g : X \longrightarrow Y$ is a fuzzy homeomorphism. Then, we have to show that.

$g\left(\overline{A}\right) = \overline{g(A)}$, where A is any fuzzy set in X.

Since g is a fuzzy homeomorphism, therefore both g and g^{-1} are fuzzy continuous. Hence, for any fuzzy set A in X, we have.

$$g\left(\overline{A}\right) \leq \overline{g(A)} \tag{5.14}$$

Again since g^{-1} is also fuzzy continuous, it follows that for any fuzzy set B in Y, we have

$$g^{-1}\left(\overline{B}\right) \leq \overline{g^{-1}(B)}$$

Let B = g(A), then $\overline{B} = \overline{g(A)}$

$$\begin{aligned} \therefore\ g^{-1}\left(\overline{g(A)}\right) &\leq \overline{g^{-1}(g(A))} \\ &= \overline{A} \\ \therefore\ g(g^{-1}\left(\overline{g(A)}\right) &\leq g\left(\overline{A}\right) \end{aligned}$$

$$\therefore\ \overline{g(A)} \leq g\left(\overline{A}\right) \tag{5.15}$$

Thus from Eqs. (5.14) and (5.15), we get

$$g\left(\overline{A}\right) = \overline{g(A)}$$

The condition is sufficient:

Let us suppose that g is a one-one onto mapping of a fuzzy topological space X onto a fuzzy topological space Y such that $g\left(\overline{A}\right) = \overline{g(A)}$, where A is any fuzzy set in X.

Then we shall show that g is a fuzzy homeomorphism.

We have, $g\left(\overline{A}\right) = \overline{g(A)}$

Therefore, $g\left(\overline{A}\right) \leq \overline{g(A)}$, where A is a fuzzy set in X and hence g is fuzzy continuous.

Now it only remains to show that g^{-1} is fuzzy continuous.

For any fuzzy set B in a fuzzy topological space Y, we have the inequality

$$g^{-1}(\overline{B}) \leq \overline{g^{-1}(B)}$$

Let $g^{-1}(B) = A$ in $g(\overline{A}) = \overline{g(A)}$
then,$g\left(\overline{g^{-1}(B)}\right) = \overline{g(g^{-1}(B))} = \overline{B}$
Since the mapping is onto Y.

$$g^{-1}\left\{g\left(\overline{g^{-1}(B)}\right)\right\} = g^{-1}(\overline{B})$$

i.e., $\left(\overline{g^{-1}(B)}\right) = g^{-1}(\overline{B})$, for any fuzzy set B in Y.

Therefore, $g^{-1}(\overline{B}) \leq \left(\overline{g^{-1}(B)}\right)$

Hence g^{-1} is F-Continuous. Thus, g and g^{-1} both are F-Continuous; hence, f is a fuzzy homeomorphism.

5.12 Compact Fuzzy Space

A fuzzy topological space (X,$\mathcal{T}$) is considered compact if and only if each open cover has a finite sub-cover.

5.13 Finite Intersection Property

We say that a family F of a fuzzy set has the finite intersection property iff the intersection of the members of each finite sub-family of F is non-empty.

C. L. Chang has proved two theorems concerning compact sets.

Theorem 5.22 A fuzzy topological space is compact if and only if each family of closed fuzzy sets with the finite intersection property has a non-empty intersection. [C. L. Chang: Fuzzy Topological Spaces, J. Mathematical Analysis, and Applications, V.24, p.189 (1968)].

Theorem 5.23 Let f be an F-continuous function carrying the compact fuzzy topological spaces X onto the fuzzy topological spaces Y. Then Y is compact.

J. A. Goguen [The Fuzzy Tychonoff Theorem, J. Mathematical Analysis, and Applications, V.43, p.734–742 (1973)] pointed out the deficiency in Chang's compactness theory by showing that Tychonoff Theorem is false for infinite product fuzzy topological spaces.

R. Lowen observed that with the definition of fuzzy topological spaces and fuzzy continuous function, constant functions between fuzzy topological spaces are not necessarily fuzzy continuous. This has led to an alternative definition of fuzzy topological space and the compact fuzzy topological space by R. Lowen.

R. Lowen proposed the notion of fuzzy topological space [Fuzzy Topological Spaces and Fuzzy Compactness, J. Mathematical Analysis, and Applications, V.56, p.621–683] as follows:

5.14 Fuzzy Topology by Lowen

A family $\delta \subseteq I^E$ of fuzzy subsets is called a fuzzy topology (in the sense of Lowen) for E if it satisfies the following:

(i) For all α (constant function) $\alpha \in \delta$
(ii) For all $\mu, \gamma \Rightarrow \mu \wedge \gamma \in \delta$
(iii) For all $(\mu_j)_{j \in J} \subset \delta \Rightarrow \underset{j \in J}{Sup}\ \mu_j \in \delta$

where I = [0, 1].

Obviously, we can associate a fuzzy topology with a given topology on a set and viceversa as follows:

Let $\mathcal{T}(E)$ be the set of all fuzzy topologies on E and $W(E)$ be the set of all fuzzy topologies on$E$. OnR, we consider a topology $\mathcal{T}_r = \{]\alpha, \infty[: \alpha \in R\} \cup \{\phi\}$, then induced topology on I is defined as I_r

We define two mappings i *and* w as follows:

(i) $i : W(E) \to \mathcal{T}(E) : \delta \to i(\delta)$

, where $i(\delta)$ is the initial topology on E for the family of "functions δ" and topological space I_r

(ii) $w : \mathcal{T}(E) \to W(E) : \mathcal{T} \to w(\mathcal{T})$

, where $w(\mathcal{T}) = C(E, I_r)$ the continuous functions from $E(\mathcal{T})$ *to* I_r.

R. Lowen has called fuzzy topological spaces, as defined by C. L. Chang as quasi-fuzzy topological space, and fuzzy compactness, as defined by C. L. Chang, as quasi-fuzzy compactness, which makes sense only in the class of quasi-fuzzy topological space.

As Chang and Lowen proposed, both definitions are being used by fuzzy topologists. One of the main advantages of R. Lowen's description over Change's meaning is that all constant functions are fuzzy and continuous.

5.15 Continuous Function Between Fuzzy Topological Spaces

A function $g : (E, i(\delta)) \to (F, i(\delta))$ is continuous. Chang's definition of compactness for (quasi) fuzzy topological space is formally the same as for the topological space. With this definition, one does not have that if $(X, \mathcal{T})$ is compact, $(X, w(\mathcal{T}))$

is compact. R. Lowen has shown this by giving a suitable example. Therefore, R. Lowen changed the definition of fuzzy compactness as follows:

5.16 Fuzzy Compactness for Fuzzy Set

Let (E, δ) be a fuzzy topological space. A fuzzy set $V \in I^E$ is said to be fuzzy compact if and only if, for all family β, $\beta \subset \delta$ *such that* $\underset{\mu \in \beta}{Sup} \mu \geq V$

and for all $\in > 0$, There exists a finite sub-family $\beta_0 \subset \beta$ such that $\underset{\mu \in \beta_0}{Sup} \mu \geq V - \in$.

5.17 Fuzzy Compactness for Quasi-Fuzzy Topological Space

The fuzzy topological space (quasi-fuzzy topological space) (E, δ) is said to be fuzzy compact if and only if each constant fuzzy set in (E, δ) is fuzzy compact.

The greatest advantage of R. Lowen's notion of fuzzy compactness is that the Tychonoff product theorem for a family of compact fuzzy topological space holds, while with Change's definition of compactness, Tychonoff product theorem is true only for a finite number of compact fuzzy topological spaces.

Lowen defines initial fuzzy topology as follows:

5.18 Initial Fuzzy Topology

The initial fuzzy topology on E for the family of fuzzy topological spaces $\left(F_j, \gamma_j\right)_{j \in J}$ and the family of functions

$$g_j : E \longrightarrow \left(F_j, \gamma_j\right)$$

is the smallest fuzzy topology on E making each function g_j fuzzy continuous, where the smallest fuzzy topology with the above property is a fuzzy topology on E, where sub-base is $\underset{j \in J}{\cup} g_j^{-1}\left(\gamma_j\right)$ where

$$g_j^{-1}\left(\gamma_j\right) = \left\{g_j^{-1}(\lambda) : \lambda \in \gamma_j\right\}$$

5.19 Fuzzy Product Topology

If $\left(E_j,\ \delta_j\right)_{j\in J}$ is a family of fuzzy topological spaces, then the fuzzy product topology on $\prod_{j\in J}E_j$ is defined as initial fuzzy topology on $\prod_{j\in J}E_j$, for the family of spaces $\left(E_j,\ \delta_j\right)_{j\in J}$ and functions $g_i : \prod_{j\in J} E_j \longrightarrow E_i$ where for all $i \in J, g_i = pr_i$, the projection on ith co-ordinate. The product fuzzy topology on $\prod_{j\in J}E_j$ is denoted as$\prod_{j\in J}\delta_j$.

Finally, R. Lowen proves the Tychonoff product theorem [Initial and Final Topology, J. of Mathematical Analysis, and Applications, V.58, Theorem 1.6, pp. 16, (1977)].

Tychonoff Product Theorem 5.24: If $\left(E_j,\ \delta_j\right)_{j\in J}$ be a family of fuzzy compact fuzzy topological spaces, then the product spaces $(\prod_{j\in J}E_j,\ \prod_{j\in J}\delta_j)$ is also fuzzy compact.

R. Lowen believes that the notion of fuzzy compactness, as he proposed, is correct. One can obtain the Tychonoff theorem for arbitrary products. However, it has been observed that with drastic changes in the definition of fuzzy topological space and compact fuzzy space, Lowen has lost the concept that fuzzy topology generalizes topology. The usual definition of a fuzzy topology given by Chang includes ordinary topology, but Lowen's definition of fuzzy topology excludes ordinary topology from being fuzzy topology.

Some More Illustrative Examples:

Examples 5.3 Let X be a non-empty set and we define a fuzzy set A_n for each n ∈ N by$A_n(x) = 1 - \frac{1}{n}$, for all x ∈ X, then $\mathcal{T} = \{A_n : n \in N\} \cup \{\tilde{1}\}$ forms a fuzzy topology on X.

Examples 5.4 Let X be a non-empty set and a ∈ X. Let$\mathcal{T} = \{a_p : p \in (0, 1]\} \cup \{U, \mathcal{T}\}$, where the fuzzy point a_p is defined by

$$\begin{aligned} a_p(x) &= 0,\ if\ x \neq a \\ &= p,\ if\ x = a \end{aligned}$$

then $\mathcal{T}$ forms a fuzzy topology on X.

Examples 5.5 Suppose that X = {a, b, c}. A fuzzy set $A : X \rightarrow I$ be defined by A(a) = 0.2, A(b) = 0.4 and A(c) = 0.6, then the collection $\mathcal{T} = \{\tilde{0}, A, \tilde{1}\}$ then $\mathcal{T}$ forms a fuzzy topology on X.

Examples 5.6 Let $(X, \mathcal{T})$ be a crisp topological space. If we consider the characteristic function of each member $A \in \mathcal{T}$ defined by

$$\begin{aligned} \chi_A(x) &= 1,\ if\ x \in A \\ &= 0,\ otherwise \end{aligned}$$

then $\mathcal{T} = \{\chi_A : A \in \mathcal{T}\}$ forms a fuzzy topology on X.

Bases

Examples 5.7 In Example 5.3 $\mathfrak{B} = \{A_n : n \in N\}$ forms a base for fuzzy topology $\mathcal{T}$.

Examples 5.8 In Example 5.1, if we consider any fuzzy point x_p then

$$x_p \notin A_n^c(x) \Rightarrow p > A_n^c(x) \Rightarrow p > \frac{1}{n} \Rightarrow n > \frac{1}{p}$$

Let $n_0 = \left[\frac{1}{p}\right] + 1$ where $\left[\frac{1}{p}\right]$ denotes the greatest integer not exceeding $\frac{1}{p}$. Then $Q(x_p) =$ the collection of all fuzzy subset of X containing A_{n_0}.

Examples 5.9 In Example 5.4, if we consider a fuzzy point a_p then $a_p \notin {a^c}_p$ if$p > 1 - r \Rightarrow r > 1 - p$. So $Q(a_p) =$ the collection of all fuzzy sets containing the fuzzy point a_r where $r > 1 - p$. Here, we note that a_r is an open set which is quasi-coincidence with $a_p, \forall\, r > 1 - p$. For other fuzzy points x_p *with* $x \neq a, \mathcal{T}$ is the only fuzzy open set quasi-coincidence witha_p.

Hence, $Q(x_p) = \{\mathcal{T}\} \forall\, x \neq a$ *and* $\forall\, p \in (0, 1]$.

Examples 5.10 In Example 5.5, we have X = {a, b, c}and the fuzzy set A is defined as

$A = \{(a, 0.2), (b, 0.4), (c, 0.6)\}$ and $\mathcal{T} = \{\tilde{0}, A, \tilde{1}\}$. Then, find the q-neighborhood system of all the fuzzy points.

Solution: For fuzzy point a_p with support a if $p > 1 - 0.2 = 0.8$ then $a_{pq} \in A$ so (Fig. 5.1).

$Q(a_p) =$ the collection of all fuzzy sets containing A, if $p > 0.8$

$$= \{\tilde{1}\}\ if\ \ p \leq 0.8$$

Similarly, we can find $Q(b_p)$ and $Q(c_p)$ for $p > 0.6,\ p \leq 0.6$ and $p > 0.4,\ p \leq 0.4$, respectively, to get the same system as in the case of $Q(a_p)$.

Examples 5.11 In Example 5.4, if we consider the sub-collection $\mathfrak{B} = \left\{a_r : r = \frac{1}{n}, n \in N\right\}$ then $\mathfrak{B}$ is a countable base for $\mathcal{T}$.

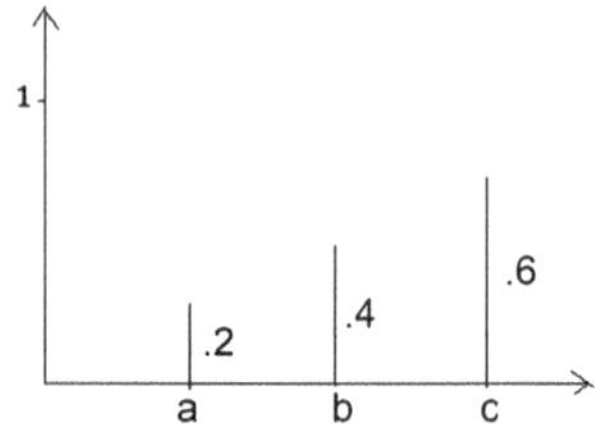

Fig. 5.1 Fuzzy set A

Examples 5.12 Let X be any non-empty set and we define a set of fuzzy subsets $\mathfrak{B}_n, \forall n \in N$ given by $\mathfrak{B}_n(x) = \frac{1}{n}, \forall x \in X$. Then, the collection $\mathcal{T} = \{\mathfrak{B}_n : n \in N\} \cup \{\tilde{0}\}$ forms a fuzzy co-topology on X. If we want to find the closure of any fuzzy point say x_p, then we have to search the smallest closed fuzzy subset containing x_p. Now we see that $x_p \in \mathfrak{B}_n \Rightarrow p \leq \frac{1}{n} \Rightarrow n \leq \frac{1}{p}$.

Let $n_0 = \left[\frac{1}{p}\right]$, where $\left[\frac{1}{p}\right]$ denotes the greatest integer not exceeding $\frac{1}{p}$ then $\overline{x_p} = \mathfrak{B}_{n_0}$.

Exercise-5

1. Define fuzzy topology as proposed by C. L. Chang and give a suitable example.
2. Show that intersection of an arbitrary family of fuzzy sets is again a fuzzy set.
3. Define the interior of a fuzzy set. Show that a fuzzy set A is open iff A = A^0.
4. Define the closure of a fuzzy set and show that A is closed iff $A = \overline{A}$.
5. If A is a fuzzy set in a fuzzy topological space$(X, \mathcal{T})$, then show that the following results hold in it:

   (i) $\overline{\overline{A}} = \overline{A}$ *and* $(ii)\left(A^0\right)^0 = A^0$

6. If A and B are fuzzy sets in a fuzzy topological space$(X, \mathcal{T})$, then establish the following results in it:

   (i) $(a)A^0 \wedge B^0 = (A \wedge B)^0 and (b)A^0 \vee B^0 = (A \vee B)^0$
   (ii) $(a)\overline{A} \wedge \overline{B} \geq \overline{A \wedge B}$ *and* $(b)\overline{A} \vee \overline{B} = \overline{A \vee B}$
   (iii) $(1 - A)^0 = 1 - \overline{A}$
   (iv) $\overline{1 - A} = 1 - A^0$

7. What is a fuzzy point, in the words of Pu and Liu? When such a fuzzy point x_λ is said to be quasi-coincident with a fuzzy set A?
8. What is a fuzzy limit point? If A is a fuzzy set in a fuzzy topological space$(X, \mathcal{T})$, then A is closed iff $A' \leq A$. Furthermore, show that $A'' \leq \overline{A} = A \vee A'$.
9. If A is a fuzzy set in a fuzzy topological space $(X, \mathcal{T})$ and $A \subset X$, then the family $\mathcal{T}_A = \{G/A : G \in \mathcal{T}\}$ is a fuzzy topology on X. What name has been assigned to this topology?
10. Define the basis and subbasis for a fuzzy topology $\mathcal{T}$. Give a suitable example.
11. If N_x *and* P_x are the neighborhood of a point $x \in X$ in a fuzzy topological space $(X, \mathcal{T})$, then show that $N_x \wedge P_x$ is also a neighborhood of x.
12. Explain the concept of fuzzy set induced by a mapping.
13. Explain the meaning of fuzzy continuous functions. Define fuzzy homeomorphism.
14. Define the cover and subcover of a fuzzy set.
15. Explain the notion of fuzzy topological space as proposed by R. Lowen.
16. Let X = [−1, 1]; we define a set of fuzzy subsets of X by

$$A_n(x) = 1, \text{ if } |x| \leq \frac{1}{n}$$
$$= 0, \text{ otherwise}$$

Then, (i) Show that $\{A_n : n \in N\}$ forms a base for a fuzzy topology.

(ii) Find the closed sets.
(iii) Find the interior of a fuzzy set U.
(iv) Find the closure of the fuzzy points $0_{0.5}$ and $0.5_{0.5}$, where $0_{0.5}$ is the fuzzy point which assigns o.5 grade to its support 0, $0.5_{0.5}$ may similarly be defined.
(v) Find $Q(1_{0.2})$.

17. Let A be a fuzzy subset of $\mathbb{R}$ defined by

$$A(x) = 0.7 \; if \; x \in Q$$
$$= 0.5 \quad otherwise$$

where Q is the set of all rational numbers. Let$\mathcal{T} = \{\tilde{0}, A, \tilde{1}\}$. Find the q-neighborhood systems of the fuzzy points $\left(\sqrt{2}\right)_{0.6}$ *and* $(2)_{0.6}$ which assign grade 0.6 at $\sqrt{2}$, 2.
18. Let X = [0, 1] and $\mathcal{T} = \{\tilde{0}, x, Sinx, \tilde{1}\}$ be a fuzzy topology on X. Find the interior and closure of the fuzzy set Cosx. Find $Q(x_{0.5}), \forall\, x \in X$.
19. Include minimum number of fuzzy sets to the collection $\{\tilde{0}, |Cosx|, |Sinx|, \tilde{1}\}$ so that it reduces to a fuzzy topology on X = [−1, 1]. Find the closure of the fuzzy subset {x} of X w.r.t. the 0.
20. Let X = [0, 1] = I and $\mathcal{T} = \left\{\tilde{0}, \tilde{1}, x, \frac{x^2}{2}, \frac{x^3}{3}, \ldots..\right\}$ be a fuzzy topology on X. Find all the open q-neighborhoods of $x_p, \forall\, x_p \in Pt(I^X)$.

Hint: $x_p\, q \frac{x_n}{n} \Rightarrow p > 1 - \frac{x_n}{n} \Rightarrow \frac{x_n}{n} > 1 - p$
So, $Q(x_p) = \left\{\frac{x_n}{n} : \frac{x_n}{n} > 1 - p\right\}$

21. Let X = [0, 1], and $A_n \in I^X$ be defined by

$$A_n(x) = 1, \; if \; x < \frac{1}{n}$$
$$= 0 \; otherwise \; \forall\, n \in N$$

Then show that $\{A_n : n \in N\}$ forms a countable base for a fuzzy topology$\mathcal{T}$. Find. A° *and* $\overline{A}$ *in*$(X, \mathcal{T})$, where A is the constant fuzzy set defined by A(x) = 0.6,$\forall\, x \in X$.

Hint: As $A_1 \supseteq A_2 \supseteq A_3 \supseteq \ldots\ldots\ldots\ldots..$ so the collection $B = \{A_n : n \in N\}$ is closed under finite intersection and $A_1 = \mathcal{T}_X$, so B forms a countable base for a fuzzy topology. Clearly, A_2 is (Fig. 5.2).
the largest open set contained in A so $A^{\circ} = A_2$

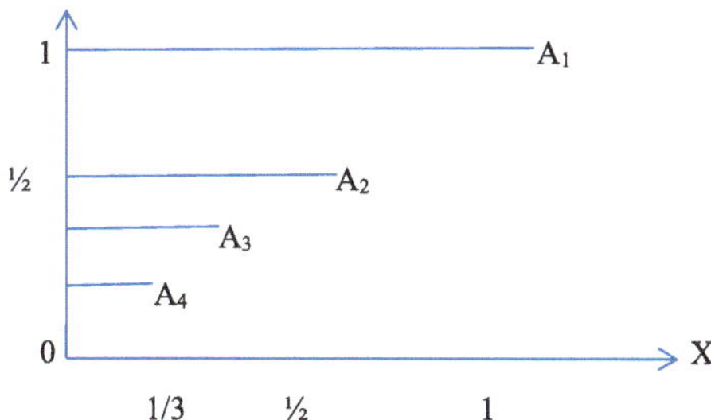

Fig. 5.2 Representing $A_n \in I^X$

Objective-Type Questions and Answers

1. The notion of fuzzy topology was first introduced by
   Pu Pao Ming (b) R. Lowen (c) Lao Mao Kang (d) C. L. Chang
2. According to C. L. Chang, a fuzzy topology is
   (a) A collection of fuzzy sets (b) A fuzzy collection of crisp sets
   (b) A fuzzy family of fuzzy sets (d) Both (a) and (b)
3. The main drawback of the fuzzy topology, introduced by C. L. Chang was
   (a) The properties of neighborhood systems deviate remarkably from crisp topology
   (b) Crisp topology cannot be obtained from it as a particular case
   (c) Fuzziness in the concept of openness is absent
   (d) All the above
4. Intuitionistic fuzzy topology was introduced by
   (a) U Hohle (b) Atanassov (c) C. L. Chang (d) Ramadan
5. If A° be the interior of a fuzzy set A in a fuzzy topological space$(X, \mathcal{T})$, then which one of the following statement is correct?
   (a) $(A^\circ)^\circ \subset A^\circ$ (b) $(A^\circ)^\circ \supset A^\circ$ (c) $(A^\circ)^\circ = A^\circ$ (d) $(A^\circ)^\circ \neq A^\circ$
6. Let X be a non-empty set, and for each $n \in N$, $A_n \in I^X$, be defined by $A_n = 1-\frac{1}{n}$ for some $x \in X$. Then $\mathcal{U} = \{A_n : n \in N\}$
   (a) $\mathcal{U}$ is a fuzzy topology on X (b) $\mathcal{U}$ is a base for a fuzzy topology but not a fuzzy topology on X
   (c) $\mathcal{U}$ is a sub-base but not a base for a fuzzy topology on X (d) None of these
7. Let X be a non-empty set and $x \in X$. Let $\mathcal{U} = \{x_\lambda : \lambda < 0.5\} \cup \{\tilde{0}, \tilde{1}\}$ where x_λ is a fuzzy point with support x and grade λ. Then
   (a) $\mathcal{U}$ is a fuzzy topology on X (b) $\mathcal{U}$ is a base for a fuzzy topology but not a fuzzy topology on X
   (c) $\mathcal{U}$ is a sub-base but not a base for a fuzzy topology on X (d) None of these
8. If A and B are fuzzy sets in a fuzzy topological space$(X, \mathcal{T})$, which of the following statements is correct?
   (a) $\overline{A} \vee \overline{B} \geq \overline{A \vee B}$ (b) $\overline{A} \vee \overline{B} \leq \overline{A \vee B}$ (c) $\overline{A} \vee \overline{B} \neq \overline{A \vee B}$ (d) $\overline{A} \vee \overline{B} = \overline{A \vee B}$
9. If A and B are fuzzy sets in a fuzzy topological space$(X, \mathcal{T})$, which of the following statements is correct?
   (b) $\overline{A} \wedge \overline{B} \geq \overline{A \wedge B}$ (b) $\overline{A} \wedge \overline{B} \leq \overline{A \wedge B}$ (c) $\overline{A} \wedge \overline{B} \neq \overline{A \wedge B}$ (d) $\overline{A} \wedge \overline{B} = \overline{A \wedge B}$

10. If A° be the interior of a fuzzy set A in a fuzzy topological space$(X, \mathcal{T})$, which of the following statements is correct?
   (a)$(A \wedge B)^\circ \neq A^\circ \wedge B^\circ$ (b) $(A \wedge B)^\circ \leq A^\circ \wedge B^\circ$
   (c) $(A \wedge B)^\circ = A^\circ \wedge B^\circ$ (d) $(A \wedge B)^\circ \geq A^\circ \wedge B^\circ$
11. If A° be the interior of a fuzzy set A in a fuzzy topological space$(X, \mathcal{T})$, which of the following statement is correct?
   (a)$(A \vee B)^\circ = A^\circ \vee B^\circ$ (b) $(A \vee B)^\circ \leq A^\circ \vee B^\circ$
   (c) $(A \vee B)^\circ \neq A^\circ \vee B^\circ$ (d) $(A \vee B)^\circ \geq A^\circ \vee B^\circ$
12. If A and B are fuzzy sets in a fuzzy topological space $(X, \mathcal{T})$ such that $A \leq B$, then which of the following statements is correct?
   $(a)\overline{A} \leq \overline{B}$ (b) $\overline{A} \geq \overline{B}$ (c) $\overline{A} \neq \overline{B}$ (d) $\overline{A} = \overline{B}$
13. If A and B are fuzzy sets in a fuzzy topological space $(X, \mathcal{T})$ such that$A \leq B$, then which of the following statements is correct?
   (a)$A^\circ \geq B^\circ$ (b) $A^\circ \neq B^\circ$ (b) $A^\circ = B^\circ$ (d) $A^\circ \leq B^\circ$
14. If $\overline{A}$ be the closure of a fuzzy set A in a fuzzy topological space $(X, \mathcal{T})$, which of the following statements is correct?
   (a) $\overline{\overline{A}} \supset \overline{A}$ (b) $\overline{\overline{A}} \subset \overline{A}$ (c) $\overline{\overline{A}} = \overline{A}$ (d) $\overline{\overline{A}} \neq \overline{A}$
15. A discrete fuzzy topology contains
   (a) ϕ *and X only* (b) Contains all fuzzy sets on X.
   (c) Contains some fuzzy sets on X along with ϕ *and X* (d) None of these.
16. If $(X, \mathcal{T})$ is a fuzzy topological space and $A_i \in \mathcal{T}, \forall i \in I$, then which statement is correct?
   (a) $\cap A_i \in \mathcal{T}$ (b) $\cup A_i \in \mathcal{T}$ (c) ϕ *and X* $\notin \mathcal{T}$ (d) None of these
17. Which statement is correct if $(X, \mathcal{T})$ be a fuzzy topological space and $A, B, A_i \in \mathcal{T}, \forall i \in I$ are closed fuzzy sets?
   (a) $A \vee B$ is a closed fuzzy set (b) $A \wedge B$ is a closed fuzzy set
   (c) $\wedge A_i$ is a closed fuzzy set (d) None of these.
18. Indiscrete fuzzy topology on X contains
   (a) All fuzzy sets on X (b) Contains some fuzzy sets on X along with ϕ *and X*
   (c) ϕ *and X only* (d) None of these.
19. If A is a fuzzy set in a fuzzy topological space $(X, \mathcal{T})$ and A° is the interior of A, then which of the following statements is correct?
   (a) A° is open (b) A° is open and the smallest open fuzzy set less than or equal to A.
   (c) A° is open and is the largest open fuzzy set less than or equal to A (d) None of these.
20. Which of the following statement is correct if A is a fuzzy set in a fuzzy topological space $(X, \mathcal{T})$ and $\overline{A}$ is the closure of A?
   (a) $\overline{A}$ is open (b) $\overline{A}$ is closed.
   (c) $\overline{A}$ is closed and is the least closed fuzzy set greater than or equal to A (d) None of these.
21. If A and B are fuzzy set in a fuzzy topological space $(X, \mathcal{T})$, $\overline{A}$ is the closure of A and A° the Interior of A, then which one of the following statements is correct?

(a) $(1-A)^\circ = 1-\overline{A}$ (b) $\overline{(1-A)} = 1-A^\circ$ (c) $1-\overline{A} = 1-A^\circ$ (d) None of these

22. If A and B are fuzzy set in a fuzzy topological space $(X, \mathcal{T})$, A° is the interior of A, and $\overline{A}$ is the closure of A, then which one of the following statements is correct?
    (a) $1-A^\circ = 1-\overline{A}$ (b) $(1-A)^\circ = \overline{A}-1$ (c) $\overline{(1-A)} = 1-A^\circ$ (d) None of these
23. The convergence of fuzzy nets and filters was studied first by introducing a q-neighborhood system by
    (a) Pu Pao Ming (b) A. S. Ramadan (c) Lao Mao Kang (d) C. L. Chang.
24. Which neighborhood system works better for the study of Moor- Smith convergence in fuzzy topology?
    (a) Neighborhood system built with the help of $\hat{a}$ belongs to $\hat{a}$ relation
    (b) q-neighborhood system
    (c) Remote neighborhood system
    (d) None of these
25. Let $x \in X, \mathcal{T} = \left\{X, \tilde{0}, x_{\frac{1}{2}}\right\}; e = x_{\frac{3}{4}}$; then the open q- neighborhood of e in $(X, \mathcal{T})$ are
    (a) X *and* $x_{\frac{1}{2}}$ (b) $\left\{x_\lambda; \lambda \geq \frac{1}{2}\right\}$ (c) $X \cup \left\{x_\lambda; \lambda \geq \frac{1}{2}\right\}$ (d) $X \cup \left\{x_\lambda; \lambda \geq \frac{3}{4}\right\}$
26. If $\mathcal{F}$ is the closure operator on a fuzzy topological space$(X, \mathcal{T})$, then $\left\{A \in I^X : \mathcal{F}(A) = A\right\}$
    (a) Forms a local base of $\mathcal{T}$ (b) Forms a fuzzy topology coarser than $\mathcal{T}$
    (c) Forms a base for $\mathcal{T}$ (d) Forms a co-topology on X.
27. If I is the interior operator in a fuzzy topological space$(X, \mathcal{T})$, then $\left\{A \in I^X : I(A) = A\right\}$
    (a) Generates $\mathcal{T}$ (b) Forms a fuzzy topology coarser than $\mathcal{T}$
    (c) Forms a base for $\mathcal{T}$ (d) Forms a co-topology on X.
28. Let A be a fuzzy subset of R (the set of real numbers) be defined by

$$\begin{aligned} A(x) &= 0.7, \forall x \in Q \\ &= 0.5, \textit{otherwise} \end{aligned}$$

    Let $\mathcal{T} = \left\{\tilde{0}, A, X\right\}$, then the q-neighborhood system of the fuzzy point $\left(\sqrt{2}\right)_{.6}$ with support $.\sqrt{2}$ and grade 0.6 is.

    (a) $\{A, X\}$ (b) The collection of all fuzzy sets containing A
    (b) $\left\{\tilde{0}, X\right\}$ (d) The collection of all fuzzy points of X with irrational supports.

29. Let A be a fuzzy subset of R (the set of real numbers) be defined by

$$\begin{aligned} A(x) &= 0.5, \forall x \in Q \\ &= 0.7, \textit{otherwise} \end{aligned}$$

Let $\mathcal{T} = \{\tilde{0}, A, X\}$, then the neighborhood system (represented by the belongingness relation) of the fuzzy point $2_{0.6}$ with support *two* and a grade of 0.6 is

(a) $\{A, X\}$ (b) The collection of all fuzzy sets containing A.
(c) $\{X\}$ (d) The collection of all fuzzy points of X with rational supports.

30. Let X = [0, 1], and $\mathcal{T} = \{\tilde{0}, x, Cosx, X\}$, then the interior of the fuzzy set *Sinx* is

    (a) $\tilde{0}$ (b) x (c) *Cosx* (d) X.

31. Let X = [0, 1], and $\mathcal{T} = \left\{\tilde{0}, \frac{x^2}{2}, Sinx, X\right\}$, then the closure of the fuzzy set *Cosx* is

    (a) $\tilde{0}$ (b) $1 - \frac{x^2}{2}$ (c) $1 - Sinx$ (d) X.

32. Let X = [0, 1], for each $n \in N$, the fuzzy subset A_n be defined as

$$A_n(x) = \frac{1}{n}, \, if \; x < \frac{1}{n}$$
$$= 0, \; otherwise$$

Let $\mathcal{T}$ be the fuzzy topology on X generated by the collection$\{A_n, n \in N\}$. Let A be the constant fuzzy subset of X defined by $A(x) = 0.6, \forall\, x \in X$, then A° is

(a) A_1 (b) A_2 (c) A_3 (d) A_4.

33. Let X be a non-empty set, $I = [0, 1]$, *and* $\mathcal{T} = \{A \in I^X; Supp(A^c) = finite\} \cup \{\tilde{0}\}$ where $Supp(A^c)$, is the support of A^c, then $\mathcal{T}$ is

    (a) $\mathcal{T}_0$ *but not* $\mathcal{T}_1$ (b) $\mathcal{T}_1$ *but not* $\mathcal{T}_2$ (c) $\mathcal{T}_2$ (d) $\mathcal{T}_3$.

34. A sequence $\{A_n\}$ of fuzzy sets in a fuzzy topological space (X, T) is said to converge to a fuzzy set *A* if it is

    (i) Frequently contained in each neighborhood of *A*
    (ii) Eventually contained in each neighborhood of *A*
    (iii) Cluster fuzzy set of A (iv) None of these

35. Let X = [0, 1], for each $n \in N$, the fuzzy subset A_n be defined as

$$A_n(x) = \frac{1}{n}, \; if \; x < \frac{1}{n}$$
$$= 0, \; otherwise$$

Let $\mathcal{T}$ be the fuzzy topology on X generated by the collection$\{A_n, n \in N\}$. Let A be the constant fuzzy subset of X defined by $A(x) = 0.6, \forall\, x \in X$, then $\overline{A}$ is.

(a) A_1^c (b) A_2^c (c) A_3^c (d) A_4^c

36. If A is a fuzzy set and A' is the derived fuzzy set in a fuzzy topological space $(X, \mathcal{T})$, then A is Closed if and only if
    (a) $A' \geq A$ (b) $A' = A$ (c) $A' \neq A$ (d) $A' \leq A$
37. If A is a fuzzy set in a fuzzy topological space $(X, \mathcal{T})$, A' is the derived fuzzy set, and $\overline{A}$ is the closure of A, then which one of the following statements is correct?
    (a) $\overline{A} \leq A''$ (b) $\overline{A} \geq A''$ (c) $\overline{A} = A''$ (d) $\overline{A} > A''$
38. If A is a fuzzy set in a fuzzy topological space $(X, \mathcal{T})$, $\overline{A}$ is the closure, and A' is the derived fuzzy set of A, then which one of the following statements is correct?

    (a) $\overline{A} = A \vee A'$ (b) $\overline{A} \subset A \vee A'$ (c) $\overline{A} \supset A \vee A'$ (d) $\overline{A} \subseteq A \vee A'$

39. Let $(X, \mathcal{T})$be a fuzzy topological space and N_x, P_x are neighborhoods of $x \in X$, then which one of the following statements is correct?

    (a) $N_x \vee P_x$ is a neighborhood's of x (b) $N_x \vee P_x = \phi$
    (c) $N_x \wedge P_x$ is a neighborhood's of x (d) None of these

40. Do you know which one of the following statements is correct?

    (a) The identity function is not fuzzy continuous.
    (b) Composition of fuzzy continuous functions is fuzzy continuous.
    (c) If $f : (X, \mathcal{T}) \to (Y, U)$ is fuzzy continuous, then the inverse of each U-open fuzzy set is not a $\mathcal{T}$-open fuzzy set.
    (d) None of these.

Answers

1. (d) 2. (a) 3. (c) 4. (b) 5. (c) 6. (a) 7. (a) 8. (d) 9. (a) 10. (c) 11. (d) 12. (a) 13. (d) 14. (c) 15. (b) 16. (b) 17. (a) and (c) 18. (c) 19. (a) and (c) 20. (b) and (c) 21. (a) 22. (c) 23. (a) 24. (b) 25. (a) 26. (d) 27. (a) 28. (b) 29. (c) 30. (a) 31. (b) 32. (b) 33. (b) 34. (b) 35. (c) 36. (d) 37. (b) 38. (a) 39. (c) 40. (b)

True/False Type Statements and Answers

1. The notion of fuzzy topology was initiated by C. L. Chang in 1968.
2. Fuzzy topology extends the concept of ordinary or classical topological space.
3. A fuzzy set in X is called a fuzzy point iff it takes the value 0 for all $y \in X$ except one $x \in X$.
4. A fuzzy set is called topologically closed or $\mathcal{T}$-closed if and only if its complement is topologically open or $\mathcal{T}$-open or open fuzzy set.
5. If $\mathcal{T}$ is a fuzzy topology on X, then ϕ *and* $X \notin \mathcal{T}$.
6. The pair $(X, \mathcal{T})$ is called fuzzy topological space.
7. The fuzzy topology $\mathcal{T}$ is termed "discrete" if it does not contain all of the fuzzy sets on X.
8. The fuzzy topology $\mathcal{T}$ is termed "indiscrete" if it contains only ϕ and X.
9. The members of $\mathcal{T}$ are called $\mathcal{T}$ -closed fuzzy sets (or closed fuzzy sets).

10. The pair set $\{X, P(X)\}$ is a fuzzy topological space called the discrete fuzzy topology of X.
11. We expect more robust results in the case of fuzzy topology.
12. On X the indiscrete fuzzy topology is the finest.
13. On X the discrete fuzzy topology is the coarser.
14. If the inclusion relation U $\subset \mathcal{T}$ holds, we say that U is coarser, weaker, or smaller than $\mathcal{T}$.
15. If $\{\mathcal{T}_i$: i $\in$ I}, I being an arbitrary set, be a collection of fuzzy topologies on X, then $\underset{i \in I}{\cap} \mathcal{T}_i$ is also a fuzzy topology on X.
16. If A and $\{A_i : i \in I\}$ are fuzzy sets in X, then $\{A_i : i \in I\}$ be called a cover of A if and only if$\cup\{A_i : i \in I\} \subset A$.
17. Two fuzzy sets A and B in X are said to be intersecting if and only if there exists a point x $\in$ X such that $(A \cap B)(x) > 0$.
18. The closure of the fuzzy set A in a fuzzy topological space $(X, \mathcal{T})$ is denoted by $\overline{A}$.
19. The interior of a fuzzy subset A in a fuzzy topological space $(X, \mathcal{T})$ is denoted byA°.
20. We always have $A^\circ = IntA = sup\{B : B \subseteq A, B \in \mathcal{T}\}$.
21. The interior of the fuzzy set A, i.e., A° is closed, and is the smallest closed fuzzy set contained in A.
22. A fuzzy point x_λ is said to be contained in a fuzzy set A if and only if f(λ) < A(x), and is denoted by $x_\lambda \in A$.
23. If A and B are fuzzy sets in a fuzzy topological space $(X, \mathcal{T})$, then $A^\circ \vee B^\circ = (A \wedge B)^\circ$.
24. For any fuzzy sets A and B in a fuzzy topological space $(X, \mathcal{T})$, we have $\overline{A} \vee \overline{B} = \overline{A \vee B}$.
25. If A is a fuzzy set in a fuzzy topological space $(X, \mathcal{T})$, then $\overline{1 - A} = 1 - A^\circ$.
26. A fuzzy topological space $(X, \mathcal{T})$ is said to be compact iff each open cover has an infinite subcover.
27. The derived fuzzy set of a fuzzy set A in a fuzzy topological space $(X, \mathcal{T})$ is denoted by A'.
28. A fuzzy set A in a fuzzy topological space $(X, \mathcal{T})$ considered closed iff $A' \geq A$.
29. Any one–one mapping $g : X \longrightarrow Y$ from a fuzzy topological space $(X, \mathcal{T})$ onto a fuzzy topological space $(Y, \mathcal{T}_1)$ considered fuzzy homeomorphic if $g\left(\overline{A}\right) \neq \overline{g(A)}$.
30. R. Lowen observed that with the definition of fuzzy topological spaces and fuzzy continuous function, constant functions between fuzzy topological spaces are not necessarily fuzzy continuous. This compels him to give another notion of fuzzy topology.
31. Any finite topological space, including the null set, is compact.
32. Any space carrying the cofinite topology is not compact.
33. The identity mapping $I : (X, \mathcal{T}) \longrightarrow (Y, \mathcal{F})$ is fuzzy continuous.

34. The composition of two fuzzy continuous functions, f *and* g, *i.e.*,$f \circ g$ is fuzzy continuous.
35. The notion of intuitionistic fuzzy topological space was proposed by Coker and Demirci in 1996 in Sostak's sense.
36. For a fuzzy set A in a fuzzy topological space $(X, \mathcal{T})$,$1 - IntA = Cl(1 - A)$, and $1 - ClA = Int(1 - A)$ holds good.
37. Let $\mathcal{T}$ be a fuzzy topology. A subfamily B of $\mathcal{T}$ is a base for $\mathcal{T}$ iff each member of $\mathcal{T}$ can be expressed as the union of some members of B.
38. A fuzzy set A in $(X, \mathcal{T})$ is called a neighborhood of a fuzzy point p and x_p iff there exists a $B \in \mathcal{T}$ such that $x_p \in B \geq A$.
39. In point set theory, we know that the closure of the product of the sets is the product of closures, i.e., $Cl(A \times B) = ClA \times ClB$. Is this result true in the fuzzy setting if A *and* B are two fuzzy sets?
40. If A is a fuzzy set of a fuzzy topological space $(X, \mathcal{T})$ and B is a fuzzy set of a fuzzy topological space $(Y, \mathcal{F})$, then $ClA \times ClB \leq Cl(A \times B)$.
41. If A is a fuzzy set of a fuzzy topological space $(X, \mathcal{T})$ and B is a fuzzy set of a fuzzy topological space $(Y, \mathcal{F})$, then $IntA \times IntB \leq Int(A \times B)$.
42. R. Lowen introduced a new definition of fuzzy topology, replacing the first condition of Chang for all a (Constant function), $a \in \mathcal{T}$.
43. One of the main advantages of Lowen's definition over Chang's definition is that all constant functions are fuzzy continuous.
44. Let $g : (X, \mathcal{T}) \longrightarrow (Y, \mathcal{F})$, where $(X, \mathcal{T})$ and $(Y, \mathcal{F})$ are fuzzy topological spaces, then the inverse image of every closed fuzzy set is open.
45. If $(X, \mathcal{T})$ and $(Y, \mathcal{F})$ are two fuzzy topological spaces and $g : (X, \mathcal{T}) \longrightarrow (Y, \mathcal{F})$, the function g is F-continuous.

Answers

1.(T); 2. (T); 3. (T); 4. (T); 5. (F); 6. (T); 7. (F); 8. (T); 9. (F); 10. (T); 11. (F); 12. (F); 13. (F); 14. (T); 15. (T); 16. (F); 17. (T); 18. (T); 19. (T); 20. (T); 21. (F); 22. (T); 23. (F); 24. (T); 25. (T); 26. (F); 27. (T); 28. (F); 29. (F); 30. (T); 31. (T); 32. (F); 33. (T); 34. (T); 35. (T). 36. (T); 37. (T); 38. (F); 39. (F); 40. (F); 41. (T); 42. (T); 43. (T); 44. (F); 45. (T).

Chapter 6
Fuzzy Subgroups and Fuzzy Normal Subgroups

In this chapter, we will introduce some fuzzy analogs of the conception of a subgroupoid of a groupoid, a subgroup of a group, a normal subgroup of a group, a level subgroup of a group, and cosets of a subgroup of a group. Further, it has been shown how some basic notions of group theory should be extended in an elementary way to develop the theory of fuzzy subgroupoids, fuzzy subgroups, and homomorphism between fuzzy groups, sup property, fuzzy groups redefined, fuzzy level subgroups, fuzzy normal subgroups, and fuzzy cosets. Also, we have here the fuzzy version of the famous "Lagrange's theorem" and the fundamental homomorphism theorem. We should be familiar with the fuzzy subgroup and the fuzzy normal subgroup are generalizations of their classical counterparts.

6.1 Introduction

Azriel Rosenfeld of the University of Maryland was the first mathematician who did fantabulous work by applying fuzzy sets in the realm of group theory and propounded the novelties of fuzzy subgroups of a group G w.r.t. t-norm min. Since then, mathematicians have tried to fuzzify various notions and results of abstract algebra to the broader framework of the fuzzy setting. In 1971, when fuzzy set theory was in its infancy, Rosenfeld hypothesized the "Fuzzy Groups" notion that appeared in the (V.35 of the Journal of Mathematical Analysis and Applications). In this chapter, the fundamental abstraction revolves around fuzzy subgroups and fuzzy normal subgroups, which are fuzzifications of their respective counterparts in abstract algebra. Fuzzy set theory (FST) provides incredible richness in application than the ordinary set theory. FST is regarded as more general than the classical set theory; in the same way, Fuzzy Groups are more general than the classical groups. We will focus on this aspect in the chapter. Apart from this, several results of classical group

M. K. Singh, *Applied Fuzzy Mathematics*, Forum for Interdisciplinary Mathematics,
https://doi.org/10.1007/978-981-97-3257-9_6

theory can be derived in FST, and some new results can also be established in the usual way.

6.2 Fuzzy Subgroupoids and Fuzzy Ideals

Let G be a groupoid, i.e., a set closed under a binary operation denoted multiplicatively. Further, we assume that $\lambda : G \to [0, 1]$ is a fuzzy set in G.

6.2.1 Fuzzy Subgroupoid

A mapping $\lambda : G \to [0, 1]$ be called a fuzzy subgroupoid of G if

$$\lambda(ab) \geq min\{\lambda(a), \lambda(b)\}, \forall\, a, b \in G \tag{6.1}$$

6.2.2 Fuzzy Ideal

A fuzzy set λ in G be called a **fuzzy left ideal** if,

$$\lambda(ab) \geq \lambda(b), \forall\, a , b \in G; \tag{6.2}$$

and a fuzzy set λ in G be called a **fuzzy right ideal** if,

$$\lambda(ab) \geq \lambda(a), \forall\, a, b \in G \tag{6.3}$$

and on combining (6.1) and (6.2), we can say that λ is a **fuzzy ideal** if,

$$\lambda(ab) \geq max\{\lambda(a), \lambda(b)\}, \forall\, a , b \in G$$

That is, if λ is both a fuzzy left ideal and a fuzzy right ideal.

Theorem 6.1 If λ is a fuzzy subgroupoid of G, then $\lambda(a^n) \geq \lambda(a), \forall\, a \in G$.

Proof Let λ be a fuzzy subgroupoid in G, and a^n is any composite of a's. Then, we have

$$\begin{aligned}\lambda\left(a^n\right) &= \lambda(a \cdot a \cdot a \cdot \text{.............up to n times}), \ \forall\, a \in G\\ &\geq min\{\lambda(a), \lambda(a), \lambda(a), \text{...............up to n times}\}, \forall\, a \in G, \text{ using (6.1)}\\ &\geq \lambda(a), \forall\, a \in G\end{aligned}$$

$\therefore \lambda(a^n) \geq \lambda(a), \forall a \in G$.

Theorem 6.2 Show that, for any $\alpha \in [0, 1]$, $\{t : t \in G, \lambda(t) \geq \alpha\}$ is a subgroupoid or (left, right) ideal if λ is a fuzzy subgroupoid or (left, right) ideal.

Proof Let λ be a fuzzy subgroupoid of the groupoid G and $\alpha \in [0, 1]$. Let $A = \{t : t \in G, \lambda(t) \geq \alpha\}$; then we claim that A is a subgroupoid of λ. Obviously, A is a subset of G. Let $t_1, t_2 \in A$. Then,

$$\lambda(t_1) \geq \alpha \text{ and } \lambda(t_2) \geq \alpha$$

We have $\lambda(t_1.t_2) \geq min\{\lambda(t_1), \lambda(t_2)\} \geq min\{\alpha, \alpha\} = \alpha$
$\therefore \lambda(t_1.t_2) \geq \alpha \Rightarrow t_1.t_2 \in A \Rightarrow A$ is a subgroupoid of λ.

Next, we assume that λ is a fuzzy left ideal of G. Let $t \in A$ and $g \in G$ and λ is a fuzzy left ideal, therefore

$$\lambda(gt) \geq \lambda(t) \geq \alpha \Rightarrow \lambda(g.t) \geq \alpha \Rightarrow g.t \in A$$

This implies that $t \in A$ and $g \in G \Rightarrow g.t \in A$. It shows that A is a left ideal of the fuzzy subgroupoid λ. Similarly, we can show that A is a right ideal of λ.

Theorem 6.3 If λ is the characteristic function of a subset $A \subseteq G$, then λ is a fuzzy subgroupoid or (left, right) ideal of the groupoid G if and only if A is a subgroupoid or (left, right) ideal, respectively.

Proof Let $\lambda : G \to \{0, 1\}$ is such that

$$\lambda_T(x) = 1 \text{ if } x \in T$$
$$= 0 \text{ if } x \notin T$$

So that λ_T is the characteristic function of a subset T of G.

Let T be the subgroupoid of G and let $t_1, t_2 \in T$. Since T is a subgroupoid of G, therefore $t_1.t_2 \in T$. Hence, $\lambda_T(t_1) = 1, \lambda_T(t_2) = 1$ and $\lambda_T(t_1.t_2) = 1$.

We have,

$$\lambda_T(t_1.t_2) = 1 \geq min\{1\,, 1\} = min\{\lambda_T(t_1), \lambda_T(t_2)\}$$

i.e., $\lambda_T(t_1.t_2) \geq min\,\{\lambda_T(t_1), \lambda_T(t_2)\} \Rightarrow \lambda_T$ is a fuzzy subgroupoid.

Conversely, assume that λ_T is a fuzzy subgroupoid of the groupoid G. Now, we claim that λ_T is a subgroupoid of G. Now, for $t_1, t_2 \in T$

$$\lambda_T(t_1 \cdot t_2) \geq min\,\{\lambda_T(t_1), \lambda_T(t_2)\} \geq min\{1, 1\} = 1$$

i.e., $\lambda_T(t_1.t_2) \geq 1$. So $\lambda_T(t_1 \cdot t_2) = 1$ as $\lambda_T(t_1 \cdot t_2) > 1$ *is not true*. and consequently $t_1.t_2 \in \lambda_T \Rightarrow \lambda_T$ is a subgroupoid of G.

Let T is a left ideal of G. Hence, $t \in T$ and $g \in G \Rightarrow gt \in T$. This implies that $\lambda_T(gt) = 1 \geq \lambda_T(t)$.

It follows that λ_T is a fuzzy left ideal.

Conversely, we assume that λ_T is a fuzzy left ideal. Therefore,

$$\lambda_T(gt) \geq \lambda_T(t) = 1$$

Thus, we have $\lambda_T(gt) = 1 \Rightarrow gt \in \lambda_T \Rightarrow \lambda_T$ is a fuzzy left ideal. Similarly, we can establish the result for a fuzzy right ideal.

Note 1: The characteristic function of T is denoted by φ_T.

6.2.3 The Lattices of Fuzzy Subgroupoids and Ideals

Let (G, ·) be a groupoid and $\tilde{P}(G)$ be the set of all fuzzy sets on G. Let λ and μ are two fuzzy subsets of G. If we define the relation $\leq$ on the set $\tilde{P}(G)$ of all fuzzy sets in G as $\lambda \leq \mu$ means $\lambda \subseteq \mu, \forall\, a \in G$ then, it can be inferred that $\left(\tilde{P}(G), \leq\right)$ is a partially ordered set and $\leq$ is an ordered relation on G, consequently G is an ordered set under "$\leq$." If we define supremum and infimum of a family of fuzzy sets on G as the union ($\cup$) and intersection ($\cap$) respectively of the family of fuzzy sets, then we observe that $\left(\tilde{P}(G), \leq\right)$ forms a complete distributive lattice L. The least and the greatest elements of L are the constant functions 0 and 1. These functions are nothing but just the φ_ϕ and φ_G, so they are fuzzy ideals, particularly fuzzy subgroupoids.

Theorem 6.4 The intersection of any set of fuzzy subgroupoids is a fuzzy subgroupoid.

Proof Let G be a groupoid and $\{\lambda_i : i \in I\}$ be a family of fuzzy subgroupoids of the groupoid G. Then, we have for all $a, b \in G$.

$$\begin{aligned}
\left[\bigcap_{i \in I} \lambda_i\right](ab) &= \inf_{i \in I} [\lambda_i(ab)] \\
&\geq \inf_{i \in I} [\min\{\lambda_i(a)\ ,\ \lambda_i(b)\}] \\
&\geq \min\left\{\inf_{i \in I} \lambda_i(a),\ \inf_{i \in I} \lambda_i(b)\right\} \\
&\geq \min\left\{\bigcap_{i \in I} \lambda_i(a),\ \bigcap_{i \in I} \lambda_i(b)\right\} \\
&\Rightarrow \bigcap_{i \in I} \lambda_i \text{ is a fuzzy subgroupoid of G.}
\end{aligned}$$

Theorem 6.5 The intersection and union of any family of fuzzy (left, right) ideals is a fuzzy (left, right) ideal.

Proof Let $\{\lambda_i : i \in I\}$ be a family of fuzzy (left, right) ideals of G. Now, for all $a, b \in G$, we have
$\left[\underset{i \in I}{\cap} \lambda_i\right](ab) = \underset{i \in I}{inf}[\lambda_i(ab)] \geq \underset{i \in I}{inf}[\lambda_i(b)] = \left[\underset{i \in I}{\cap} \lambda_i\right](b) \Rightarrow \underset{i \in I}{\cap} \lambda_i$ is a fuzzy left ideal.

Further, we also have
$\left[\underset{i \in I}{\cup} \lambda_i\right](ab) = \underset{i \in I}{sup}[\lambda_i(ab)] \geq \underset{i \in I}{sup}[\lambda_i(b)] = \left[\underset{i \in I}{\cup} \lambda_i\right](b) \Rightarrow \underset{i \in I}{\cup} \lambda_i$ is a fuzzy left ideal.

Similarly, we can establish the results for fuzzy right ideal.

6.3 Homomorphism in Fuzzy Subgroupoids and Fuzzy Ideals

6.3.1 Image of a Fuzzy Set Under a Mapping

Let f be a function defined from A onto B, i.e., $f : A \to B. Let\ \lambda$ be a fuzzy set in A and μ is a fuzzy set in $f(A) = B$, then the fuzzy set $\lambda = f \circ \mu$ in A defined by the relation

$$\mu(b) = \underset{a \in f^{-1}(b)}{sup} \lambda(a), \forall\, b \in f(A) = B$$

is called the image of the fuzzy set λ under the mapping f.

6.3.2 Preimage of a Fuzzy Set Under a Mapping

Let μ be a fuzzy set in $f(A) = B$, then the fuzzy set $f \circ \mu = \lambda$ in A defined by the relation

$$\lambda(a) = \mu(f(a)), \forall\, a \in A$$

is called the preimage of the fuzzy set μ under the mapping f.

Theorem 6.6 To show that a homomorphic preimage of a fuzzy subgroupoid or (left, right) ideal is a fuzzy subgroupoid or (left, right) ideal, respectively.

Proof Let $f : A \rightarrow B$ be a homomorphism from a groupoid A onto a groupoid B. Let λ be the preimage of the fuzzy subgroupoid μ in B. Now, for $a, b \in A$, we have $f(a), f(b) \in f(A) = B$. Then,

$$\begin{aligned}\lambda(ab) &= \mu(f(ab)), \text{ using def. 6.3.2}\\ &= \mu\{f(a) \cdot f(b)\}, \text{ since is a homomorphism}\\ &\geq min\big[\mu\{f(a)\}, \mu\{f(b)\}\big], \text{ since } \mu \text{ is a fuzzy subgroupoid}\\ &= min\{\lambda(a), \lambda(b)\}, \text{ using def. 6.3.2}\end{aligned}$$

Thus, we get that $\lambda(ab) \geq min\{\lambda(a), \lambda(b)\}, \forall\, a, b \in A \Rightarrow \lambda$ the homomorphic preimage of a fuzzy subgroupoid μ in B is also a fuzzy subgroupoid in A.

Next, we consider the case when λ is the preimage of the fuzzy left ideal μ in B. Then, we have

$$\begin{aligned}\lambda(ab) &= \mu(f(ab)), \text{ using def. 6.3.2}\\ &= \mu\{f(a) \cdot f(b)\}, \text{ since is a homomorphism}\\ &\geq \mu\{f(b)\}, \text{ since } \mu \text{ is a fuzzy left ideal}\\ &= \lambda(a), \text{ using def. 6.3.2}\end{aligned}$$

$\therefore\ \lambda(ab) \geq \lambda(a), \forall\, a, b \in A \Rightarrow \lambda$, the homomorphic preimage of a fuzzy left ideal is again is again a fuzzy left ideal and similar to right ideals.

6.3.3 Sup Property

A fuzzy subset λ of a group G is said to have the sup property if, for any subset, A of G, there exists $a_0 \in A$ such that $\lambda(a_0) = \sup_{a \in A} \lambda(a)$.

Theorem 6.7 To show that a homomorphic image of a fuzzy subgroupoid or (left, right) ideal with the sup property is a fuzzy subgroupoid or (left, right) ideal, respectively.

Proof Suppose that $f : A \rightarrow B$ is a homomorphism from a groupoid A onto a groupoid B. Let λ be the fuzzy subgroupoid of A, which has the sup property, and let the fuzzy set μ is the homomorphic image of λ under f in B. Now, for given $a, b \in A$, we have $f(a), f(b) \in f(A) = B$.

Let $a_0 \in f^{-1}(f(a))$, and $b_0 \in f^{-1}(f(b))$ be such that

$\lambda(a_0) = \sup_{t \in f^{-1}(f(a))} \lambda(t)$ and $\lambda(b_0) = \sup_{t \in f^{-1}(f(b))} \lambda(t)$ respectively. We also have

$$f(a_0 b_0) = f(a_0) f(b_0) = f(a) f(b) = f(ab)$$

$\therefore a_0 b_0 \in f^{-1}(f(ab))$.

Then, we have

$$\begin{aligned}
\mu\{f(a), f(b)\} &= \sup_{z \in f^{-1}\{f(a) \cdot f(b)\}} \lambda(z) \\
&= \sup_{f(z) = \{f(a).f(b)\}} \lambda(z) \\
&= \sup_{f(z) = f(ab)} \lambda(z) \\
&= \lambda(ab) \\
&= \lambda(a_0 b_0) \\
&\geq \min\{\lambda(a_0), \lambda(b_0)\}, \text{ since } \lambda \text{ is a fuzzy subgroupoid} \\
&= min\left\{\sup_{t \in f^{-1}(f(a))} \lambda(t), \sup_{t \in f^{-1}(f(b))} \lambda(t)\right\} \\
&= min\{\mu(f(a)), \mu(f(b))\}
\end{aligned}$$

Thus, we have $\mu\{f(a), f(b)\} = min\{\mu(f(a)), \mu(f(b))\} \Rightarrow \mu$ is a fuzzy subgroupoid of $f(A) = B$.

Similarly, we can establish the result for the fuzzy (left, right) ideal.

6.4 Fuzzy Subgroups

As fuzzy sets are regarded as more general than the classical sets in the same vein, fuzzy subgroups are more prevalent than the classical groups or crisp groups.

6.4.1 Fuzzy Subgroups

A fuzzy subset λ of a group G is called a fuzzy subgroup (FSG) of G if for all $a, b \in G$, the following axioms are satisfied:

(i) $\lambda(ab) \geq min\{\lambda(a), \lambda(b)\}$

and

(ii) $\lambda(a^{-1}) \geq \lambda(a)$

In other words, a fuzzy subgroupoid λ of a group G will be called a fuzzy subgroup of G if

$$\lambda(a^{-1}) \geq \lambda(a), \forall\, a \in G$$

Theorem 6.8 If λ is a fuzzy subgroup of the group G, then $\lambda(a^{-1}) = \lambda(a)$ and $\lambda(a) \leq \lambda(e), \forall\, a \in G$, where *e* is the identity of G.

Proof Let λ be a fuzzy subgroup of group G. Then, from the condition (ii) of def. (6.4.1), we have

$$\lambda(a^{-1}) \geq \lambda(a), \forall\, a \in G$$

On replacing a by a^{-1} in the above equation, we get

$$\lambda\left(\left(a^{-1}\right)^{-1}\right) \geq \lambda(a^{-1}) \Rightarrow \lambda(a) \geq \lambda(a^{-1})$$

Hence,

$$\lambda(a^{-1}) = \lambda(a) \tag{6.4}$$

Let e be the identity element of the group G. Then, for all $a \in G$, we have $aa^{-1} = e$.

∴

$$\begin{aligned}\lambda(e) &= \lambda(aa^{-1}) \geq \min\{\lambda(a), \lambda(a^{-1})\}\\ &= \min\{\lambda(a), \lambda(a)\}, \text{ using (6.4)}\\ &= \lambda(a)\end{aligned} \tag{6.5}$$

Thus, we get that $\lambda(a) \leq \lambda(e), \forall\, a \in G$. If $\lambda(a) = 1$ for at-least one $a \in G$. Then, we get that $\lambda(e) \geq 1$. But $\lambda(e) > 1$ is not true, therefore $\lambda(e) = 1$ if $\lambda(a) = 1$ for at-least one $a \in G$.

Hence, we may consider the following as a modified definition of the fuzzy subgroup:

Let $(G, \cdot)$ be a group in the crisp sense. Then, a fuzzy subset λ of G will be called a fuzzy subgroup of G if

(i) $\lambda(ab) \geq min\{\lambda(a), \lambda(b)\}, \forall\, a, b \in G$
(ii) $\lambda(a^{-1}) = \lambda(a), \forall\, a \in G$
(iii) $\lambda(e) = 1$, if $\lambda(a) = 1$ for at-least one $a \in G$

Let us consider the case when λ is an ordinary subset of G, then $\lambda(a) = 1$ and $\lambda(b) = 1$ for all $a, b \in G$. In consequence, from Definition 6.4 of FSG, we have

$$\begin{aligned}\lambda(ab^{-1}) &\geq min\{\lambda(a), \lambda(b^{-1})\}\\ &= min\{\lambda(a), \lambda(b)\}, \text{ using (6.4)}\\ &= min\{1, 1\} = 1\end{aligned}$$

Thus, we get
$\lambda(ab^{-1}) \geq 1$. But $\lambda(ab^{-1}) > 1$ *is not true.*
Hence, $\lambda(ab^{-1}) = 1 \Rightarrow ab^{-1} \in \lambda$
In this way, we observe that for all $a, b \in \lambda \Rightarrow ab^{-1} \in \lambda$
Thus, from above, it can be deduced that the notion of fuzzy subgroup has been considered so that when λ is considered an ordinary subset of G under the notion proposed by Rosenfeld, λ turns out to be a subgroup of G.

Example 6.1 Let G be the Klein's four group named after Felix Klein (1849–1925), where the set G is $G = \{a, b : a^2 = b^2 = (ab)^2 = e\}$. Thus, the elements of G are e, a, b, ab. Let $t_i, 0 \leq i \leq 2$ be three numbers lying in the closed interval [0, 1] and that $t_0 > t_1 > t_2$. Let λ be a fuzzy subset of G, i.e., a mapping $\lambda : G \rightarrow [0, 1]$ defined by $\lambda(e) = t_0, \lambda(a) = t_1, \lambda(b) = t_2, \lambda(ab) = t_2$. Then λ is a fuzzy subgroup of G.

Example 6.2 Let G be the Klein's four group. Therefore, we have $G = \{e, a, b, ab\}$, where $a^2 = e = b^2$ and $ab = ba$.

For $0 \leq i \leq 5$, let t_i be the numbers in the closed interval [0, 1] such that $t_0 > t_1 > t_2 > t_3 > t_4 > t_5$. Now, let us define fuzzy subsets λ and $\gamma : G \rightarrow [0, 1]$ as follows:

$$\lambda(e) = t_1, \lambda(a) = t_3 \text{ and } \lambda(b) = \lambda(ab) = t_4$$

$$\gamma(e) = t_0, \gamma(a) = t_5, \gamma(b) = t_2 \text{ and } \gamma(ab) = t_5$$

Let λ and γ be fuzzy subgroups of G, according to Rosenfeld.
We have,

$$(\lambda \vee \gamma)(e) = max\{t_1, t_0\} = t_0, \ (\lambda \vee \gamma)(a) = max\{t_3, t_5\} = t_3,$$
$$(\lambda \vee \gamma)(b) = max\{t_4, t_2\} = t_2, \ (\lambda \vee \gamma)(ab) = max\{t_4, t_5\} = t_4$$

But $\lambda \vee \gamma = \delta$(say) is not a fuzzy subgroup of G. Since, we get

$$(\lambda \vee \gamma)(ab) = \delta(ab) \geq min\{\delta(a), \delta(b)\} = min\{(\lambda \vee \gamma)(a), (\lambda \vee \gamma)(b)\} = min\{t_3, t_2\} = t_3$$

Thus, $(\lambda \vee \gamma)(ab) = \delta(ab) \geq t_3$. But, we have $(\lambda \vee \gamma)(ab) = t_4$.
Hence, $(\lambda \vee \gamma)(ab) \geq min\{\delta(a), \delta(b)\}$ is not true in this case.

Theorem 6.9 If φ_T is the characteristic function of a subset T of G, then φ_T is a fuzzy subgroup if and only if T is a subgroup.

Proof Let G be a group and $T \subseteq G$. Then the characteristic function of T is denoted by φ_T and is defined as

$$\varphi_T(a) = \begin{cases} 1, \text{ iff } & a \in T \\ 0, \text{ iff } & a \notin T \end{cases}$$

Firstly, we assume that φ_T is a fuzzy subgroup of G, and we claim that T is a subgroup of G. Let $a, b \in T$. Then $\varphi_T(a) = 1$ and $\varphi_T(b) = 1$.

Now, we have

$$\begin{aligned} \varphi_T\left(ab^{-1}\right) &\geq min\left\{\varphi_T(a), \varphi_T\left(b^{-1}\right)\right\}, \forall\, a, b \in G \\ &= min\{\varphi_T(a), \varphi_T(b)\} = min\{1, 1\} = 1 \end{aligned}$$

Thus, $\varphi_T\left(ab^{-1}\right) = 1 \Rightarrow ab^{-1} \in T \Rightarrow T$ is a subgroup of G.

Conversely, we assume that T is a subgroup of G and $a, b \in T \Rightarrow ab \in T$ and $a^{-1} \in T$. Therefore,

$\varphi_T(a) = 1, \varphi_T(b) = 1, \varphi_T\left(a^{-1}\right) = 1$ and $\varphi_T(ab) = 1$.

We have, $\varphi_T(ab) = 1 \Rightarrow \varphi_T(ab) \geq min\{1, 1\} = min\{\varphi_T(a), \varphi_T(b)\}$

$\therefore \varphi_T(ab) \geq min\{\varphi_T(a), \varphi_T(b)\}$.

We also have $\varphi_T\left(a^{-1}\right) = \varphi_T(a) = 1$.

So φ_T, the characteristic function of T, is a fuzzy subgroup of G.

Theorem 6.10 The intersection of any family of fuzzy subgroups of a group G is a fuzzy subgroup of G.

Proof Let us assume that $\{\lambda_i : i \in I\}$ be a family of fuzzy subgroups of the group G, where I is an index set and is such that $\forall\, i \in I$, λ_i is a fuzzy subgroup of G.

Let $\lambda = \bigcap_{i \in I} \lambda_i = \{a \in G : a \in \lambda_i, \forall\, i \in I\}$. Now, we will claim that λ is a fuzzy subgroup of G. Let $a, b \in G$. Then, we have

$$a \in \bigcap_{i \in I} \lambda_i \Rightarrow a \in \lambda_i, \forall\, i \in I \text{ and } b \in \bigcap_{i \in I} \lambda_i \Rightarrow b \in \lambda_i, \forall\, i \in I$$

$$\begin{aligned} \lambda(ab) &= \left[\bigcap_{i \in I} \lambda_i\right](ab) = \inf_{i \in I} [\lambda_i(ab)] \\ &\geq \inf_{i \in I} [\min\{\lambda_i(a)\,,\, \lambda_i(b)\}] \\ &\geq min\left\{\inf_{i \in I} \lambda_i(a), \inf_{i \in I} \lambda_i(b)\right\} \\ &\geq min\left\{\bigcap_{i \in I} \lambda_i(a), \bigcap_{i \in I} \lambda_i(b)\right\} = min\{\lambda(a), \lambda(b)\} \end{aligned}$$

and $\lambda\left(a^{-1}\right) = \left[\bigcap_{i \in I} \lambda_i\right]\left(a^{-1}\right) = \inf_{i \in I} \lambda_i\left(a^{-1}\right) = \inf_{i \in I} \lambda_i(a) = \left[\bigcap_{i \in I} \lambda_i\right](a) = \lambda(a)$

Thus, the intersection of any set of fuzzy subgroups of G is again a fuzzy subgroup of G.

Theorem 6.11 If λ is a fuzzy subgroup of G, then $\lambda(ab^{-1}) = \lambda(e) \Rightarrow \lambda(a) = \lambda(b)$.

Proof Let λ be a fuzzy subgroup of the group G, and e is the identity of G. Then, $b^{-1}b = e$. Further, we assume that $\lambda(ab^{-1}) = \lambda(e)$.

$$\lambda(a) = \lambda(ae) = \lambda(a(b^{-1}b)) = \lambda\{(ab^{-1})b\}, \text{ by associativity in G}$$

We have,
$$\geq min\{\lambda(ab^{-1}), \lambda(b)\}$$
$$= min\{\lambda(e), \lambda(b)\} = \lambda(b), \text{ using equation (6.5)}$$

$\therefore \lambda(a) \geq \lambda(b)$.

$$\lambda(b) = \lambda(be) = \lambda(b(a^{-1}a)) = \lambda\{(ba^{-1})a\}$$
$$\geq min\{\lambda(ba^{-1}), \lambda(a)\}$$

and
$$= min\left\{\lambda(ab^{-1})^{-1}, \lambda(a)\right\}$$
$$= min\{\lambda(ab^{-1}), \lambda(a)\}, \text{ using equation (6.4)}$$
$$= min\{\lambda(e), \lambda(a)\} = \lambda(a)$$

$\therefore \lambda(b) \geq \lambda(a)$.

So, we have $\lambda(a) = \lambda(b)$.

Theorem 6.12 If λ is a fuzzy subgroup of G, then $\{a \in G : \lambda(a) = \lambda(e)\}$ is a subgroup of G.

Proof Let G be a group, and λ is a fuzzy subgroup of G. Let $H = \{a \in G : \lambda(a) = \lambda(e)\}$. Let $a, b \in H$. Then

$$\lambda(a) = \lambda(e) \text{ and } \lambda(b) = \lambda(e) \forall\, a, b \in G \tag{6.6}$$

Since λ is a fuzzy subgroup, therefore

$$\lambda(a^{-1}) = \lambda(a), \forall\, a \in G$$

So $\lambda(a^{-1}) = \lambda(a) = \lambda(e) \Rightarrow a^{-1} \in H$, using equation (6.6).
Thus, we get that $a \in H \Rightarrow a^{-1} \in H$ and

$$\lambda(ab^{-1}) \geq \min\{\lambda(a), \lambda(b^{-1})\}$$
$$= min\{\lambda(a), \lambda(b)\}, \text{ using equation (6.4)}$$
$$= min\{\lambda(e), \lambda(e)\}, \text{ using equation (6.6)}$$
$$= \lambda(e)$$

Thus, $\lambda(ab^{-1}) \geq \lambda(e)$.
However, by Theorem 6.7, we have $\lambda(e) \geq \lambda(ab^{-1})$ for all $ab^{-1} \in G$.

Therefore, $\lambda(ab^{-1}) = \lambda(e) \Rightarrow ab^{-1} \in H$. Thus, $a \in H, b \in H \Rightarrow ab^{-1} \in H \Rightarrow$ H is a subgroup of G. Hence $H = \{x \in G : \lambda(a) = \lambda(e)\}$ is a subgroup of G and is usually denoted by G_λ. Thus, we have

$$G_\lambda = \{a \in G : \lambda(a) = \lambda(e)\}$$

Theorem 6.13 If λ is a fuzzy subgroup of a group G, then λ is constant on each coset of G_λ.

Proof Let λ be a fuzzy subgroup of a group G. Then, we have

$$G_\lambda = \{a \in G : \lambda(a) = \lambda(e)\} \tag{6.7}$$

Let $b \in G$. Then the right coset of G_λ be $G_\lambda b$ and $b \in G_\lambda b$ and is defined as $G_\lambda b = \{ab : \lambda(ab) = \lambda(e)\}$, $ab \in G_\lambda$ as G_λ is a subgroup.
Now, we have

$$\lambda\left((ab)b^{-1}\right) = \lambda\left(a\left(\left(bb^{-1}\right)\right)\right) = \lambda(ae) = \lambda(a) = \lambda(e), \text{ using equation (6.7)}$$

Thus, $\lambda\left((ab)b^{-1}\right) = \lambda(e) \Rightarrow \lambda(ab) = \lambda(b)$, by Theorem 6.11.
So λ is constant on each right coset of G_λ. If bG_λ is the left coset of G_λ, then

$$bG_\lambda = \{ba : \lambda(ba) = \lambda(e)\}$$

and

$$\lambda\left(b^{-1}(ba)\right) = \lambda\left(\left(b^{-1}b\right)a\right) = \lambda(ea) = \lambda(a) = \lambda(e), \text{ using equation (6.7)}$$
$$\Rightarrow \lambda\left(b^{-1}(ba)\right) = \lambda(e) \Rightarrow \lambda\left\{b^{-1}\left(\left((ba)^{-1}\right)^{-1}\right)\right\} = \lambda(e)$$
$$\Rightarrow \lambda\left(b^{-1}\right) = \lambda(ba)^{-1} \Rightarrow \lambda(b) = \lambda(ba), \text{ by theorem 6.11 since } \lambda \text{ is a FSG}$$

So λ is constant on each left coset of G_λ, and consequently λ is constant on each coset (left or right) of G_λ.

Theorem 6.14 λ is a fuzzy subgroup of G if and only if $\lambda\left(ab^{-1}\right) \geq min\{\lambda(a), \lambda(b)\}, \forall\, a, b \in G$.

Proof Suppose that λ is a fuzzy subgroup of G and $a, b \in G$. Then, we have

$$\lambda\left(ab^{-1}\right) \geq min\left\{\lambda(a), \lambda\left(b^{-1}\right)\right\} \geq \min\{\lambda(a), \lambda(b)\}$$

Conversely, we assume that $\lambda\left(ab^{-1}\right) \geq min\{\lambda(a),\ \lambda(b)\}$.
On putting $b = a$, we get $\lambda\left(aa^{-1}\right) \geq min\{\lambda(a), \lambda(a)\} \Rightarrow \lambda(e) \geq \lambda(a), \forall\, a \in G$
We also have,
$\lambda(ab) = \lambda\left\{a\left(b^{-1}\right)^{-1}\right\} \geq min\left\{\lambda(a), \lambda\left(b^{-1}\right)\right\} \geq min\{\lambda(a), \lambda(b)\} \Rightarrow \lambda$ is a FSG.

Theorem 6.15 If λ is a fuzzy subgroup of a group G, then the following conditions are equivalent:

(i) $\lambda(ab) = \lambda(ba), \forall\, a, b \in G$
(ii) $\lambda(aba^{-1}) = \lambda(b), \forall\, a, b \in G$

Proof Let λ is a fuzzy subgroup of G, and $a, b \in G$. Firstly, we claim that (i) $\Rightarrow$ *(ii)*.
We have,

$$\begin{aligned}\lambda(aba^{-1}) &= \lambda\{(ab)a^{-1}\} = \lambda\{a^{-1}(ab)\}, \text{ using (i)} \\ &= \lambda\{(a^{-1}a)b\}, \textit{by associativity in } G \\ &= \lambda\{eb\} = \lambda(b)\end{aligned}$$

$\therefore\ \lambda(aba^{-1}) = \lambda(b), \forall\, a, b \in G.$
Now, we will show that *(ii)* $\Rightarrow$ *(i)*.
We have,

$$ab = (ab)e = (ab)(aa^{-1}) = a(ba)a^{-1} \tag{6.7}$$

$$\therefore \begin{aligned}\lambda(ab) &= \lambda\{a(ba)a^{-1}\}, \text{ using equation (6.7)} \\ &= \lambda(ba), \text{ using condition (ii)}\end{aligned}.$$

Theorem 6.16 If λ is a fuzzy subgroup of a group G and if for all $a, b \in G$, $\lambda(a) < \lambda(b)$, then $\lambda(ab) = \lambda(a) = \lambda(ba)$.

Proof Let G be a group, and λ is a fuzzy subgroup of G. Let for all $a, b \in G$, $\lambda(a) < \lambda(b)$, then it follows that

$$\lambda(ab) \geq min\{\lambda(a), \lambda(b)\} = \lambda(a) \tag{6.8}$$

Now

$$\begin{aligned}\lambda(a) = \lambda(ae) = \lambda(a(bb^{-1})) = \lambda((ab)b^{-1})) \geq \min\{\lambda(ab), \lambda(b^{-1})\} \\ = min\{\lambda(ab), \lambda(b)\} \\ = \lambda(ab), \text{ since } \lambda(ab) \geq \lambda(a)\end{aligned}$$

It gives us

$$\lambda(a) \geq \lambda(ab) \tag{6.9}$$

On combining (6.8) and (6.9), we get that

$$\lambda(ab) = \lambda(a)$$

Similarly, we can show that $\lambda(ba) = \lambda(a)$.

Theorem 6.17 If λ is a fuzzy subgroup of a group G and $a \in G$, then

$$\lambda(ab) = \lambda(b), \forall\, b \in G \Leftrightarrow \lambda(a) = \lambda(e)$$

Proof Let λ be a fuzzy subgroup of a group G and be such that $\lambda(ab) = \lambda(b), \forall\, b \in G$. Then on choosing $b = e$, we will obtain $\lambda(ae) = \lambda(e) \Rightarrow \lambda(a) = \lambda(e)$.

Conversely, let us assume that

$$\lambda(a) = \lambda(e) \tag{6.10}$$

Then

$$\lambda(e) = \lambda\left(bb^{-1}\right) \geq min\left\{\lambda(b), \lambda\left(b^{-1}\right)\right\} = min\{\lambda(b), \lambda(b)\} = \lambda(b)$$

$\therefore \lambda(b) \leq \lambda(e) = \lambda(a) \Rightarrow \lambda(b) \leq \lambda(a), \forall\, b \in G,$ using equation (6.10).

Now $\lambda(ab) \geq min\{\lambda(a), \lambda(b)\} = \lambda(b) \Rightarrow \lambda(ab) \geq \lambda(b), \forall\, b \in G$

But $\begin{aligned} \lambda(b) &= \lambda(eb) = \lambda\left(\left(a^{-1}a\right)b\right) = \lambda\left(a^{-1}(ab)\right) \geq min\left\{\lambda\left(a^{-1}\right), \lambda(ab)\right\} \\ &= min\{\lambda(a), \lambda(ab)\} \end{aligned}$.

But $\lambda(a) \geq \lambda(b), \forall\, b \in G$ and $\lambda(b) \leq \lambda(ab) \Rightarrow min\{\lambda(a), \lambda(ab)\} = \lambda(ab)$.

Therefore, we have that

$$\lambda(b) \geq \lambda(ab),\ \forall\, b \in G$$

Consequently, the result $\lambda(ab) = \lambda(b)$ follows.

Remark 6.1 It is easy to verify that if $\lambda(a) = \lambda(e)$, then $\lambda(ab) = \lambda(b), \forall\, b \in G$.

Theorem 6.18 If λ is a fuzzy subgroup of a group G, then $\lambda(ab) = min\{\lambda(a), \lambda(ab)\}, \forall\, a, b \in G$.

with $\lambda(a) \neq \lambda(b)$.

Proof Let G is a group, and λ is a fuzzy subgroup of G. For all a, b in G, we have that $\lambda(a) \neq \lambda(b)$. Let e be the identity of the group G.

So, let us assume that

$$\lambda(a) > \lambda(b) \tag{6.11}$$

Then, we have

$$\begin{aligned} \lambda(b) = \lambda(eb) = \lambda\left(\left(a^{-1}a\right)b\right) &= \lambda\left\{a^{-1}(ab)\right\} \\ &\geq min\left\{\lambda\left(a^{-1}\right), \lambda(ab)\right\} \\ &= min\{\lambda(a), \lambda(ab)\} \\ &= \lambda(ab), \text{ using equation (6.11)} \end{aligned}$$

$\therefore \lambda(b) \geq \lambda(ab)$

Since λ is a fuzzy subgroup of G, therefore

$$\lambda(ab) \geq min\{\lambda(a), \lambda(b)\} = \lambda(b) \Rightarrow \lambda(ab) \geq \lambda(b)$$

Consequently, we get $\lambda(ab) = \lambda(b) = min\{\lambda(a), \lambda(b)\}$, using (6.11).
$\therefore \lambda(ab) = min\{\lambda(a), \lambda(b)\}$.

6.4.2 Proper Fuzzy Subgroups

The fuzzy subgroup λ of a group G is said to be a proper fuzzy subgroup of G if λ(a) $\neq 1$ for at least one $a \in G$, i.e., $\lambda \neq G$.

Theorem 6.19 A group cannot be the union of two proper fuzzy subgroups.

Proof Let λ *and* μ be two proper subgroups of the group G and are such that $\lambda \cup \mu = G$. Suppose that
$\lambda(a) = 1$ or $\mu(a) = 1, \forall\, a \in G$. Again, let $a_1, a_2 \in G$ and are such that $\lambda(a_1) = 1, \mu(a_1) < 1$, and $\lambda(a_2) < 1, \mu(a_2) = 1$.
Since, $a_1, a_2 \in G \Rightarrow a_1a_2 \in G$. Then, we claim that

$$\lambda(a_1a_2) \neq 1 \text{ and } \mu(a_1a_2) \neq 1$$

For otherwise, let us assume that

$$\lambda(a_1a_2) = 1 \tag{6.12}$$

We have

$$\begin{aligned} \lambda(a_2) = \lambda(ea_2) &= \lambda\big((a_1^{-1}a_1)a_2\big) \\ &= \lambda\big\{a_1^{-1}(a_1a_2)\big\}, \text{ by associativity in G} \\ &\geq min\big\{\lambda\big(a_1^{-1}\big), \lambda(a_1a_2)\big\}, \text{ since } \lambda \text{ is a FSG} \\ &= min\{\lambda(a_1), \lambda(a_1a_2)\}, \text{ using equation (6.4)} \\ &= min\{1, 1\} = 1, \text{ using equation (6.12)} \end{aligned}$$

The above result is contrary to our assumption.
Let us now assume that $\mu(a_1a_2) = 1$. Then

$$\begin{aligned} \mu(a_1) = \mu(a_1e) &= \mu\big\{a_1\big(a_2a_2^{-1}\big)\big\} \\ &= \mu\big\{(a_1a_2)a_2^{-1}\big\} \\ &\geq min\big\{\mu(a_1a_2), \mu\big(a_2^{-1}\big)\big\} \\ &= min\{\mu(a_1a_2), \mu(a_2)\} \\ &= min\{1, 1\} = 1 \end{aligned}$$

Again, this is a contradiction to our assumption. Hence, $\lambda \cup \mu \neq G$.

Theorem 6.20 A homomorphic preimage of a fuzzy subgroup is a fuzzy subgroup.

Proof Let f be a homomorphism from a group A onto a group B. Further, we assume that λ is the preimage of a fuzzy subgroup μ in B. Let $a, b \in A$, then $f(a), f(b) \in B$.Now, we have

$$\lambda(ab) = \mu(f(ab)) = \mu\{f(a) \cdot f(b)\} \geq min\left[\mu\{f(a)\}, \mu\{f(b)\}\right] = min\{\lambda(a), \lambda(b)\}$$

We also have, $\lambda\left(a^{-1}\right) = \mu\left(f\left(a^{-1}\right)\right) = \mu\{f(a)\}^{-1} \geq \mu\{f(a)\} = \lambda(a)$.

From above, it follows that λ is a fuzzy subgroup of A, i.e., the homomorphic preimage λ of a fuzzy subgroup μ in B is a fuzzy subgroup.

Theorem 6.21 A homomorphic image of a fuzzy subgroup having sup property is also a fuzzy subgroup.

Proof Let f be a homomorphism from group A onto a group B. Next, we assume that λ is a fuzzy subgroup in A, which has the sup property, and μ is the homomorphic image of λ in B. For $a, b \in A$, then $f(a), f(b) \in B$. We have seen in Theorem 6.7 that the first condition holds for subgroupoids. Therefore, it is required to verify the second condition only. Then

$$\begin{aligned} \mu\{f(a)\}^{-1} &= \mu\left\{f(a)^{-1}\right\} = \sup_{z \in f^{-1}\left\{f(a)^{-1}\right\}} \lambda(z) \\ &= \sup_{f(z) = \{f(a)\}^{-1}} \lambda(z) = \sup_{f(z) = f\left(a^{-1}\right)} \lambda(z) \\ &= \lambda\left(a_0^{-1}\right) \geq \lambda(a_0) = \sup_{t \in f^{-1}\left(f(a)^{-1}\right)} \lambda(t) = \mu(f(a)) \end{aligned}$$

Hence, homomorphic image of a fuzzy subgroup λ having sup property is also a fuzzy subgroup.

6.4.3 *f*-Invariant

Let f be any function from a set S to a set T, and let λ be a fuzzy subset of S. Then λ is called f-invariant if

$$f(a) = f(b) \Rightarrow \lambda(a) = \lambda(b), \forall\, a, b \in S.$$

Theorem 6.22 Fuzzy (left, right) ideals in a group G are just the constant function.

Proof Let G be a group and λ be a fuzzy ideal in G. Therefore, λ is a fuzzy left ideal and a fuzzy right ideal. Let λ be a fuzzy left ideal so that for all $a, b \in G$.

$$\lambda(ab) \geq \lambda(\mathrm{b}) \tag{6.13}$$

On putting $b = e$ in (6.13), we get

$$\lambda(ae) \geq \lambda(\mathrm{e}) \Rightarrow \lambda(a) \geq \lambda(\mathrm{e}), \forall\, a \in G$$

And on putting $a = b^{-1}$ in (6.13), we have

$$\lambda\left(b^{-1}b\right) \geq \lambda(\mathrm{b}) \Rightarrow \lambda(e) \geq \lambda(\mathrm{b}), \forall\, b \in G$$

Thus, we get that $\lambda(a) \geq \lambda(\mathrm{e}), \forall\, a \in G$ and $\lambda(e) \geq \lambda(\mathrm{b}), \forall\, b \in G \Rightarrow \lambda = \lambda(e)$ is a constant function and, similarly, for fuzzy right ideals.

Conversely, we assume that λ is a constant function on the group G. For $a, b \in G \Rightarrow ab \in G$ and hence $\lambda(a) = \lambda(\mathrm{e}), \lambda(b) = \lambda(\mathrm{e})$, and $\lambda(ab) = \lambda(\mathrm{e})$. Then, we have

$$\lambda(ab) = \lambda(a) = \lambda(b) = \lambda(\mathrm{e}), \forall\, a, b \in G$$

So $\lambda(ab) \geq \lambda(\mathrm{b})$ and $\lambda(ab) \geq \lambda(\mathrm{a})$ which implies that λ is a fuzzy ideal of G.

Theorem 6.23 If λ is a fuzzy subgroupoid of a finite group G, then λ is a fuzzy subgroup of G.

Proof Let λ be a fuzzy subgroupoid of a finite group G and a be any element of G. Since $a \in G$ and G is finite, the order of the element a is finite. Let $o(a) = n \Rightarrow a^n = e$, e being the identity of G. We know that $O(a) = O\left(a^{-1}\right)$. Hence

$$a^n = e = \left(a^{-1}\right)^n \Rightarrow a^n = e = a^{-n}$$

We have $a^n = e \Rightarrow a^{n-1}.a = e.$

Operating, both sides by a^{-1} in right, we get

$$\Rightarrow a^{n-1}.\left(a.a^{-1}\right) = e.a^{-1} \Rightarrow a^{n-1}.e = a^{-1} \Rightarrow a^{n-1} = a^{-1} \tag{6.14}$$

Hence, we have

$$\begin{aligned}\lambda\left(a^{-1}\right) &= \lambda\left(a^{n-1}\right), \text{ using equation (6.14)}\\ &= \lambda\left(a^{n-2}.a\right)\\ &\geq min\left\{\lambda\left(a^{n-2}\right), \lambda(a)\right\} \geq min\left\{\lambda\left(a^{n-3}\right), \lambda(a), \lambda(a)\right\}\end{aligned}$$

Now, on applying the definition of fuzzy subgroupoid repeatedly n-3 times, we will get

$$\lambda\left(a^{-1}\right) \geq \lambda(a)$$

We also have $a^{-n} = e \Rightarrow a^{-n+1}.a^{-1} = e$.
Operating, both sides by a in right, we get

$$a^{-n+1}.\left(a^{-1}.a\right) = e.a \Rightarrow a^{-n+1}.e = a \Rightarrow a^{-n+1} = a \tag{6.15}$$

Therefore,

$$\begin{aligned}\lambda(a) &= \lambda\left(a^{-n+1}\right), \text{ using equation (6.15)}\\ &= \lambda\left(a^{-n+2}.a^{-1}\right) \geq min\left\{\lambda\left(a^{-n+2}\right), \lambda\left(a^{-1}\right)\right\} \geq min\left\{\lambda\left(a^{-n+3}\right), \lambda\left(a^{-1}\right), \lambda\left(a^{-1}\right)\right\}\end{aligned}$$

Again on applying the definition of fuzzy subgroupoid repeatedly n-3 times, we will get

$$\lambda(a) \geq \lambda\left(a^{-1}\right)$$

Thus, $\lambda\left(a^{-1}\right) \geq \lambda(a)$ and $\lambda(a) \geq \lambda\left(a^{-1}\right) \Rightarrow \lambda\left(a^{-1}\right) = \lambda(a)$, and consequently, we have the result that λ is a fuzzy subgroup of the finite group G.

Theorem 6.24 If λ is a fuzzy subgroup of a group G having e as its identity element such that λ(e) = 1, then the function η defined as $\eta(a) = \frac{\lambda(a)}{\lambda(e)}, \forall\, a \in G$ is a fuzzy subgroup of G such that $\eta(e) = 1$.

Proof Let λ be a fuzzy subgroup of the group G, and a fuzzy set η is defined on G as follows:

$$\eta(a) = \frac{\lambda(a)}{\lambda(e)}, \forall\, a \in G \tag{6.16}$$

$$\begin{aligned}\Rightarrow \eta(e) &= \frac{\lambda(e)}{\lambda(e)} = 1\\ &= \frac{\lambda\left(aa^{-1}\right)}{\lambda(e)} \geq \frac{\min\left\{\lambda(a),\ \lambda\left(a^{-1}\right)\right\}}{\lambda(e)}\\ &= \frac{\min\{\lambda(a)\ ,\ \lambda(a)\}}{\lambda(e)} = \frac{\lambda(a)}{\lambda(e)} = \eta(a)\end{aligned}$$

Thus, $\eta(e) \geq \eta(a)$. From above it follows that $\eta(e) = 1$ and that $0 \leq \eta(a) \leq 1$. Further, we have

$$\begin{aligned}\eta(ab) &= \frac{\lambda(ab)}{\lambda(e)}, \forall\, a, b \in G\\ &\geq \frac{\min\{\lambda(a), \lambda(b)\}}{\lambda(e)} = \min\{\lambda(a), \lambda(b)\}, \text{ as } \lambda(e) = 1\\ &= min\left\{\frac{\lambda(a)}{1}, \frac{\lambda(b)}{1}\right\} = min\left\{\frac{\lambda(a)}{\lambda(e)}, \frac{\lambda(b)}{\lambda(e)}\right\}\end{aligned}$$

$= min\{\eta(a), \eta(b)\}$, using equation (6.16)

$\therefore \eta(ab) \geq min\{\eta(a), \eta(b)\}$
Moreover, $\eta(a^{-1}) = \frac{\lambda(a^{-1})}{\lambda(e)} = \frac{\lambda(a)}{\lambda(e)} = \eta(a)$, i.e.,$\eta(a^{-1}) = \eta(a)$.
This establishes the claim that η is a fuzzy subgroup of G w. r. t. min.

Theorem 6.25 If X is the set of all fuzzy subgroups of G and " ~ " is a relation in X defined by $\alpha \sim \beta$ if and only if for all $a, b \in G, \alpha(a) > \alpha(b) \Leftrightarrow \beta(a) > \beta(b)$ then, " ~ " is an equivalence relation.

Proof Let X be the set of all fuzzy subgroups of group G. Let α , β, and γ be fuzzy subgroups in X and a, b, and $c \in G$. The relation " ~ " in X is defined by the setting

$$\alpha \sim \beta \text{ iff } \alpha(a) > \alpha(b) \Leftrightarrow \beta(a) > \beta(b), \forall\, a, b \in G$$

Now, we will claim that " ~ " is reflexive, symmetric, and transitive.
We have $\alpha \sim \alpha$.
$\therefore \alpha(a) > \alpha(b) \Leftrightarrow \alpha(a) > \alpha(b) \Rightarrow$ $' \sim '$is reflexive.
We have $\alpha \sim \beta$.
$\therefore \alpha(a) > \alpha(b) \Leftrightarrow \beta(a) > \beta(b)$.
It follows that, $\beta(a) > \beta(b) \Leftrightarrow \alpha(a) > \alpha(b) \Rightarrow$ $'sim'$ is symmetric.
We have $\alpha \sim \beta$ and $\beta \sim \gamma$. Therefore,
$\alpha(a) > \alpha(b) \Leftrightarrow \beta(a) > \beta(b)$ and $\beta(a) > \beta(b) \Leftrightarrow \gamma(a) > \gamma(b)$.
Combining these two results, we at once get $\alpha(a) > \alpha(b) \Leftrightarrow \gamma(a) > \gamma(b)$.
Hence, " ~ " is transitive, and consequently, " ~ " is an equivalence relation.

6.5 Rosenfeld Notion of Fuzzy Subgroup of a Group (G, +)

Rosenfeld proposed the conception of a fuzzy subgroup of a group (G,·) under a binary composition multiplication. Here, we (**Singh**) will consider the binary operation "+" and wish to see the behavior of the notion of the fuzzy group introduced by Rosenfeld.

Let $(G, +)$ be a group and λ be a fuzzy subset of G. Then, λ will be called a fuzzy subgroup of G if, for all $a, b \in G$, the following conditions are satisfied:-

(i) $\lambda(a + b) \geq min\{\lambda(a), \lambda(b)\}(6.17)$
(ii) $\lambda(-a) \geq \lambda(a)(6.18)$

Now, we will analyze the above definition of the fuzzy subgroup.
On replacing a by (-a), we get

$$\lambda\{-(-a)\} \geq \lambda(-a) \Rightarrow \lambda(a) \geq \lambda(-a)$$

And consequently, we will have $\lambda(-a) = \lambda(a)$.

Let us consider the circumstances when λ is an ordinary subset of G. Then, we have $\lambda(a) = \lambda(b) = 1$ for all $a, b \in G$. Now, from Eq. (6.17), we get $\lambda(a+b) \geq min\{1, 1\}$, i.e., $\lambda(a+b) \geq 1$. But $\lambda(a+b) \ngtr 1$ *is not true*. Hence, we must have $\lambda(a+b) = 1$, which in turn implies that $a+b \in G$. Thus, we get that for all $a, b \in G \Rightarrow a+b \in G$, and from Eq. (6.18), we get that $\lambda(-a) = 1 = \lambda(a), \forall\, a \in G$. Thus, $\forall\, a \in G \Rightarrow -a \in G$.

From the above discussion, it can be inferred that the notion of fuzzy subgroups λ of a group $(G, +)$ has been considered in such a way that when λ is considered to be an ordinary subset of G under the definition proposed by Rosenfeld, λ turns out to be a subgroup of the group $(G, +)$.

6.6 Redefined Notion of Fuzzy Subgroups

The fuzzy subgroup discussed above is simply with respect to the t-norm "min." **J. M. Anthony** and **H. Sherwood** (Journal of Mathematical Analysis and Applications; V.69; 1979) have redefined the notion of fuzzy subgroups with respect to a general t-norm. Subsequently, among others, **Sherwood** (1983), **Abuosman** (1987), **P. Das** (1984), and **Singh** (2003) have studied fuzzy subgroups and showed that most of the results obtained by **Rosenfeld** with respect to the t-norm "min" are valid in their case. In this section, we will discuss the results obtained by **Anthony and Sherwood**.

According to them, some mathematical structures that intuitively seem "fuzzy" do not satisfy the definition as proposed by **Rosenfeld**. So they attempted to modify the definition of **Rosenfeld** to include those objects which Rosenfeld has excluded. In order to do this, they proposed two classes of examples.

Example 6.3 Let G be a groupoid and be the class of all random variables on the probability space (Ω, Ψ, P). Again, let B be the Borel subset of real numbers, which is a subgroup of real numbers under the binary composition "addition of real numbers." Obviously, G is a groupoid with point-wise addition. Let us define a mapping $\delta_B : G \to [0, 1]$ by

$$\delta_B(X) = P\{\omega \in \Omega; X(\omega) \in B\} = P\left[X^{-1}(B)\right]$$

Here $\delta_B(X)$ is the probability that X belongs to the subgroup B. The function δ_B seems to be a fuzzy subgroupoid, but it does not satisfy the condition of a fuzzy subgroupoid. In order to see this, let us suppose that $\Omega = [0, 1]$, Ψ be the Borel subset of [0, 1] and P be Lebesgue measure with B as the set of integers. Let us define X and Y by the setting

$$X(\omega) = \begin{cases} 1, \omega \in \left[0, \frac{1}{2}\right] \\ \frac{1}{2}, \omega \in \left(\frac{1}{2}, 1\right] \end{cases} \text{ and } Y(\omega) = \begin{cases} \frac{1}{2}, \omega \in \left[0, \frac{1}{3}\right] \\ 1, \omega \in \left(\frac{1}{3}, 1\right] \end{cases}$$

Since X and Y are random variables, $X + Y$ will be defined as

$$(X+Y)(\omega) = X(\omega) + Y(\omega) = \begin{cases} 1 + \frac{1}{2} = \frac{3}{2},\ \omega \in \left[0, \frac{1}{3}\right] \\ 1 + 1 = 2,\ \omega \in \left(\frac{1}{3}, \frac{1}{2}\right] \\ \frac{1}{2} + 1 = \frac{3}{2},\ \omega \in \left(\frac{1}{2}, 1\right] \end{cases}$$

Evidently, we have

$$\begin{aligned} \delta_B(X) &= P\left[X^{-1}(B)r\right] = P\left[0, \frac{1}{2}\right] = \frac{1}{2} - 0 = \frac{1}{2} \\ \delta_B(Y) &= P\left[Y^{-1}(B)\right] = P\left(\frac{1}{3}, 1\right] = P\left]\frac{1}{3}, 1\right] = 1 - \frac{1}{3} = \frac{2}{3} \\ \delta_B(X+Y) &= P\left[(X+Y)^{-1}(B)\right] = P\left(\frac{1}{3}, \frac{1}{2}\right] = P\left]\frac{1}{3}, \frac{1}{2}\right] = \frac{1}{2} - \frac{1}{3} = \frac{1}{6} \end{aligned}$$

Since $\delta_B(X+Y) = \frac{1}{6} < min\left\{\frac{1}{2}, \frac{2}{3}\right\} = min\{\delta_B(X), \delta_B(Y)\}$.
i.e., $\delta_B(X+Y) < min\{\delta_B(X), \delta_B(Y)\}$, this clearly shows that δ_B is not a fuzzy subgroupoid, according to Rosenfeld.

Example 6.4 Let us suppose that Ω is the collection of subgroups of G. For all $a \in G$, let us define γ_a by the setting

$$\gamma_a = \{H \in \Omega : \mathrm{a} \in \mathrm{H}\}$$

, i.e., γ_a is defined as the collection of all those subsets of Ω of which a is an element. Let B be a σ-algebra of subsets of Ω such that $\gamma_a \in B$ for all $a \in G$. We also assume that P is a probability measure on the measure space (Ω, B). Let us define a function

$$m_\Omega(a) = P(\gamma_a), \forall\, \mathrm{a} \in \mathrm{G}$$

We may interpret the number $m_\Omega(a)$ as the probability that a subgroup chosen randomly from the collection Ω will contain a as an element. In this particular example, we are unsure which "subgroup" of G has been chosen, so it is "fuzzy." According to Rosenfeld, the function m_Ω is a reasonable candidate for a fuzzy subgroup of G, if Ω is linearly ordered by set inclusion. If Ω is not linearly ordered by set inclusion, then m_Ω will not satisfy the conditions of a fuzzy subgroup of Rosenfeld.

For instance, let $(Z, +)$ be the additive group of integers. Further, we also assume that $\forall\, i \in Z^+$, let S_i be the subgroup of integers multiple of i, that is, it is a subgroup generated by i. Hence

$$S_i = \{mi : m \in Z \text{ and } i \text{ is a} + \text{ive integer}\}$$

If $\Omega = \{S_2, S_3, S_5\}$ and B is the family of sets or set of sets or power set of Ω, that is

$$B = \{\emptyset, \{S_2\}, \{S_3\}, \{S_5\}, \{S_2, S_3\}, \{S_3, S_5\}, \{S_2, S_5\}, \Omega\}$$

Let the probability measure P is defined on the measure space (Ω, B) by the setting

$$P(S_2) = P(S_3) = P(S_5) = \frac{1}{3}$$

Under the mapping, $m_\Omega : Z \to [0, 1]$, we have $m_\Omega(a) = P(\gamma_a), \forall \text{ a} \in \text{G} = P\{S \in \Omega : \text{a} \in \text{S}\}$

we have $P(\gamma_6) = P\{S \in \Omega : \sigma \in \text{S}\} = P\{S_2, S_3\}$. Then,

$$\begin{aligned} m_\Omega(6) = P(\gamma_6) &= P(S_2) + P(S_3), \text{ as } \{S_2\} \cap \{S_3\} = \phi \\ &= \frac{1}{3} + \frac{1}{3} = \frac{2}{3} \end{aligned}$$

$$\begin{aligned} m_\Omega(15) = P(\gamma_{15}) &= P\{S \in \Omega : 15 \in \text{S}\} = P\{S_3, S_5\} \\ &= P(S_3) + P(S_5), \text{ as } \{S_3\} \cap \{S_5\} = \phi \\ &= \frac{1}{3} + \frac{1}{3} = \frac{2}{3} \end{aligned}$$

$$m_\Omega(21) = P(\gamma_{21}) = P\{S \in \Omega : 21 \in \text{S}\} = P\{S_3\} = \frac{1}{3}$$

However, $m_\Omega(21) = \frac{1}{3} < min\{\frac{2}{3}, \frac{2}{3}\} = min\{m_\Omega(6), m_\Omega(15)\}$

i.e., $m_\Omega(21) < min\{m_\Omega(6), m_\Omega(15)\} \Rightarrow$ that m_Ω is not even a fuzzy subgroupoid of Z, according to the notion of Rosenfeld.

A class of semi-groups on the closed interval [0, 1] called t-norms was proposed by *Schweizer* and *Sklar* (Statistical Metric Spaces in P. J. Math., V.10 in 1960) to generalize the triangular inequality in a metric space to the more general probabilistic metric spaces. They introduced the concept of t-norm as follows:

6.6.1 T-Norm

A function $T : [0, 1] \times [0, 1] \to [0, 1], \forall \alpha, \beta, \gamma, \delta \in [0, 1]$ be called a t-norm if the following conditions are satisfied:

(i) $T(0, 0) = 0, T(\beta, 1) = \beta = T(1, \beta)$
(ii) $T(\alpha, \beta) \leq T(\gamma, \delta)$ if $\alpha \leq \gamma$ and $\beta \leq \delta$
(iii) $T(\alpha, \beta) = T(\beta, \alpha)$

(iv) $T\{\alpha, T(\beta, \gamma)\} = T\{T(\alpha, \beta), \gamma\}$

The function "min" defined on $[0, 1] \times [0, 1] \to [0, 1]$ is a t-norm. Some other t-norms that are very interesting and helpful during the study of probabilistic metric spaces are "Prod" and T_m defined by:

Prod $(\alpha, \beta) = \alpha \cdot \beta, \forall \alpha, \beta \in [0, 1]$.

and

$$T_m(\alpha, \beta) = max(\alpha + \beta - 1, 0), \forall \alpha, \beta \in [0, 1]$$

From the condition $T(0, 0) = 0$, we infer that if we have no individual information regarding α *and* β being members of the subgroupoid produces no information about $\alpha\beta$ being in the subgroupoid. From the condition $T(\beta, 1) = \beta = T(1, \beta)$, we conclude that if it is definite that one of α *and* β, say α, belongs to the subgroupoid. The amount of information derived about $\alpha\beta$ belonging to the subgroupoid is at least as much as the amount of information about β belonging to it. From the second condition, we infer that as much information we have regarding α *and* β individually being in the subgroupoid, we can derive more information regarding $\alpha\beta$ belonging to the subgroupoid. The third and fourth conditions are only required when dealing with commutative and associative subgroupoids. Keeping the above facts in their mind, **Anthony & Sherwood** introduced the notion of fuzzy subgroupoids in the following way:

6.6.2 *Fuzzy Subgroupoid (Redefined)*

Let G is a groupoid. A mapping $\lambda : G \to [0, 1]$ is a fuzzy subgroupoid of G with respect to a t-norm T, if and only if

$$\lambda(ab) \geq T\{\lambda(a), \lambda(b)\}, \forall a, b \in G$$

6.6.3 *Fuzzy Subgroup (Redefined)*

Let G is a group, then a fuzzy subgroupoid λ of G with respect to a t-norm T, is called a fuzzy subgroup of G, if and only if

$$\lambda\left(a^{-1}\right) = \lambda(a), \forall a \in G$$

Alternatively, if G is a group, then a mapping $\lambda : G \to [0, 1]$ is a fuzzy subgroup with respect to a t-norm T, if and only if for all $a, b \in G$.

(i) $\lambda(ab) \geq T\{\lambda(a), \lambda(b)\}$

(ii) $\lambda\left(a^{-1}\right) = \lambda(a)$
(iii) $\lambda(e) = 1$, e is the identity of G.

In this section, we will show some results that will exemplify the above definitions so that it is possible to eliminate the shortcomings that occurred in Examples 6.3 and 6.4, with the help of these definitions.

Theorem 6.26 The functions δ_B defined in Example 6.3. are fuzzy subgroups with respect to the t-norm,T_m.

Proof Let $\omega \in X^{-1}(B) \cap Y^{-1}(B)$. Then $\omega \in X^{-1}(B)$ and $\omega \in Y^{-1}(B)$ which implies that $X(\omega) \in B$ and $Y(\omega) \in B$. Since B is a subgroup, it is closed under addition, and so

$$(X + Y)(\omega) = X(\omega) + Y(\omega) \in B \Rightarrow \omega \in (X + Y)^{-1}(B)$$

In this way, we have that

$$\omega \in X^{-1}(B) \cap Y^{-1}(B) \Rightarrow \omega \in (X + Y)^{-1}(B) \Rightarrow X^{-1}(B) \cap Y^{-1}(B) \subset (X + Y)^{-1}(B)$$

Consequently, we will have

$$P\left[X^{-1}(B) \cap Y^{-1}(B)\right] \leq P\left[(X + Y)^{-1}(B)\right]$$

$$\begin{aligned}
\delta_B(X + Y) &= P\left[(X + Y)^{-1}(B)\right] \\
&\geq P\left[X^{-1}(B) \cap Y^{-1}(B)\right] \\
&= P\left[X^{-1}(B)\right] + P\left[Y^{-1}(B)\right] - P\left[X^{-1}(B) \cup Y^{-1}(B)\right], \text{ using the property of probability} \\
&= \delta_B(X) + \delta_B(Y) - 1
\end{aligned}$$

$\therefore \delta_B(X + Y) \geq \delta_B(X) + \delta_B(Y) - 1$, since $P\left[X^{-1}(B) \cup Y^{-1}(B)\right] = 1$.
Since $\delta_B(X + Y) \geq 0$, therefore, we at once get that

$$\begin{aligned}
\delta_B(X + Y) &\geq max\{\delta_B(X) + \delta_B(Y) - 1, 0\} \\
&= T_m\{\delta_B(X), \delta_B(Y)\}
\end{aligned}$$

i.e., $\delta_B(X + Y) \geq T_m\{\delta_B(X), \delta_B(Y)\} \Rightarrow T_m$ is a fuzzy subgroupoid of G with respect to the t-norm, T_m.

Moreover, since B is a group, therefore, if $X \in G$, then

$$\begin{aligned}
X^{-1}(B) &= \{\omega \in \Omega : X(\omega) \in B\} \\
&= \{\omega \in \Omega : -X(\omega) \in B\} \\
&= (-X)^{-1}(B)
\end{aligned}$$

$$\delta_B(X) = P\left[X^{-1}(B)\right]$$

Therefore, $= P\left[(-X)^{-1}(B)\right]$

$= \delta_B(-X)$

Moreover, hence δ_B is a fuzzy subgroup of G with respect to the t-norm T_m.

Theorem 6.27 The functions m_Ω defined in Example 6.4. are all fuzzy subgroups with respect to the t-norm, T_m.

Proof Let us assume that for each $a, b \in G$, $\gamma_a, \gamma_b \in B$.

$H \in \gamma_a \cap \gamma_b \Rightarrow H \in \gamma_a \; and \; H \in \gamma_b$

Let $\Rightarrow a \in H \; and \; b \in H$.

$\Rightarrow ab \in H \Rightarrow H \in \gamma_{ab}$

Thus, we have that $H \in \gamma_a \cap \gamma_b \Rightarrow H \in \gamma_{ab} \Rightarrow \gamma_a \cap \gamma_b \subset \gamma_{ab}$.

We have $m_\Omega(ab) \geq 0$, therefore

$$\begin{aligned} m_\Omega(ab) &= P(\gamma_{ab}) \geq P(\gamma_a \cap \gamma_b) \\ &= P(\gamma_a) + P(\gamma_b) - P(\gamma_a \cup \gamma_b) \\ &= m_\Omega(a) + m_\Omega(b) - 1 \\ &= max\{m_\Omega(a) + m_\Omega(b) - 1, 0\} \\ &= T_m\{m_\Omega(a), m_\Omega(b)\} \end{aligned}$$

$\therefore m_\Omega(ab) \geq T_m\{m_\Omega(a), m_\Omega(b)\}$.

It implies that m_Ω is a fuzzy subgroupoid of G with respect to the t-norm T_m. Further, we also observe that if $H \in \gamma_a$ then $a \in H$ and hence $a^{-1} \in H$. This implies that $H \in \gamma_{a^{-1}}$.

Thus, we have $H \in \gamma_a \Rightarrow H \in \gamma_{a^{-1}} \Rightarrow \gamma_a \subset \gamma_{a^{-1}}$.

Similarly, if $H \in \gamma_{a^{-1}}$ then $a^{-1} \in H$ and hence $\left(a^{-1}\right)^{-1} = a \in H$, which implies that $H \in \gamma_a$. Thus,

$$H \in \gamma_{a^{-1}} \Rightarrow H \in \gamma_a \Rightarrow \gamma_{a^{-1}} \subset \gamma_a$$

Consequently, $\gamma_{a^{-1}} = \gamma_a$.

Thus, we have

$$m_\Omega(a) = P\left[\gamma_a\right] = P\left[\gamma_{a^{-1}}\right] = m_\Omega\left(a^{-1}\right)$$

It follows that m_Ω is a fuzzy subgroup of G with respect to the t-norm T_m.

Theorem 6.28 The function m_Ω defined in Example 6.4. are all fuzzy subgroups with respect to the t-norm min if Ω is linearly ordered by set inclusion.

Proof Let G be a group, and let $a, b \in G$. Let $\gamma_a, \gamma_b \in \Omega$. Since Ω is linearly ordered by set inclusion, therefore, either $\gamma_a \subset \gamma_b$ or $\gamma_b \subset \gamma_a$. Without losing any generality, let us assume that $\gamma_a \subset \gamma_b$. Hence, $P(\gamma_a) \leq P(\gamma_b)$.

If $a \in H$ then $H \in \gamma_a$. But $\gamma_a \subset \gamma_b$, hence $H \in \gamma_b$, and consequently, $b \in H$. Thus, $a \in H$, $b \in H$ and H is a group, it follows that $ab \in H \Rightarrow H \in \gamma_{ab}$. In this way, we have $H \in \gamma_a \Rightarrow H \in \gamma_{ab}$. It implies that

$$\gamma_a \subset \gamma_{ab} \Rightarrow P[\gamma_{ab}] \geq P[\gamma_a] \geq min\{P(\gamma_a), P(\gamma_b)\}$$

Therefore,
$$\begin{aligned} m_\Omega(ab) &= P(\gamma_{ab}) \\ &\geq min\{P(\gamma_a),\ P(\gamma_b)\} \\ &= min\{m_\Omega(a), m_\Omega(b)\} \end{aligned}$$

$\therefore m_\Omega(ab) \geq min\{m_\Omega(a), m_\Omega(b)\} \Rightarrow m_\Omega$ is a fuzzy subgroupoid of G with respect to min.

Since $\gamma_{a^{-1}} = \gamma_a$ (Theorem 6.27), we have $m_\Omega(a) = m_\Omega(a^{-1}) \Rightarrow m_\Omega$ is a fuzzy subgroup of G with respect to min.

Most of the results proved for fuzzy subgroups with respect to "min" have slight variation for t-norms.

In order to exemplify this, some results are discussed below.

Theorem 6.29 If λ is a fuzzy subgroup of a group G with respect to a t-norm T and if $\lambda(ab^{-1}) = 1$, then $\lambda(a) = \lambda(b)$.

Proof Suppose that λ is a fuzzy subgroup of a group G with respect to a t-norm T. Hence, for all $a, b \in G$, we have

$$\lambda(ab) \geq T\{\lambda(a), \lambda(b)\} \text{ and } \lambda(a^{-1}) = \lambda(a)$$

According to the hypothesis, we have $\lambda(ab^{-1}) = 1$. Now, we have

$$\begin{aligned} \lambda(a) = \lambda(ae) = \lambda\{a(b^{-1}b)\} &= \lambda\{(ab^{-1})b\} \\ &\geq T\{\lambda(ab^{-1}), \lambda(b)\}, \text{ since } \lambda \text{ is a FSG w.r.t. T} \\ &= T\{1, \lambda(b)\} \\ &= \lambda(b), \text{ by definition of t} - \text{norm} \end{aligned}$$

$\therefore \lambda(a) \geq \lambda(b)$.

Again, we have

$$\begin{aligned} \lambda(b) = \lambda(b^{-1}) = \lambda(eb^{-1}) = \lambda\{(a^{-1}a)b^{-1}\} &= \lambda\{a^{-1}(ab^{-1})\} \\ &\geq T\{\lambda(a^{-1}), \lambda(ab^{-1})\}, \text{ since } \lambda \text{ is a FSG w.r.t. T} \\ &= T\{\lambda(a^{-1}), 1\} \\ &= T\{\lambda(a), 1\} \\ &= \lambda(a), \text{ by definition of t} - \text{norm} \end{aligned}$$

$\therefore \lambda(b) \geq \lambda(a)$.

Consequently, we will get $\lambda(a) = \lambda(b)$.

Theorem 6.30 If λ is a fuzzy subgroup of a group G with respect to a t-norm T, then $H = \{a \in G : \lambda(a) = 1\}$ is either empty or is a subgroup of G.

Proof Let λ be a fuzzy subgroup of a group G with respect to a t-norm T. Hence, we have

$$\lambda(ab) \geq T\{\lambda(a), \lambda(b)\}, \forall\, a, b \in G \text{ and } \lambda\left(a^{-1}\right) = \lambda(a), \forall\, a \in G$$

Let H be a subset of G defined by $H = \{a \in G : \lambda(a) = 1\}$. Then, we claim that H is a subgroup of G. Let $a, b \in H$, then by definition of H, we have

$$\lambda(a) = 1 \text{ and } \lambda(b) = 1$$

We have, $\lambda\left(ab^{-1}\right) \geq T\left\{\lambda(a),\ \lambda\left(b^{-1}\right)\right\} = T\{\lambda(a),\ \lambda(b)\} = T\{1, 1\} = 1$.

That is, $\lambda\left(ab^{-1}\right) \geq 1 \Rightarrow ab^{-1} \in H$. Thus, $a, b \in H \Rightarrow ab^{-1} \in H \Rightarrow H$ is a subgroup of G.

Theorem 6.31 If λ is a fuzzy set of a group G and T be any given t-norm such that $\lambda(e) = 1$ and

$\lambda\left(ab^{-1}\right) \geq T\{\lambda(a), \lambda(b)\}, \forall\, a, b \in G$, then λ is a fuzzy subgroup of G with respect to "T."

Proof Let G be a group and λ is a fuzzy subset of G, i.e., $\lambda : G \to [0, 1]$. Let T be the given t-norm such that $\lambda\left(ab^{-1}\right) \geq T\{\lambda(a), \lambda(b)\}, \forall\, a, b \in G$ and $\lambda(e) = 1$, e being the identity of G. We have

$$\begin{aligned}\lambda\left(b^{-1}\right) &= \lambda\left(eb^{-1}\right) \geq T\{\lambda(e), \lambda(b)\}, \text{ using the given condition}\\ &= T\{1, \lambda(b)\}\\ &= \lambda(b)\end{aligned}$$

$\therefore \lambda\left(b^{-1}\right) \geq \lambda(b)$.

Also, we have
$$\begin{aligned}\lambda(b) &= \lambda(eb) = \lambda\left\{e\left(b^{-1}\right)^{-1}\right\}\\ &\geq T\left\{\lambda(e), \lambda\left(b^{-1}\right)\right\}\\ &= T\left\{1, \lambda\left(b^{-1}\right)\right\}\\ &= \lambda\left(b^{-1}\right), \text{ by definition of t} - \text{norm}\end{aligned}$$

$\therefore \lambda(b) \geq \lambda\left(b^{-1}\right)$.

Consequently, we will have $\lambda\left(b^{-1}\right) = \lambda(b)$.

Moreover,

$$\begin{aligned}\lambda(ab) &= \lambda\left\{a\left(b^{-1}\right)^{-1}\right\}\\ &\geq T\left\{\lambda(a), \lambda\left(b^{-1}\right)\right\}\end{aligned}$$

$$= T\{\lambda(a), \lambda(b)\}$$

$\therefore \lambda(ab) \geq T\{\lambda(a), \lambda(b)\}$.

Hence, λ is a fuzzy subgroup of G w.r.t. t-norm "T."

Theorem 6.32 If λ is a fuzzy subgroup of a group G with respect to a t-norm T, and if there is a sequence $\{a_n\}$ in G such that $\lim_{n \to \infty} T\{\lambda(a_n), \lambda(a_n)\} = 1$ then λ(e) = 1, where e is the identity of G.

Proof Let G is a group and λ be a fuzzy subgroup of G with respect to a given t-norm T. Let a ∈ G. then $aa^{-1} = e$. Hence, we have

$$\lambda(e) = \lambda\left(aa^{-1}\right) \geq T\left\{\lambda(a), \lambda\left(a^{-1}\right)\right\}$$
$$= T\{\lambda(a), \lambda(a)\}$$

Further, there exists a sequence $\{a_n\}$ in G such that $\lim_{n \to \infty} T\{\lambda(a_n), \lambda(a_n)\} = 1$. Therefore, for all n, we will have

$$\lambda(e) \geq \lim_{n \to \infty} T\{\lambda(a_n), \lambda(a_n)\} = 1, \text{ i.e., } \lambda(e) \geq T\{\lambda(a_n), \lambda(a_n)\}$$

Since $1 \geq \lambda(e) \geq \lim_{n \to \infty} T\{\lambda(a_n), \lambda(a_n)\} = 1$

It implies that $\lambda(e) = 1$.

6.6.4 Level Subset

If λ is a fuzzy subset of G, then for any $t \in [0, 1]$, the set $\lambda_t = \{a \in G : \lambda(a) \geq t\}$ is called a level subset of the fuzzy set λ. The set λ_t is a subset of G in the classical sense. *Zadeh* was the first who introduced the notion of level subsets of a fuzzy set.

If λ is a fuzzy subgroup of G, then we have $\lambda(a) \leq \lambda(e), \forall\, a \in G$.

Theorem 6.33 If λ is a fuzzy subgroup of a group G, then the level subset λ_t, for $t \in [0, 1]$ and $\lambda(e) \geq t$ is a subgroup of G, where e is the identity of G.

Proof Let G be a group, and λ is a fuzzy subgroup of G. Let $\lambda_t = \{a \in G : \lambda(a) \geq t\}$ be the fuzzy subset of G. Obviously, λ_t is non-empty, as $e \in G$, and $\lambda(e) \geq t \Rightarrow e \in \lambda_t$.

Let $a, b \in \lambda_t \Rightarrow \lambda(a) \geq t$ and $\lambda(b) \geq t$. Since λ is a fuzzy subgroup of G, therefore

$$\lambda(ab) \geq min\{\lambda(a), \lambda(b)\} \geq min\{t, t\} = t$$

It follows that $\lambda(ab) \geq t \Rightarrow ab \in \lambda_t$.

Again $a \in \lambda_t \Rightarrow \lambda(a) \geq t$. Since λ is a fuzzy subgroup, we have $\lambda\left(a^{-1}\right) \geq \lambda(a)$ and hence $\lambda\left(a^{-1}\right) \geq \lambda(a) \geq t \Rightarrow a^{-1} \in \lambda_t$. Therefore λ_t is a fuzzy subgroup of G.

Theorem 6.34 If λ is a fuzzy subset of a group G such that λ_t is a subgroup for all $t \in [0, 1]$ and $\lambda(e) \geq t$, then λ is a fuzzy subgroup of G.

Proof Let λ be a fuzzy subset of the group G and for all $t \in [0, 1]$ and $\lambda(e) \geq t$, λ_t is a subgroup where $\lambda_t = \{a \in G : \lambda(a) \geq t\}$.Let $a, b \in G$ and let $\lambda(a) = t_1$ and $\lambda(b) = t_2$. Then $a \in \lambda_{t_1}$ and $b \in \lambda_{t_2}$. Now let us assume that $t_1 < t_2$, then it implies that $\lambda_{t_2} \subseteq \lambda_{t_1}$. Thus, $b \in \lambda_{t_2} \Rightarrow b \in \lambda_{t_1}$. Then it follows that $a, b \in \lambda_{t_1}$. By supposition, λ_t is a subgroup that is λ_{t_1} is a subgroup and hence $ab \in \lambda_{t_1}$. Therefore, we have

$$\lambda(ab) \geq t_1 = min\{t_1, t_2\} = min\{\lambda(a), \lambda(b)\}, \text{ i.e., } \lambda(ab) \geq min\{\lambda(a), \lambda(b)\}$$

Next, we assume that $a \in G$ and $\lambda(a) = t$. This gives $a \in \lambda_t$. But λ_t is a subgroup of G, hence $a \in \lambda_t \Rightarrow a^{-1} \in \lambda_t \Rightarrow \lambda\left(a^{-1}\right) \geq t = \lambda(a)$, i.e., $\lambda\left(a^{-1}\right) \geq \lambda(a)$.

Thus, λ is a fuzzy subgroup of G.

The notion of the level subgroup of a fuzzy subgroup was introduced by **P. S. Das** in 1981(Journal of Mathematical Analysis and Applications; V.84, pp. 264–269).

6.7 Level Subgroup of Fuzzy Subgroup

If λ is a fuzzy subgroup of G, then for any $t \in [0, 1]$ with $\lambda(e) \geq t$, the subgroups λ_t of G in the usual sense of Das are called level subgroups of λ.

Example 6.5 Let $G = \{e, a, b, ab\}$ be Klein's four group. Let t_i for $0 \leq i \leq 2$ be three numbers lying in the closed interval [0, 1] such that $t_0 > t_1 > t_2$. Let us define a fuzzy set by the mapping $\lambda : G \to [0, 1]$ by the setting $\lambda(e) = t_0, \lambda(a) = t_1, \lambda(b) = t_2, \lambda(ab) = t_2$. Then, it follows that λ is a fuzzy subgroup of G. In this case, we have the image set of λ as $\{t_0, t_1, t_2\}$. Therefore, the level subgroups of λ are given by

$$\lambda_{t_0} = \{e\}, \lambda_{t_1} = \{e, a\}, \lambda_{t_2} = \{e, a, b, ab\} = G$$

Now let us consider another set of three numbers s_0, s_1, and s_2 such that $s_0 > s_1 > s_2$, where $s_0, s_1, s_2 \in [0, 1]$ and that the two sets $\{s_i\}$and$\{t_i\}$ are disjoint, i.e., have empty intersection. Further, we define a mapping $\mu : G \to [0, 1]$ by the setting

$$\mu(e) = s_0, \mu(a) = s_1, \mu(b) = s_2, \ \mu(ab) = s_2$$

Again, we can see that μ is a fuzzy subgroup of G and the family of level subgroups of $\mu$are given by

$$\mu_{t_0} = \{e\}, \mu_{t_1} = \{e,\ a\},\ \mu_{t_2} = \{e,\ a,\ b,\ ab\} = G$$

In this way, we observe that the two fuzzy subgroups λ and μ have the same family of level subgroups, but λ and μ are unequal.

6.7.1 Image Set of the Fuzzy Subgroup

Let λ be a fuzzy subgroup of a group G. Then the image set of fuzzy subgroup λ consists of $\{t_0, t_1, t_2, \ldots, t_n\}$, and the family of level subgroups $\left\{\lambda_{t_i} : 0 \leq i \leq n\right\}$ constitutes the complete list of level subgroups of λ. The image set of the fuzzy subgroup on a finite group G is denoted by $Im(\lambda) = \{t_0, t_1, t_2, \ldots, t_n\}$ where $t_0 > t_1 > t_2 > \cdots > t_n$, then the level subgroups of λ form a chain

$$\lambda_{t_0} \subseteq \lambda_{t_1} \subseteq \lambda_{t_2} \subseteq \cdots \subseteq \lambda_{t_n} = G$$

where $\lambda(\mathrm{e}) = t_0$.

Theorem 6.35 If λ be a fuzzy subgroup of a group G with $t_1 < t_2$ as elements of the image set of λ, then the two level subgroups $\lambda_{t_1}$and λ_{t_2} will be equal if and only if there exists no $a \in G$ such that $t_1 < \lambda(a) < t_2$.

Proof Let λ is a fuzzy subgroup of a group G. Let t_1 and t_2 are numbers lying in the interval [0, 1] such that $t_1 < t_2$. Suppose that the two level subgroups are equal, i.e., $\lambda_{t_1} = \lambda_{t_2}$.

Now, we assume that there does not exist an element $a \in G$ such that $t_1 < \lambda(a) < t_2$, implies that $\lambda_{t_2} \subseteq \lambda_{t_1}$, since $a \in \lambda_{t_1}$, but $\notin \lambda_{t_2}$,which is against the supposition.

Conversely, we assume that $\nexists\ a \in G$ such that $t_1 < \lambda(a) < t_2$, since $t_1 < t_2$ we have $\lambda_{t_2} \subseteq \lambda_{t_1}$. Let $a \in \lambda_{t_1}$, then $\lambda(a) \geq t_1$ and hence $\lambda(a) \geq t_2$, since $\lambda(\mathrm{a})$ does not lie between $t_1$and t_2. Hence $a \in \lambda_{t_2} \Rightarrow \lambda_{t_1} \subseteq \lambda_{t_2}$ consequently, we will get $\lambda_{t_1} = \lambda_{t_2}$.

Theorem 6.36 Any subgroup H of group G can be realized as a level subgroup of some fuzzy subgroup of G.

Proof Let G be a group and H be a non-trivial subgroup of G. Obviously, H $\neq$ e or H $\neq$ G. Let us define a mapping $\lambda : G \to [0, 1]$ by

$$\lambda(a) = \begin{cases} t \ \mathit{if}\ a\ \in H \\ 0\ \mathit{if}\ a \notin H \end{cases}$$

where t is an arbitrary number in [0, 1]. Now, we claim that λ is a fuzzy subgroup of G. For, let $a, b \in G$. If $a, b \in H$ then $ab \in H$ as H is a subgroup of G. Now, from the definition of fuzzy subset λ, we have

$$\lambda(ab) = t \text{ and } \lambda(a) = \lambda(b) = t$$

Therefore, we will have

$$\lambda(ab) = t \geq min\{t, t\} = min\{\lambda(a), \lambda(b)\}$$

Since H is a subgroup of G, therefore $a \in H \Rightarrow a^{-1} \in H$. Now, we get that $\lambda(a^{-1}) = t$ and $\lambda(a) = t$. So $\lambda(a^{-1}) \geq \lambda(a)$. So $\lambda(a^{-1}) \geq \lambda(a)$.

Next, we suppose that $a \in H$ and $b \notin H$, then $ab \notin H$, this gives that $\lambda(a) = t$, $\lambda(b) = 0$

and $\lambda(ab) = 0$. Therefore,

$$\lambda(ab) = 0 \geq min\{t, 0\} = min\{\lambda(a), \lambda(b)\}$$

$\therefore \lambda(ab) \geq min\{\lambda(a), \lambda(b)\}$.

If $a \notin H$ then $b^{-1} \notin H$, then $\lambda(a) = 0$, $\lambda(a^{-1}) = 0$ and consequently $\lambda(a^{-1}) = \lambda(a)$.

Further, we assume that $a, b \notin H$ then ab may or may not belong to H. In any case, as above we will get $\lambda(ab) \geq min\{\lambda(a), \lambda(b)\}$ and $\lambda(a^{-1}) \geq \lambda(a)$.

It follows that, in all the cases, λ is a fuzzy subgroup of G, and in this case, we have $\lambda_t = H$.

6.8 Fuzzy Normal Subgroups

This section, will discuss about the concept of fuzzy normal subgroups and fuzzy cosets. Fuzzy normal subgroups and fuzzy cosets were introduced by *N. P. Mukherjee* and *Prabir Bhattacharya* in 1984. It was also studied by *Mustafa Akgul* (1988), *M. K. Singh* (2006) and others.

6.8.1 Fuzzy Normal Subgroup

A fuzzy subgroup λ of group G is called a normal fuzzy subgroup (according to *Mukherjee & Bhattacharya*) if and only if

$$\lambda(ab) = \lambda(ba), \ \forall\, a, b \in G \tag{6.19}$$

Alternatively, a fuzzy subgroup λ of a group G be called a fuzzy normal subgroup (FNSG) (according to *Akgul*) if and only if for all $a, b \in G$

$$\lambda\left(aba^{-1}\right) = \lambda(b) \text{ or } \lambda\left(a^{-1}ba\right) = \lambda(b) \tag{6.20}$$

We may (according to *Singh*) also call a fuzzy subgroup λ of a group G to be a fuzzy normal subgroup of G if and only if

$$\lambda\left(a^{-1}ba\right) \geq \lambda(b) \,\forall\, a, b \in G \tag{6.21}$$

Now, we consider the situation when λ is an ordinary subset of G. Then, we have $\lambda(b) = 1, \forall\, b \in G$. From any equations from (6.20) and (6.21), we will have $\lambda\left(a^{-1}ba\right) = 1$ as the value is never greater than 1. Therefore, $\lambda\left(a^{-1}ba\right) = 1 \Rightarrow a^{-1}ba \in \lambda$.

Thus, we conclude that $\forall\, a, b \in G \Rightarrow a^{-1}ba \in \lambda$.

From the above discussion, we infer that the notion of fuzzy normal subgroup has been considered so that when λ is considered an ordinary subset of G under the above definition, it turns out to be a normal subgroup of G.

N. P. Mukherjee & Prabir Bhattacharya (1984) and *Mustafa Akgul* (1988) proposed the notion of FNSG of a group (G, ·) Here, in place of in place of (G, ·), (G, +) (Singh) has been considered.

6.9 Notion of Fuzzy Normal Subgroup of a Group (G, +)

Let (G, +) be a group and λ be a fuzzy subset of G. Then λ be called a fuzzy normal subgroup of G if and only if the following conditions are satisfied:

(i) λ is a fuzzy subgroup of G
(ii) $\lambda(-a + b + a) \geq \lambda(b), \forall\, a, b \in G$

Alternatively, a fuzzy subgroup λ of a group (G, +) be called a fuzzy normal subgroup of (G, +) iff

$$\lambda(-a + b + a) \geq \lambda(b),\ \forall\, a, b \in G$$

We assume that λ is an ordinary subset of (G, +). Then we have $\lambda(b) = 1, \forall b \in G$.
Therefore, $\lambda(-a + b + a) \geq 1$. But $\lambda(-a + b + a) \not> 1$.
Hence $\lambda(-a + b + a) = 1 \Rightarrow (-a + b + a) \in \lambda$.
Thus, we observe that $\forall\, a, b \in \lambda \Rightarrow (-a + b + a) \in \lambda$.

So, we can say that the notion of FNSG has been considered so that when λ is considered an ordinary subset of (G, +), λ turns out to be a normal subgroup of (G, +).

Theorem 6.37 If λ is a fuzzy subgroup of a group G, then the following conditions are equivalent:

(i) $\lambda(ab) = \lambda(ba), \forall\, a, b \in G$

(ii) $\lambda(aba^{-1}) = \lambda(b), \forall\, a, b \in G$

Now, we will claim that both conditions (i) & (ii) are equivalent. In order to establish this, let us assume that $a, b \in G$.

Firstly, we establish that (i)⇒ (ii).

We have

$$\begin{aligned}\lambda(aba^{-1}) &= \lambda\{(ab)a^{-1}\}, \text{ by associativity in G} \\ &= \lambda\{a^{-1}(ab)\}, \text{ using equation (6.19)} \\ &= \lambda\{(a^{-1}a)b\}, \text{ by associativity in G} \\ &= \lambda(eb) = \lambda(b)\end{aligned}$$

Thus,$\lambda(aba^{-1}) = \lambda(b), \forall\, a, b \in G$

(ii)⇒ (i)

We have $ab = (ab)e = (ab)(aa^{-1}) = a(ba)a^{-1}$.

Therefore

$$\begin{aligned}\lambda(ab) &= \lambda\{a(ba)a^{-1}\} \\ &= \lambda(ba), \text{ as equation (6.20) holds}\end{aligned}$$

$\therefore \lambda(ab) = \lambda(ba), \forall\, a, b \in G.$

Thus, it follows that both definitions are equivalent.

A fuzzy subgroup λ of G, which satisfies the above equivalent conditions, is considered a fuzzy normal subgroup of the group G.

Theorem 6.38 If λ and μ are fuzzy normal subgroups of a group G, then $\lambda \cap \mu$ is also a fuzzy normal subgroup of G.

Proof Let G be a group, and λ *and* μ are fuzzy normal subgroups of the group G. We will claim that $\lambda \cap \mu$ is a fuzzy normal subgroup of G. Since λ and μ are fuzzy normal subgroups of G, therefore

$$\lambda(aba^{-1}) = \lambda(b), \forall\, a, b \in G$$

and $\mu(aba^{-1}) = \mu(b), \forall\, a, b \in G$

$$\begin{aligned}(\lambda \cap \mu)(aba^{-1}) &= min\{\lambda(aba^{-1}), \mu(aba^{-1})\} \\ \text{Then} \qquad &= min\{\lambda(b), \mu(b)\} \\ &= (\lambda \cap \mu)(b)\end{aligned}$$

Thus $(\lambda \cap \mu)(aba^{-1}) = (\lambda \cap \mu)(b) \Rightarrow \lambda \cap \mu$ is a fuzzy normal subgroup of the group G.

Theorem 6.39 The intersection of any family of fuzzy normal subgroups of group G is again a fuzzy normal subgroup.

Proof Let G be a group and $\{\lambda_i : i \in I\}$ be any family of fuzzy normal subgroups of G, where I is an index set such that $\forall\, i \in I$, λ_i is a fuzzy normal subgroup.

Let $\lambda = \underset{i \in I}{\cap} \lambda_i = \{a \in G : a \in \lambda_i, \forall\, i \in I\}$. Now we claim that λ is a fuzzy normal subgroup of G.

$\therefore \lambda\,\left(aba^{-1}\right) = \left[\underset{i \in I}{\cap} \lambda_i\right]\left(aba^{-1}\right) = \left[\underset{i \in I}{inf}\, \lambda_i\right]\left(aba^{-1}\right) = \underset{i \in I}{inf}\, \lambda_i(b) = \left[\underset{i \in I}{\cap} \lambda_i\right](b) = \lambda(b)$.

$\therefore \lambda\,\left(aba^{-1}\right) = \lambda(b) \Rightarrow \lambda$ is a fuzzy normal subgroup of G.

Theorem 6.40 If G is an abelian group, then every fuzzy subgroup of G is a fuzzy normal subgroup.

Proof Let λ be any fuzzy subgroup of an abelian group G. Then $\forall\, a, b \in G$, we have

$$\begin{aligned}\lambda\left(aba^{-1}\right) &= \lambda\left\{a\left(ba^{-1}\right)\right\}, \text{ by associativity in G}\\ &= \lambda\left\{a\left(a^{-1}b\right)\right\}, \text{ G is abelian} \Rightarrow ba^{-1} = a^{-1}b\\ &= \lambda\{\left(aa^{-1}\right)b)\}, \text{ by associativity in G}\\ &= \lambda\{eb\} = \lambda(b)\end{aligned}$$

$\therefore \lambda\,\left(aba^{-1}\right) = \lambda(b) \Rightarrow \lambda$ is a fuzzy normal subgroup of G.

Theorem 6.41 If f is a homomorphism of group G onto another group H, and μ is a fuzzy (normal) subgroup of group H, then $\vartheta = \mu \circ f$, the pre-image of μ is a fuzzy (normal) subgroup of G.

Proof Let G and H be two groups, and let $f : G \to H$ be a homomorphism. Let μ be a fuzzy normal subgroup of the group $f(G) = H$, and ϑ is a fuzzy set in G defined by

$$\begin{aligned}\vartheta(a) &= (\mu \circ f)(a)\\ &= \mu\{f(a)\}, \forall\, a \in G\end{aligned}$$

is called the pre-image of μ under f. We now claim that ϑ is a fuzzy normal subgroup of group G. Let $a, b \in G$. Then $f(a), f(b)$, and $f\left(a^{-1}\right) \in f(G) = H$. We have

$$\begin{aligned}\vartheta\left(aba^{-1}\right) &= (\mu \circ f)\left(aba^{-1}\right)\\ &= \mu\left\{f\left(aba^{-1}\right)\right\}\\ &= \mu\left\{f(a) \cdot f(b) \cdot f\left(a^{-1}\right)\right\}, \text{ since f is a homomorphism}\\ &= \mu\left\{f(a) \cdot f(b) \cdot (f(a))^{-1}\right\}, \text{ since f is a homomorphism}\\ &= \mu\{f(b)\}, \textit{since } \mu \text{ is a FNSG}\\ &= (\mu \circ f)(b)\\ &= \vartheta(b)\end{aligned}$$

$\therefore \vartheta\left(aba^{-1}\right) = \vartheta(b) \Rightarrow \vartheta$, the pre-image of a fuzzy normal subgroup is a fuzzy (normal) subgroup.

Alternatively,

$$\begin{aligned} \vartheta(ab) &= (\mu \circ f)(ab) \\ &= \mu\{f(ab)\} \\ &= \mu\{f(a) \cdot f(b)\}, \text{ since } f \text{ is a homomorphism} \\ &= \mu\{f(b) \cdot f(a)\}, \text{ since } f \text{ is a homomorphism} \\ &= \mu\{f(ba)\} \\ &= (\mu \circ f)(ba) \\ &= \vartheta(ba) \end{aligned}$$

$\therefore \vartheta(ab) = \vartheta(ba) \Rightarrow \vartheta$ is a fuzzy normal subgroup of G.

Theorem 6.42 A fuzzy subgroup λ of a group G is a fuzzy normal subgroup of G if and only if λ is a constant on the conjugate classes of G.

Proof Let G be a group and λ is a fuzzy subgroup of G. Let us assume that $C(a)$ be the collection of all elements conjugate to an element $a \in G$. Thus, we have

$$C(a) = \left\{x^{-1}ax : x \in G\right\}$$

We assume that λ is a fuzzy normal subgroup of G. Then

$$\lambda(xa) = \lambda(ax), \forall\, x,\ a \in G$$

Therefore, we have

$$\begin{aligned} \lambda\left(x^{-1}ax\right) &= \lambda\left\{x^{-1}(ax)\right\} \\ &= \lambda\left\{(ax)x^{-1}\right\} = \lambda\left\{a\left(xx^{-1}\right)\right\} = \lambda(ae) = \lambda(a) \end{aligned}$$

$\therefore \lambda\left(x^{-1}ax\right) = \lambda(a), \forall\, a, x \in G \Rightarrow \lambda\{C(a)\} = \lambda(a) \Rightarrow \lambda$ is constant on the conjugate classes of G.

Conversely, let us suppose that the fuzzy group λ is constant on each conjugate class of G. That is,

$$\lambda\left(x^{-1}ax\right) = \lambda(a), \forall\, a, x \in G$$

Then

$$\lambda(xa) = \lambda(xae) = \lambda\left(xa\left(xx^{-1}\right)\right) = \lambda\left\{x(ax)x^{-1}\right\} = \lambda(ax)$$

$\therefore \lambda(xa) = \lambda(ax), \forall\, x, a \in G \Rightarrow \lambda$ is a fuzzy normal subgroup of G.

We now provide another elaboration for the notion of fuzzy normal subgroups in terms of the "commutator" of a group. If G is any group and a and $x \in G$, the element $a^{-1}x^{-1}ax$ is usually referred to as the **commutator** of the ordered pair (a, x) and is denoted by the symbol $[a, x]$. If a and x commute with each other then

$$axa^{-1}x^{-1} = axx^{-1}a^{-1} = aea^{-1} = aa^{-1} = e$$

that is $[a, x] = e$. This is the reason behind this motivation.

6.10 Subgroup Generated

If H and K are two subgroups of a group G, then the subgroup $[H, K]$ is called as the subgroup generated by the elements a and x and is given by

$$\{[a, x] : a \in H, x \in K\}$$

Theorem 6.43 If λ is a fuzzy subgroup of a group G, then λ is a fuzzy normal subgroup of G if and only if $\lambda([a, x]) \geq \lambda(\mathrm{a}), \forall\, a, x \in G$.

Proof Let λ be a fuzzy subgroup of the group G. Let a *and* $x \in G$, and suppose that $[a, x]$ is the commutator of a and x in G, i.e., $[a, x] = axa^{-1}x^{-1}$.Now, we assume that λ is a fuzzy normal subgroup of G. Therefore, $\lambda\left(x^{-1}ax\right) = \lambda(a)$

$$\begin{aligned}\lambda\left(a^{-1}x^{-1}ax\right) &= \lambda\left\{a^{-1}\left(x^{-1}ax\right)\right\} \geq min\left\{\lambda\left(a^{-1}\right), \lambda\left(x^{-1}ax\right)\right\}, \text{ since } \lambda \text{ is a FSG} \\ &= min\{\lambda(a), \lambda(a)\}, \text{ since } \lambda \text{ is a FSG as well as FNSG} \\ &= \lambda(a)\end{aligned}$$

$$\therefore\ \lambda\left(a^{-1}x^{-1}ax\right) \geq \lambda(a) \Rightarrow \lambda([a, x]) \geq \lambda(a), \forall\, a, x \in G \tag{6.22}$$

Conversely, we assume that λ satisfies the relation (6.22), i.e., $\lambda([a, x]) \geq \lambda(a), \forall\, a, x \in G$

Then for all a, x in G, we have

$$\begin{aligned}\lambda\left(x^{-1}ax\right) &= \lambda\left(ex^{-1}ax\right) \\ &= \lambda\left(aa^{-1}x^{-1}ax\right) \\ &= \lambda\left(a\left(a^{-1}x^{-1}ax\right)\right) \\ &\geq min\left\{\lambda(a), \lambda\left(a^{-1}x^{-1}ax\right)\right\}, \text{ since } \lambda \text{ is a FSG} \\ &\geq min\{\lambda(a), \lambda([a, x])\}\end{aligned}$$

$$\geq min\{\lambda(a), \lambda(a)\} = \lambda(a)$$

$\therefore$

$$\lambda\left(x^{-1}ax\right) \geq \lambda(a), \forall\, a, x \in G \tag{6.23}$$

Again, we have that

$$\begin{aligned}\lambda(a) &= \lambda(eae)\\ &= \lambda\left(x^{-1}xax^{-1}x\right)\\ &= \lambda\left(x^{-1}\left(xax^{-1}\right)x\right)\\ &\geq min\left\{\lambda\left(x^{-1}\right),\ \lambda\left(xax^{-1}\right),\ \lambda(x)\right\}\\ &= min\left\{\lambda(x),\ \lambda\left(xax^{-1}\right),\ \lambda(x)\right\}\\ &= min\left\{\lambda(x),\ \lambda\left(xax^{-1}\right)\right\}\end{aligned} \tag{6.24}$$

Now if $min\left\{\lambda(x), \lambda\left(xax^{-1}\right), \lambda(x)\right\} = \lambda(x)$, then from Eq. (6.24), we get that

$$\lambda(a) \geq \lambda(x),\ \forall\, x,\ a \in G$$

It follows that λ is a constant function; hence, the result holds good in this case. Therefore, we now consider the case when $min\left\{\lambda(x),\ \lambda\left(xax^{-1}\right),\ \lambda(x)\right\} = \lambda\left(xax^{-1}\right)$. Then from Eq. (6.24), we will have

$$\lambda(a) \geq \lambda\left(xax^{-1}\right),\ \forall\, x,\ a \in G$$

On combining the inequalities (6.23) and (6.24), we get that

$$\lambda\left(x^{-1}ax\right) = \lambda(a), \forall\, a,\ x \in G$$

It follows that λ is constant on the conjugate classes of G.

Theorem 6.44 If λ is a fuzzy normal subgroup of a group G and $t \in [0, 1]$ such that $\lambda(e) \geq t$, where e is the identity of G, then the set $\lambda_t = \{a \in G : \lambda(a) \geq t\}$ is a normal subgroup of G.

Proof Let G is a group, and λ is a fuzzy normal subgroup of G. Let t be any number in the closed interval [0, 1], and is such that $\lambda(e) \geq t$. Let us define a set called the level subset of the fuzzy normal subgroup λ, and is defined by

$$\lambda_t = \{a \in G : \lambda(a) \geq t\}$$

Then, from theorem 6.34, λ_t is a subgroup of the group G in the sense of Das. We now claim that λ_t is a normal subgroup of G. In order to show this, let $x \in G$ *and* $a \in \lambda_t$. Since λ is a fuzzy normal subgroup of G, we have

$$\lambda\left(x^{-1}ax\right) \geq \lambda(a), \forall\, a,\ x \in G$$

So we get that

$$\lambda\left(x^{-1}ax\right) \geq t \Rightarrow x^{-1}ax\ \in \lambda_t$$

Thus, we have that $x^{-1}ax \in \lambda_t, \forall\, a, x \in G \Rightarrow \lambda_t$ is a normal subgroup of G, i.e., $\lambda_t \lhd G$. That is, the level subsets of a fuzzy normal subgroup are normal subgroups of G.

If λ is a fuzzy normal subgroup of a finite group G having $\{t_0, t_1, t_2, \ldots, t_r\}$ as its image set, where $t_0 > t_1 > t_2 > \cdots > t_r$, then it follows from the above theorem (6.43) that the level subgroups of λ form a chain of normal subgroups, i.e.,

$$\lambda_{t_0} < \lambda_{t_1} < \lambda_{t_2} < \cdots < \lambda_{t_r}$$

Theorem 6.45 If λ is a subgroup of a finite group G such that all the level subgroups of λ are normal in G, then λ is a fuzzy normal subgroup.

Proof Let G be a finite group, and λ is a fuzzy subgroup of G. Then the image set of λ, i.e., $Im(\lambda) = \{t_0, t_1, t_2, \ldots, t_r\}$ be finite as G is finite where $t_0 > t_1 > t_2 > \cdots > t_r$, then the complete set of level subgroups of G is the family given by

$$\left\{\lambda_{t_i} : 0 \leq i \leq r\right\}$$

According to our assumption, each λ_{t_i} is a normal subgroup of G. From the definition of the level subgroups; it is now apparent that

$$\lambda_{t_i} \backslash \lambda_{t_{i-1}} = \{a \in G : \lambda(a) = t_i\} \tag{6.25}$$

From the fundamental result of abstract algebra, we know that a normal subgroup of a group is a complete union of conjugate classes of G. In consequence, it follows that in the chain of normal subgroups $\lambda_{t_0} < \lambda_{t_1} < \lambda_{t_2} < \cdots < \lambda_{t_r}$, each $\lambda_{t_i} \backslash \lambda_{t_{i-1}}$ is a union of some conjugate classes of G. By Eq. (6.25), we get that λ is constant on $\lambda_{t_i} \backslash \lambda_{t_{i-1}}$, it follows that λ is constant on each conjugate classes of G. Hence by Theorem 6.46, we have that λ is a fuzzy normal subgroup.

Example 6.6 Let us consider a square whose vertices are designated by 1, 2, 3, and 4 as in the following figure.

Now we will consider all rigid motions of the square into itself. The square may be thought of as being made of some rigid material such as cardboard. These rigid motions are of the following types (Fig. 6.1).

(i) The four counter-clockwise rotations about the center O of the square (in the plane of the square) through 0^0, 90^0, 180^0, 270^0, respectively.

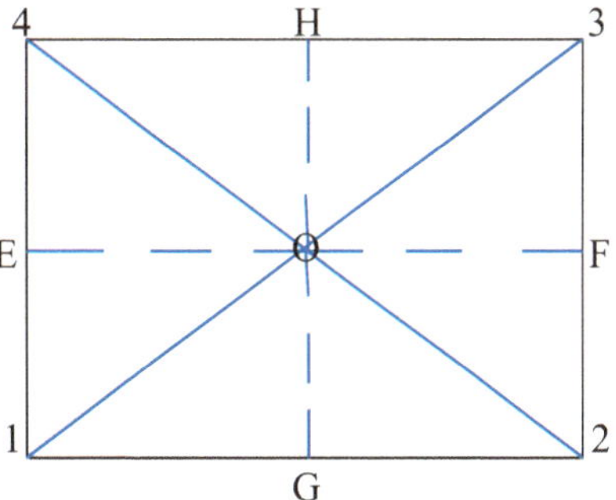

Fig. 6.1 Motions of the square

(ii) The two rotations in space through 180^0 about the diagonals 13 and 24 and the two rotations in space through 180^0 about the lines EF and GH (here, E, F, G, and H are the middle points of the sides shown in the figure). Any transformation of the above type is termed a rigid motion or symmetry of the square. The composition of symmetry is symmetry. All symmetry determines a permutation of the vertices 1, 2, 3, and 4. The four permutations corresponding to type (i) rotations are e, a, a^2 and a^3 and are given by

$$e = \begin{pmatrix} 1\,2\,3\,4 \\ 1\,2\,3\,4 \end{pmatrix},\ a = \begin{pmatrix} 1\,2\,3\,4 \\ 2\,3\,4\,1 \end{pmatrix}$$

$$a^2 = \begin{pmatrix} 1\,2\,3\,4 \\ 3\,4\,1\,2 \end{pmatrix} \text{and } a^3 = \begin{pmatrix} 1\,2\,3\,4 \\ 4\,1\,2\,3 \end{pmatrix}, \text{ respectively.}$$

The four permutations corresponding to type (ii) rotations are d, k, b and c and are given by

$$d = \begin{pmatrix} 1\,2\,3\,4 \\ 1\,4\,3\,2 \end{pmatrix},\ k = \begin{pmatrix} 1\,2\,3\,4 \\ 3\,2\,1\,4 \end{pmatrix}$$

$$b = \begin{pmatrix} 1\,2\,3\,4 \\ 4\,3\,2\,1 \end{pmatrix} \text{and } c = \begin{pmatrix} 1\,2\,3\,4 \\ 2\,1\,4\,3 \end{pmatrix}, \text{ respectively.}$$

Let G be the group of all symmetries of the square, i.e., $G = \{e, a, a^2, a^3, b, c, d, k\}$. Then G is a group of order eight generated by a rotation through $\pi/2$ and a reflection along a square's diagonal, where e is the identity, a is a rotation through $\pi/2$,and b is a reflection along a diagonal. The multiplication table for G is as shown in the following **Table 6.1.**

This particular group with eight elements is usually referred to as dihedral or the octic group. It is denoted by D_4. Since we have obtained this group by considering the rigid motions of the square, we may also call it the **group of rigid motions of**

Table 6.1 Cayley's Table for group of symmetries of the square

·	e	a	a^2	a^3	b	c	d	k
e	e	a	a^2	a^3	b	c	d	k
a	a	a^2	a^3	e	k	d	b	c
a^2	a^2	a^3	e	a	c	b	d	
a^3	a^3	e	a	a^2	d	k	c	b
b	b	d	c	k	e	a^2	a	a^3
c	c	k	b	d	a^2	e	a^3	a
d	d	c	k	b	a^3	a	e	a^2
k	k	b	d	c	a	a^3	a^2	e

the square or the *group of symmetries of a square.* This group is non-abelian as ab = k and ba = d, thus ba ≠ ab. We can easily verify that the conjugate classes of G are

$$\{e\},\ \{a^2\},\ \{b, d\},\ \{c, k\},\ \{a, a^3\}$$

Now let us assume that $H = \{e, a^2\}$ and $K = \{e, a, a^2, a^3\}$.

Then we find that H and K are normal subgroups of G. Thus, we have a chain of normal subgroups given by

$$\{e\} \subset H \subset K \subset G \tag{6.26}$$

Next, we will construct the fuzzy subgroup of G, whose level subgroups are precisely chain members given by (6.26). In order to do this, let us choose $t_i \in [0, 1]$, where $0 \le i \le 3$ such that $t_0 > t_1 > t_2 > t_3$, and define a mapping $\lambda : G \to [0, 1]$ such that

$$\lambda(e) = t_0,\ \lambda(H\backslash\{e\}) = t_1,\ \lambda(K\backslash H) = t_2, \lambda(G\backslash K) = t_3$$

From the definition of fuzzy set λ, it is evident that $\lambda(a) = \lambda(a^{-1}), \forall\, a \in G$. We can also easily verify that

$$\lambda(ab) \ge min\{\lambda(a),\ \lambda(b),\ \forall\, a, b \in G$$

Thus, it follows that λ is a fuzzy subgroup whose level subgroups are the subgroups in the chain given by (6.26). Further, we also observe that λ is constant on the conjugate classes of G, and hence by virtue of Theorem 6.46, λ is a fuzzy normal subgroup.

The notion of fuzzy cosets was introduced by ***Mukherjee and Bhattacharya*** in 1984.

6.11 Fuzzy Cosets

Let G be a finite group, and λ is a fuzzy subgroup of G. Then a mapping $\lambda : G \rightarrow [0, 1]$ defined as

$$\lambda_a(g) = \lambda\left(ga^{-1}\right), \forall\, g \in G \tag{6.27}$$

is called the "fuzzy coset" or *"fuzzy right coset"* of G determined by a and g.

Similarly,

$$a^{\lambda}(g) = \lambda\left(a^{-1}g\right), \forall\, g \in G \tag{6.28}$$

is called the "fuzzy coset" or *"fuzzy left coset"* of G determined by a and g.

Note 2: If we use the notation similar to group theory, then we may write Eqs. (6.27) & (6.28) as follows:

$$a^{\lambda}(g) = (a\lambda)(g) = \lambda\left(a^{-1}g\right),\ \forall\, g \in G$$

and

$$\lambda_a(g) = (\lambda a)(g) = \lambda\left(ga^{-1}\right), \forall\, g \in G$$

and will call $(a\lambda)$ as **left fuzzy coset** and (λa) as **right fuzzy coset**.

Now, we will see that the notion of fuzzy cosets is an extension of the notion of cosets of a group.

For this, let us assume that λ is the characteristic function of a subgroup H of a group G. That is,

$$H = \{a \in G : \lambda(a) = 1\} \text{ and } G \backslash H = \{a \in G : \lambda(a) = 0\}$$

We observe that for any $a \in G$, we have $Ga = G$. Now if $g \in H$ then

$$\lambda_a(ga) = \lambda\left(gaa^{-1}\right) = \lambda(ge) = \lambda(g) = 1$$

If $g \notin H$ then $ga \notin Ha$, and so

$$\lambda_a(ga) = \lambda\left(gaa^{-1}\right) = \lambda(ge) = \lambda(g) = 0$$

Thus, from the above discussion, it follows that

$$\lambda/Ha = 1 \text{ and } \lambda/G\backslash Ha = 0$$

It follows that λ is the characteristic function of Ha.

Group theory shows that two left (right) cosets of a subgroup H of a group G are either equal or disjoint. The following example shows that this fact is not valid in the case of fuzzy cosets.

Example 6.7 Let $G = \{e = a, b, c, d\}$ be the *Kline's* four group and λ is a fuzzy subgroup of G under "min," i.e., according to the definition of **Rosenfeld** and is defined by $\lambda(a) = 1, \lambda(b) = t_1, \lambda(c) = \lambda(d) = t_2$ such that $1 > t_1 \geq t_2 (t_2 \neq 0)$. Then $b\lambda \neq c\lambda$.

Theorem 6.46 If λ is a fuzzy normal subgroup of a group G, then for all $a, b \in G$ we have that.

$$\lambda_a(ag) = \lambda_a(ga) = \lambda(g),\ \forall\, g \in G$$

Proof Let λ be a fuzzy normal subgroup of group G. Let us define a map λ_a from G to [0, 1], *that is* $\lambda_a : G \rightarrow\ [0, 1]$ such that

$$\lambda_a(g) = \lambda\left(ga^{-1}\right),\ \forall\, g \in G$$

Let $a \in G, g \in G \Rightarrow ag \in G$. Therefore, we have

$$\begin{aligned}
\lambda_a(ag) &= \lambda_a(age) \\
&= \lambda_a\left(aga^{-1}a\right) \\
&= \lambda\left(aga^{-1}aa^{-1}\right), \text{ using the definition of } \lambda_a \\
&= \lambda\left(aga^{-1}e\right) \\
&= \lambda\left(aga^{-1}\right) \\
&= \lambda(a), \text{ since } \lambda \text{ is a FNSG}
\end{aligned}$$

Again, we have

$$\begin{aligned}
\lambda_a(ga) &= \lambda_a(ega) \\
&= \lambda_a\left(aa^{-1}ga\right) \\
&= \lambda\left(a^{-1}aa^{-1}ga\right), \text{ using equation (6.27)} \\
&= \lambda\left(ea^{-1}ga\right) \\
&= \lambda\left(a^{-1}ga\right) \\
&= \lambda(a), \text{ since } \lambda \text{ is a FNSG}
\end{aligned}$$

Thus, we have

$$\lambda_a(ag) = \lambda_a(ga) = \lambda(a),\ \forall\, g \in G$$

Remark 6.2 From group theory, we know that if G is a group and H is a normal subgroup of G, then for all $a \in G$, we have $Ha = aH$. In other words, if $H \triangleleft G$, *then* $Ha = aH, \forall\, a \in G$. Theorem 6.46 is analogous to this result of group theory.

Again, we have the result that if H is a normal subgroup of a group G, then the set G/H of all cosets of H in G is a group with respect to the multiplication of cosets. This group is called the quotient group or factor group of G by H. Next result, which we will discuss, is the analogy for the fuzzy normal subgroup.

Theorem 6.47 If λ is a fuzzy normal subgroup of a group G and if $\mathcal{F}$ is the set of all fuzzy cosets of λ in G, then $\mathcal{F}$ is a group under the composition of fuzzy coset multiplication in G. If there exists a mapping $\overline{\lambda} : \mathcal{F} \to [0, 1]$ defined by $\overline{\lambda}(\lambda_a) = \lambda(a) \forall\, a \in G$, then $\overline{\lambda}$ is a fuzzy subgroup on $\mathcal{F}$.

Proof Suppose that G is a group and λ is a fuzzy normal subgroup of G. Define a mapping $\lambda : G \to [0, 1]$ by

$$\lambda_a(g) = \lambda\left(ga^{-1}\right),\ \forall\, g \in G$$

Then λ_a is the "fuzzy coset" of G determined by a and g. Let $\mathcal{F}$ is the set of all fuzzy cosets of λ. That is $\mathcal{F} = \{\lambda_a : a \in G\}$. Let $a,\ b \in G \Rightarrow \lambda_a,\ \lambda_b \in \mathcal{F}$. Let us define a composition ∘ on $\mathcal{F}$ by the setting

$$\lambda_a \circ \lambda_b = \lambda_{ab}, \forall\, a, b \in G \tag{6.29}$$

Now, we will show that the mapping given by (6.27) is well-defined. Let $a, b, a_0, b_0 \in G$ such that

$$\lambda_a = \lambda_{a_0} \text{ and } \lambda_b = \lambda_{b_0} \tag{6.30}$$

We have, from Eq. (6.30)

$$\begin{aligned}(\lambda_a \circ \lambda_b)(g) &= \lambda_{ab}(g)\\ &= \lambda\left(g(ab)^{-1}\right),\ \text{by definition of fuzzy coset}\\ &= \lambda\left(gb^{-1}a^{-1}\right),\ \forall\, g \in G\end{aligned} \tag{6.31}$$

and

$$\begin{aligned}\left(\lambda_{a_0} \circ \lambda_{b_0}\right)(g) &= \lambda_{a_0b_0}(g)\\ &= \lambda\left(g(a_0b_0)^{-1}\right)\\ &= \lambda\left(gb_0^{-1}a_0^{-1}\right),\ \forall\, g \in G\end{aligned} \tag{6.32}$$

Now, we have

$$\begin{aligned}\lambda(gb^{-1}a^{-1}) &= \lambda(geb^{-1}a^{-1}) \\ &= \lambda(gb_0^{-1}b_0b^{-1}a^{-1}) \\ &= \lambda(gb_0^{-1}b_0eb^{-1}a^{-1}) \\ &= \lambda((gb_0^{-1}a_0^{-1})(a_0b_0b^{-1}a^{-1})) \\ &\geq min\{\lambda(gb_0^{-1}a_0^{-1}), \lambda(a_0b_0b^{-1}a^{-1})\}, \text{ since } \lambda \text{ is a FSG}\end{aligned} \tag{6.33}$$

From Eq. (6.30), we have that

$$\begin{aligned}&\lambda_a = \lambda_{a_0} \text{ and } \lambda_b = \lambda_{b_0}, \forall\, g \in G \\ &\Rightarrow \lambda_a(g) = \lambda_{a_0}(g) \text{ and } \lambda_b(g) = \lambda_{b_0}(g), \forall\, g \in G \\ &\Rightarrow \lambda(ga^{-1}) = \lambda(ga_0^{-1}) \text{ and } \lambda(gb^{-1}) = \lambda(gb_0^{-1}), \forall\, g \in G\end{aligned} \tag{6.34}$$

On substituting $g = a_0b_0b^{-1}$ in Eq. (6.34), we get that

$$\begin{aligned}\lambda(a_0b_0b^{-1}a^{-1}) &= \lambda(a_0b_0b^{-1}a_0^{-1}) \\ &= \lambda\{a_0(b_0b^{-1})a_0^{-1}\} \\ &= \lambda\{b_0b^{-1}\}, \text{ since } \lambda \text{ is a FNSG}\end{aligned} \tag{6.35}$$

and on replacing g by b_0 in $\lambda(gb^{-1}) = \lambda(gb_0^{-1})$ of Eq. (6.34), we get

$$\lambda(b_0b^{-1}) = \lambda(b_0b_0^{-1}) = \lambda(e)$$

$\therefore$

$$\lambda(a_0b_0b^{-1}a^{-1}) = \lambda\{b_0b^{-1}\} = \lambda(e), \text{ using equation (6.35)} \tag{6.36}$$

We know that, if λ is a fuzzy group of a group G having e as its identity, then $\lambda(e) \geq \lambda(a), \forall\, a \in G$. Further, we have that $g, b_0^{-1}, a_0^{-1} \in G \Rightarrow g.b_0^{-1}.a_0^{-1} \in G$, and consequently, we will have

$$\lambda(e) \geq \lambda(gb_0^{-1}a_0^{-1}) \tag{6.37}$$

Therefore, from Eq. (6.33), we get

$$\begin{aligned}\lambda(gb^{-1}a^{-1}) &\geq min\{\lambda(gb_0^{-1}a_0^{-1}), \lambda(a_0b_0b^{-1}a^{-1})\} \\ &\geq min\{\lambda(gb_0^{-1}a_0^{-1}), \lambda(e)\}, \text{ using equation (6.36)} \\ &= \lambda(gb_0^{-1}a_0^{-1}), \text{ using equation (6.37)}\end{aligned} \tag{6.38}$$

By similar arguments as discussed above, we will have

$$\lambda(gb_0^{-1}a_0^{-1}) \geq \lambda(gb^{-1}a^{-1}) \tag{6.39}$$

Thus, from Eqs. (6.37) and (6.38), we have

$$\begin{aligned}\lambda\left(gb_0^{-1}a_0^{-1}\right) = \lambda\left(gb^{-1}a^{-1}\right) &\Rightarrow \lambda_{a_0b_0}(g) = \lambda_{ab}(g)\\ &\Rightarrow \lambda_{a_0b_0} = \lambda_{ab}\\ &\Rightarrow \text{The composition is well defined.}\end{aligned}$$

Next, we will show that group postulates are satisfied in $\mathcal{F}$.

Closure: Since $\lambda_a \circ \lambda_b = \lambda_{ab}, \forall\, a\,, b \in G \Rightarrow ab \in G \Rightarrow \lambda_{ab} \in \mathcal{F} \Rightarrow \mathcal{F}$ is closed w. r. t. $\circ$.

Associative: Let $a,\ b,\ c \in G \Rightarrow \lambda_a,\ \lambda_b,\ \lambda_c \in \mathcal{F}$.

We have

$$\lambda_a \circ (\lambda_b \circ \lambda_c) = \lambda_a \circ \lambda_{bc} = \lambda_{a(bc)} = \lambda_{(ab)c} = \lambda_{ab} \circ \lambda_c = (\lambda_a \circ \lambda_b) \circ \lambda_c$$

$\Rightarrow$ the composition $\circ$ is associative in $\mathcal{F}$.

Existence of Identity: Let $a \in G$ and e be the identity of G. Therefore, $\lambda_a,\ \lambda_e \in \mathcal{F}$, we have

$$\lambda_a \circ \lambda_e = \lambda_{ae} = \lambda_a$$

and

$$\lambda_e \circ \lambda_a = \lambda_{ea} = \lambda_a$$

$\therefore\ \lambda_e$ is the identity of $\mathcal{F}$.

Existence of Inverse: Let a^{-1} is the inverse of $a \in G$. Hence, $\lambda_a\,,\, \lambda_{a^{-1}} \in \mathcal{F}$. Then we have

$$\lambda_a \circ \lambda_{a^{-1}} = \lambda_{aa^{-1}} = \lambda_e$$

$$\lambda_{a^{-1}} \circ \lambda_a = \lambda_{a^{-1}a} = \lambda_e$$

It shows that $\lambda_{a^{-1}}$ is the inverse of λ_a in $\mathcal{F}$. Hence it follows that $\mathcal{F}$ is a group.

Let us define a mapping $\overline{\lambda} : \mathcal{F} \to [0, 1]$ by the setting

$$\overline{\lambda}(\lambda_a) = \lambda(a) \forall\, a \in G \tag{6.40}$$

Then we claim that $\overline{\lambda}$ is a fuzzy subgroup on $\mathcal{F}$.

Let $a, b \in G$. Then we have that

$$\begin{aligned}\overline{\lambda}(\lambda_a \circ \lambda_b) = \overline{\lambda}(\lambda_{ab}) &= \lambda(ab),\ \text{using equation (6.40)}\\ &\geq min\{\lambda(a), \lambda(b)\},\ \text{since } \lambda \text{ is a FSG}\\ &\geq min\left\{\overline{\lambda}(\lambda_a), \overline{\lambda}(\lambda_b)\right\},\ \text{using equation (6.40)}\end{aligned}$$

Further we have that

$$\begin{aligned}\overline{\lambda}(\lambda_{a^{-1}}) &= \lambda\left(a^{-1}\right)\\ &= \lambda(a),\ \text{since}\ \lambda\ \text{is a FSG}\\ &= \overline{\lambda}(\lambda_a)\end{aligned}$$

Hence $\overline{\lambda}$ is a fuzzy subgroup on $\mathcal{F}$.

6.12 Fuzzy Quotient Group

Let λ be a fuzzy normal subgroup of a group G. Then the mapping $\overline{\lambda}$ is defined from the set of all fuzzy cosets of λ to [0, 1], i.e., $\overline{\lambda} : \mathcal{F} \to [0, 1]$ defined by $\overline{\lambda}(\lambda_a) = \lambda(a) \forall\, a \in G$ is called the fuzzy quotient group determined by λ.

Theorem 6.48 If λ is a fuzzy normal subgroup of G and $\mathcal{F}$ is the set of all fuzzy cosets of λ, then $\mathcal{F}$ is a group under the composition of fuzzy cosets multiplication. If θ is a mapping from G to $\mathcal{F}$, i.e., $\theta : G \to \mathcal{F}$ defined as.

$$\theta(a) = \lambda_a \forall\, a \in G$$

Then θ is a homomorphism with kernel G_λ, given by

$$G_\lambda = \{a \in G : \lambda(a) = \lambda(e)\}$$

where e is the identity of G.

Proof Let G be a group and λ is a fuzzy normal subgroup of G. Let $\mathcal{F}$ be the set of all fuzzy cosets of λ in G, i.e.,

$$\mathcal{F} = \{\lambda_a : a \in G\}$$

Then $(\mathcal{F}, \circ)$ forms a group under the composition $\circ$ defined as

$$\lambda_a \circ \lambda_b = \lambda_{ab} \forall\, a,\, b \in G$$

Let us define a mapping $\theta : G \to \mathcal{F}$ as follows:

$$\theta(a) = \lambda_a \forall\, a \in G$$

Let $a, b \in G$. Then

$$\begin{aligned}\theta(ab) &= \lambda_{ab}\\ &= \lambda_a \circ \lambda_b\\ &= \theta(a) \circ \theta(b)\end{aligned}$$

Hence θ is a homomorphism. Further, from group theory, we know that the kernel of a homomorphism θ of a group G into a group $\mathcal{F}$ is defined as the set G_λ of all those elements of G which are mapped onto the identity λ_e of $\mathcal{F}$, i.e.,

$$G_\lambda = \{a \in G : \theta(a) = \lambda_e\}$$

We have

$$\theta(a) = \lambda_a$$

$\therefore G_\lambda = \{a \in G : \lambda_a = \lambda_e\}$.
From the definition of fuzzy coset, we have

$$\begin{aligned}\lambda_a(a) &= \lambda\left(aa^{-1}\right) \forall\, a \in G \\ &= \lambda(e)\end{aligned}$$

and

$$\begin{aligned}\lambda_e(a) &= \lambda\left(ae^{-1}\right) \forall\, a \in G \\ &= \lambda(ae) = \lambda(a)\end{aligned}$$

$\therefore \lambda_a = \lambda_e \Rightarrow \lambda_a(a) = \lambda_e(a) \Rightarrow \lambda(e) = \lambda(a)$.
$\therefore G_\lambda = \{a \in G : \lambda(a) = \lambda(e)\}$.

Corollary 6.1 If λ is a fuzzy normal subgroup of a group G, then the set $G/\lambda = \{a\lambda : a \in G\}$ is a group with respect to the operation defined as $(a\lambda)(b\lambda) = (ab\lambda)$.

Now we wish to discuss the fuzzy version of a significant result known as the "**fundamental theorem of homomorphism of groups.**"

Theorem 6.49 If λ is a fuzzy normal subgroup of a group G and $\mathcal{F}$ is the collection of all fuzzy cosets of λ, then each fuzzy normal subgroup of $\mathcal{F}$ corresponds naturally to a fuzzy normal subgroup of G.

Proof Let G be a group and λ is a fuzzy normal subgroup of G. Let $\mathcal{F}$ be the set of all fuzzy cosets of λ, i.e.,

$$\mathcal{F} = \{\lambda_a : a \in G\}$$

Suppose that λ^* is a fuzzy normal subgroup of $\mathcal{F}$. Let us define a mapping μ from G to the closed interval [0, 1], i.e., $\mu : G \to [0, 1]$ as follows

$$\mu(a) = \lambda^*(\lambda_a) \forall\, a \in G, \text{ where } \lambda_a \text{ is a fuzzy coset of } \lambda \tag{6.41}$$

We now show that μ is a fuzzy subgroup of G. Let $a, b \in G$. Then, we have

$$\mu(ab) = \lambda^*(\lambda_{ab}) = \lambda^*(\lambda_a \circ \lambda_b) = \lambda^*(\lambda_a\lambda_b)$$

$$\geq min\{\lambda^*(\lambda_a), \lambda^*(\lambda_b)\}, \text{ since } \lambda^* \text{ is a FSG}$$
$$= min\{\mu(a), \mu(b)\}, \text{ using equation (6.41)}$$

Also, we have that

$$\mu(a) = \lambda^*(\lambda_a) = \lambda^*(\lambda_{a^{-1}}), \text{ since } \lambda^* \text{ is a FSG}$$
$$= \mu(a^{-1}), \text{ using equation (6.41)}$$

Thus μ is a fuzzy subgroup of G.
Further, we have

$$\begin{aligned}\mu(ab) &= \lambda^*(\lambda_{ab}) = \lambda^*(\lambda_a \circ \lambda_b)\\ &= \lambda^*(\lambda_b \circ \lambda_a)\\ &= \lambda^*(\lambda_{ba})\\ &= \mu(ba)\end{aligned}$$

Hence μ is a fuzzy normal subgroup.

6.13 Cardinality of the Set of All Fuzzy Cosets

Let λ be a fuzzy subgroup of a finite group G and $\mathcal{F}$ be the set of all fuzzy cosets of λ. Then the number of distinct fuzzy cosets in $\mathcal{F}$ is known as $\mathcal{F}'s$ cardinality, and $\mathcal{F}'s$ cardinality is called the **index of the fuzzy subgroup** λ.

If λ is a fuzzy subgroup of G, then the cardinality of $^{G}/_{\lambda}$ is called the index of λ in G and is denoted by $[G : \lambda]$.

Finally, this chapter will explain the fuzzy version of the famous Lagrange's theorem for finite groups.

Fuzzy Lagrange's Theorem

Theorem 6.50 The index of each fuzzy subgroup λ of a group G divides the order of the group G.

Proof Let λ be a fuzzy subgroup of a finite group G and $\mathcal{F}$ is the set of all fuzzy cosets of λ in G. Then

$$\mathcal{F} = \{\lambda_a : a \in G\}$$

where $\lambda_a(g) = \lambda(ga^{-1}) \forall\, a \in G$.

Since G is finite, it implies that $\mathcal{F}$ will also be a finite set. Then there exists a homomorphism θ from G to $\mathcal{F}$, i.e., $\theta : G \to \mathcal{F}$ defined as

$$\theta(a) = \lambda_a \,\forall\, a \in G \tag{6.42}$$

Let H be the subgroup of the group G defined by

$$H = \{h \in G : \lambda_h = \lambda_e\} \tag{6.43}$$

where e denotes the identity of G. Let $g \in G$ and $h \in H$. Then

$$\begin{aligned} \lambda_h(g) &= \lambda_e(g) \\ \Rightarrow \lambda\left(gh^{-1}\right) &= \lambda\left(ge^{-1}\right) \\ &= \lambda(ge) \\ &= \lambda(g) \end{aligned}$$

$\therefore \lambda\left(gh^{-1}\right) = \lambda(g)$.
On replacing g by e, we get

$$\begin{aligned} \lambda\left(eh^{-1}\right) &= \lambda(e) \\ \Rightarrow \lambda\left(h^{-1}\right) &= \lambda(e) \\ \Rightarrow \lambda(h) &= \lambda(e), \text{ since } \lambda \text{ is a FSG} \end{aligned}$$

On the other hand, if $\lambda(h) = \lambda(e)$.
Then, we have

$$\begin{aligned} \lambda\left(h^{-1}\right) &= \lambda(e) \\ \Rightarrow \lambda\left(eh^{-1}\right) &= \lambda(e) \end{aligned}$$

On replacing e by g, we get

$$\begin{aligned} \lambda\left(gh^{-1}\right) &= \lambda(g) \\ \Rightarrow \lambda\left(gh^{-1}\right) &= \lambda(ge) \\ \Rightarrow \lambda\left(gh^{-1}\right) &= \lambda\left(ge^{-1}\right) \\ \Rightarrow \lambda_h(g) &= \lambda_e(g) \\ \Rightarrow \lambda_h &= \lambda_e \end{aligned}$$

Thus, it follows that

$$H = \{h \in G : \lambda_h = \lambda_e\}$$

From group theory, we know that if H is a subgroup of a group G, then G is equal to the disjoint union of the distinct cosets of H in G. That is, we can decompose G as a disjoint union of the cosets of G w. r. t. H. Thus, we will have

$$G = Ha_1 \dot{\bigcup} Ha_2 \dot{\bigcup} Ha_3 \dot{\bigcup} \cdots \dot{\bigcup} Ha_k \tag{6.44}$$

where $Ha_1 = H$. Then we shall show that corresponding to each coset Ha_1 given in Eq. (6.44), there exists a fuzzy coset in $\mathcal{F}$, and this correspondence is one-to-one. In order to verify this, let us consider any coset Ha_i. For any $h \in H$, we have that

$$\begin{aligned}\theta(ha_i) = \lambda_{ha_i} &= \lambda_h \lambda_{a_i}, \text{ using equation (6.42)} \\ &= \lambda_e \lambda_{a_i}, \text{ using equation (6.43)} \\ &= \lambda_{a_i}, \text{ since } \lambda_e \text{ is the identity of } \mathcal{F}\end{aligned}$$

Thus θ maps each element of Ha_i into the fuzzy coset λ_{a_i}. We now wish to establish a natural correspondence $\overline{\theta}$ between the set $\{Ha_i : 1 \le i \le k\}$ and the set $\mathcal{F}$ by the setting,

$$\overline{\theta}(Ha_i) = \lambda_{a_i}, 1 \le i \le k$$

Then we claim that the mapping $\overline{\theta}$ is one–one. For if

$$\overline{\theta}(Ha_i) = \overline{\theta}(Ha_j) \Rightarrow \lambda_{a_i} = \lambda_{a_j}$$

Operating both sides by $\lambda_{a_j^{-1}}$, we get

$$\Rightarrow \lambda_{a_i} \lambda_{a_j^{-1}} = \lambda_{a_j} \lambda_{a_j^{-1}}$$

$$\Rightarrow \lambda_{a_i} \lambda_{a_j^{-1}} = \lambda_{a_j . a_j^{-1}}$$

$$\Rightarrow \lambda_{a_i . a_j^{-1}} = \lambda_e$$

$$\Rightarrow a_i . a_j^{-1} \in H, \text{ by the definition of H given by equation (6.43)}$$

$$\Rightarrow Ha_i = Ha_j, \text{ since H is a subgroup of G}$$

$\Rightarrow \theta$ is a one–one mapping.

It is now apparent from the above discussion that the number of distinct cosets of H in G, known as an index of H in G, is a divisor of the group's order. It follows that index of λ also divides the order of G.

Exercise-6

1. Explain the notion of fuzzy subgroupoid as proposed by Rosenfeld and give a suitable example.
2. Explain the concept of fuzzy left ideal and fuzzy right ideal. Also, define fuzzy ideal.

3. If λ is a fuzzy subgroupoid of a groupoid G, show that $\lambda(x^n) > \lambda(x)$ for all $x \in G$.
4. Show that the intersection of any set of fuzzy subgroupoids is a fuzzy subgroupoid.
5. Describe the lattice of fuzzy subgroupoid.
6. Prove that the intersection of any set of fuzzy (left, right) ideals is a fuzzy (left, right) ideal.
7. Prove or disprove that the union of any set of fuzzy (left, right) ideals is a fuzzy (left, right) ideal.
8. Define the image of a fuzzy subgroupoid under a mapping.
9. Define the pre-image of a fuzzy subgroupoid under a mapping.
10. Show that the homomorphic pre-image of a fuzzy subgroupoid is a fuzzy subgroupoid.
11. Show that the homomorphic image of a fuzzy subgroupoid having sup property is a fuzzy subgroupoid.
12. Define sup property of a group G and fuzzy subgroup of a group G and give proper examples of it.
13. Prove that the intersection of any set of fuzzy groups is a fuzzy subgroup.
14. Show that $\gamma\left(xy^{-1}\right) = \gamma(e) \Rightarrow \gamma(x) = \gamma(y), \forall\, x, y \in G$.
15. Establish the claim that a group cannot be the union of two proper subgroups.
16. Explain the conception of fuzzy subgroups of a Group (G, +) and show that the conception of fuzzy subgroups is more general than classical subgroups.
17. Who redefined the notion of fuzzy subgroups? What are the reasons for doing this?
18. Define T-norm, fuzzy subgroupoid, and fuzzy groups as proposed by Anthony & Sherwood.
19. Discuss any two good results that are valid in Rosenfeld and Anthony & Sherwood's theory with little variations.
20. Define level subgroups of a fuzzy group λ. If λ_{t_1} and λ_{t_2} are level subgroups of λ with $t_1 < t_2$,then show that any two level subgroups of a fuzzy subgroup λ will be equal if and only if $t_1 < \lambda(x) < t_2$.
21. Show that the level set λ_t, for $t \in [0, 1], t \leq \lambda(e)$ is a subgroup of G, where e is the identity of G.
22. Give an example of a level subgroup of a fuzzy subgroup.
23. Explain the conception of fuzzy normal subgroups proposed by Mukherjee & Bhattacharya and M. Akgul and show that both definitions are equivalent.
24. Discuss an example of a fuzzy normal subgroup of a group G.
25. Show that intersection of two fuzzy normal subgroups is again a fuzzy normal subgroup.
26. Prove that intersection of any set of fuzzy normal subgroups of a group is again a fuzzy normal subgroup.
27. If G and H are two normal subgroups and $\varphi : G \to H$ be a homomorphism with μ as a fuzzy normal subgroup of H. Then show that $\lambda = \mu \circ \varphi$ is a fuzzy (normal) subgroup of G.

28. Define fuzzy cosets of the group G determined by x and further explain that fuzzy cosets extend the notion of cosets of group theory.
29. Show that the homomorphic pre-image of a fuzzy normal subgroup is a fuzzy normal subgroup.
30. Discuss and prove the fundamental theorem of homomorphism of fuzzy groups.

Objective-Type Questions and Answers

1. The concept of fuzzy subgroups of a group was introduced by

   (a) Bhattacharya (b) Anthony & Sherwood (c) Das (d) Rosenfeld
2. Fuzzy subgroups were defined in terms of t-norm T by

   (a) Abuosama (b) M. Akgul (c) Anthony & Sherwood (d) Singh
3. Every fuzzy (left, right) ideal is a

   (a) Fuzzy Subgroup (b)Fuzzy subgroupoid (c) Characteristic function (d) None of these
4. The union of any set of fuzzy subgroupoids is a

   (a) Fuzzy subgroupoid (b) Fuzzy ideal (c) Fuzzy subgroup (d) None of these
5. The intersection of any set of fuzzy subgroupoids is a

   (a) Fuzzy ideal (b) Fuzzy subgroup (c) Fuzzy subgroupoid (d) None of these
6. The intersection of any set of fuzzy left ideals is a

   (a) Fuzzy right ideal (b) Fuzzy ideal (c) Fuzzy subgroupoid (d) fuzzy left ideal
7. The union of any set of fuzzy right ideals is a

   (a) Fuzzy ideal (b) Fuzzy right ideal (c) Fuzzy subgroupoid (d) fuzzy left ideal
8. A homomorphic pre-image of a fuzzy ideal is a

   (a) Fuzzy ideal (b) Fuzzy subgroup (c) Fuzzy subgroupoid (d) fuzzy left ideal
9. A homomorphic pre-image of a fuzzy subgroupoid is a

   (a) Fuzzy ideal (b) Fuzzy subgroup (c) Fuzzy subgroupoid (d) None of these
10. A fuzzy set λ in a groupoid G be called a fuzzy ideal if

    (a) $\lambda(xy) \le max\{\lambda(x), \lambda(y)\}, \forall x, y \in G$ (b) $\lambda(xy) \ge min\{\lambda(x), \lambda(y)\}, \forall x, y \in G$

    (c) $\lambda(xy) < min\{\lambda(x), \lambda(y)\}, \forall x, y \in G$ (d) $\lambda(xy) \ge max\{\lambda(x), \lambda(y)\}, \forall x, y \in G$
11. A fuzzy set λ in a groupoid G be called a fuzzy subgroupoid of G w. r. t. a t-norm T if and only if for every x, y in G

    (a) $\lambda(xy) \le T\{\lambda(x), \lambda(y)\}$ (b) $\lambda(xy) > T\{\lambda(x), \lambda(y)\}$

    (c) $\lambda(xy) < T\{\lambda(x), \lambda(y)\}$ (d) $\lambda(xy) \ge T\{\lambda(x), \lambda(y)\}$
12. The fuzzy (left, right) ideals of a groupoid G are a…of the set of all fuzzy sets in G.

    (a) Lattice $\mathcal{L}$ (b) Complemented lattice $\mathcal{L}$ (c) Complete sub lattice $\mathcal{L}$ (d) None of these

13. The least and the greatest elements of $\mathcal{L}$ are the constant functions …

    (a) 0 and 1 (b) 1 and ϕ (c) 0 and ϕ (d) None of these
14. If G is a group, a fuzzy subgroupoid λ of G will be called a fuzzy subgroup of G if

    (a) $\lambda(x^{-1}) = \lambda(x)$ (b) $\lambda(x^{-1}) \leq \lambda(x)$ (c) $\lambda(x^{-1}) \geq \lambda(x)$ (d)$\lambda(x^{-1}) > \lambda(x)$
15. The union of any family of fuzzy subgroups is a

    (a) Fuzzy subgroup (b) Fuzzy subgroupoid (c) Proper fuzzy subgroup (d) None of these
16. The intersection of any family of fuzzy subgroups is a

    (a) Fuzzy subgroup (b) Fuzzy subgroupoid (c) Proper fuzzy subgroup (d) None of these
17. The fuzzy (left, right) ideals in a group are just the

    (a) Characteristic function (b) fuzzy subgroup (c) Constant function (d) None of these
18. A homomorphic image of a fuzzy subgroup is a

    (a) Fuzzy subgroupoid (b) Fuzzy ideal (c) Proper fuzzy subgroup (d) Fuzzy subgroup
19. If λ is a fuzzy subgroup of group G, then

    (a) $\lambda(\mathrm{x}) \leq \lambda(\mathrm{e})$ (b) $\lambda(\mathrm{x}) \geq \lambda(\mathrm{e})$(c) $\lambda(\mathrm{x}) < \lambda(\mathrm{e})$ (d) $\lambda(\mathrm{x}) > \lambda(\mathrm{e})$
20. The union of two proper fuzzy subgroups is a

    (a) Group (b) Fuzzy subgroup (c) Not a group (d) Not a fuzzy subgroup
21. If λ is a fuzzy subset of a set G, and $t \in [0, 1]$, then the set $\lambda_t = \{x \in G : \lambda(x) \geq t\}$ is called

    (a) Level subgroup of λ (b) Level subset of λ (c) λ level set (d) None of these
22. If λ is a fuzzy subgroup of a group G, then for any $t \in [0, 1]$ with $\lambda(e) \geq t$, the level set λ_t is a subgroup of G in the usual sense. In this situation, the level set λ_t is called

    (a) Image set (b) Fuzzy set (c) Level subgroup (d) None of these
23. Let λ be a fuzzy subgroup of a group G. Then two level subgroups λ_{t_1}, λ_{t_2} with $t_1 < t_2$ of λ are equal if and only if there does not exist $x \in G$ such that

    (a) $t_1 < \lambda(\mathrm{x}) < t_2$ (b) $t_1 \leq \lambda(\mathrm{x}) \leq t_2$ (c) $t_1 > \lambda(\mathrm{x}) > t_2$ (d) $t_1 \geq \lambda(\mathrm{x}) \geq t_2$
24. A fuzzy subgroup λ of group G is called a normal fuzzy subgroup if

    (a) $\lambda\left(xyx^{-1}\right) \geq \lambda(y)$ (b) $\lambda\left(xyx^{-1}\right) = \lambda(y)$.

    (c) $\lambda\left(x^{-1}yx\right) \geq \lambda(y)$ (d) $\lambda\left(x^{-1}yx\right) \geq \lambda(y)$.
25. If λ is a fuzzy subgroup of G, then the conditions … are equivalent for all x, y in G.

    (a) $\lambda(\mathrm{xy}) = \lambda(\mathrm{yx})$ (b) $\lambda\left(xyx^{-1}\right) \leq \lambda(y)$.

(c) $\lambda\left(xyx^{-1}\right) \geq \lambda(y)$(d) $\lambda\left(xyx^{-1}\right) = \lambda(y)$.

26. If λ is a fuzzy subgroup of a group G and x, y, g in G, then $\lambda\left(x^{-1}g\right)$ is called a ………… of λ in G.

    (a) Fuzzy coset (b) Right fuzzy coset (c) Left fuzzy coset (d) None of these
27. If λ is a fuzzy subgroup of a group G, then $\lambda\left(gx^{-1}\right) = (\lambda x)g$ is called a …………of λ in G for all x, y, and g in G.

    (a) Left fuzzy coset (b) Right fuzzy coset (c) Fuzzy coset (d) None of these
28. Any two left (right) fuzzy cosets of a group G are

    (a) Equal (b) Disjoint (c) Neither equal nor disjoint (d) None of these
29. Let G be a group and λ_1, λ_2 are two fuzzy subgroups of G, then λ_1 is said to be conjugate to λ_2 if for all x in G, we have

    (a) $\lambda_1(g) = \lambda_2\left(x^{-1}gx\right)\forall\, g \in G$ (b)$\lambda_2(g) = \lambda_1\left(x^{-1}gx\right)\forall\, g \in G$

    (c) $\lambda_1(g) = \lambda_2\left(xgx^{-1}\right)\forall\, g \in G$ (d) None of these
30. If λ is a fuzzy subgroup of a group G, then the cardinality of G/λ is called the index of λ in G and is denoted by

    (a) [λ : G] (b) [G : λ] (c) [λ; G] (d)[G : λ]

Answers

1. (d); 2. (c); 3. (b); 4. (d); 5. (c); 6. (d); 7. (b); 8. (a); 9. (c); 10. (d); 11. (d); 12. (c); 13. (a); 14. (a) & (c); 15. (d); 16. (a); 17. (c); 18. (d); 19. (a); 20. (c); 21. (b); 22. (c); 23. (a); 24. (a to d all); 25. (a) & (d); 26. (a) & (c); 27. (b) & (c); 28. (c); 29. (a); 30. (b)

True/False Type Statements and Answers

1. A fuzzy subgroupoid is one of the smallest structures in the area of fuzzy algebraic structure.
2. The notion of fuzzy subgroupoid was proposed by A. Rosenfeld in 1971.
3. A fuzzy set λ in a groupoid G is called a fuzzy left ideal if $\lambda(ab) \geq \lambda(a), \forall\, a, b \in G$.
4. If λ is a fuzzy subgroupoid of G, then $\lambda(a^n) \geq \lambda(a), \forall\, a \in G$ and $n \in \mathbb{N}$.
5. If λ is a fuzzy right ideal of G, then $\lambda(ab) \geq \lambda(a), \forall\, a, b \in G$.
6. The intersection of any two fuzzy subgroupoids is a fuzzy subgroupoid.
7. The union of any family of fuzzy right ideals is a fuzzy right ideal.
8. A homomorphic preimage of a fuzzy subgroupoid is not a fuzzy subgroupoid.
9. Let μ be a fuzzy set in $f(A) = B$, then the fuzzy set $f \circ \mu = \lambda$ in A, defined by $\lambda(a) = \mu(f(a)), \forall\, a \in A$ is the preimage of the fuzzy set μ under the mapping f.
10. If $\lambda(ab) \leq max\{\lambda(a), \lambda(b)\}, \forall\, a, b \in G$, then λ is called the fuzzy ideal of G.
11. The intersection of any family of fuzzy left ideals is a fuzzy left ideal.
12. The homomorphic preimage of a fuzzy left ideal is a fuzzy right ideal.

13. A fuzzy subset λ of a group G is said to have the sup property if, for any subset A of G, there exists $a_0 \in A$ such that $\lambda(a_0) = \sup_{a \in A} \lambda(a)$.
14. Every homomorphic image of a fuzzy subgroupoid with the sup property is a fuzzy subgroupoid.
15. A homomorphic image of a fuzzy left ideal with the sup property is not a fuzzy left ideal.
16. A fuzzy subgroupoid λ of a group G will be called a fuzzy subgroup of G if $\lambda\left(a^{-1}\right) \geq \lambda(a), \forall\, a \in G$
17. For the fuzzy subgroup λ of group G, we always have $\lambda\left(a^{-1}\right) = \lambda(a)$ and $\lambda(a) \leq \lambda(e), \forall\, a \in G$, where *e* is the identity of G.
18. The arbitrary intersection of any family of fuzzy subgroups of a group G is a fuzzy subgroup of G.
19. $\lambda\left(ab^{-1}\right) \geq min\{\lambda(a), \lambda(b)\}, \forall\, a, b \in G$, if λ is a fuzzy subgroup of G.
20. If $\lambda(a) = \lambda(e)$,then $\lambda(ab) = \lambda(b), \forall\, b \in G$, where λ is a fuzzy subgroup of G.
21. A fuzzy subgroup λ of a group G is said to be a proper fuzzy subgroup of G if $\lambda(\mathrm{a}) = 1$ for at least one $a \in G$, i.e., $\lambda = G$.
22. A fuzzy subgroup cannot be the union of two proper fuzzy subgroups.
23. Every homomorphic preimage of a fuzzy subgroup is a fuzzy subgroup.
24. The homomorphic image of a fuzzy subgroup having sup property is not a fuzzy subgroup.
25. A group's fuzzy ideals are just the constant function.
26. If $f(a) = f(b) \Rightarrow \lambda(a) = \lambda(b), \forall\, a, b \in S$, then λ is called *f*-invariant if *f* is any function from a set S to a set T.
27. If λ is a fuzzy subset of $(G, +)$,then λ will be called a fuzzy subgroup of G if, for all $a, b \in G$,$\lambda(a + b) \geq min\{\lambda(a), \lambda(b)\}$.
28. A fuzzy subgroupoid λ of a group $(G, +)$ will be called a fuzzy subgroup of G if $\lambda(-a) \leq \lambda(a)$.
29. Anthony and Sherwood have redefined the notion of a fuzzy subgroup with respect to a general t-norm.
30. A fuzzy subgroupoid λ of $(G, .)$ with respect to a t-norm T, is a fuzzy subgroup of G, if and only if $\lambda\left(a^{-1}\right) = \lambda(a), \forall\, a \in G$.
31. $H = \{a \in G : \lambda(a) = 1\}$ is a subgroup of $(G, .)$, if λ is a fuzzy subgroup of a group $(G, .)$ with respect to a t-norm T.
32. The set $\lambda_t = \{a \in G : \lambda(a) \geq t\}$ is called a level subset of the fuzzy set λ in $(G, .)$.
33. Any subgroup H of group G cannot be realized as a level subgroup of some fuzzy subgroup of G.
34. Fuzzy normal subgroups and fuzzy cosets were defined by Mukherjee and Bhattacharya in 1984.
35. A fuzzy group λ be called a fuzzy normal subgroup of $(G, .)$ if and only if $\lambda(-a + b + a) \geq \lambda(b), \forall\, a, b \in G$.

Answers

1. (T); 2. (T); 3. (F); 4. (T); 5. (F); 6. (T); 7. (T); 8. (F); 9. (T); 10. (F); 11. (T); 12. (F); 13. (T); 14. (T); 15. (F); 16. (T); 17. (T); 18. (T); 19. (T); 20. (T); 21. (F); 22. (T); 23. (T); 24. (F); 25. (T); 26. (T); 27. (T); 28. (F); 29. (T); 30. (T); 31. (T): 32. (T); 33. (F); 34. (T): 35. (T).

Chapter 7
Application of Fuzzy Mathematics in Other Disciplines (Relevancy of Fuzzy Mathematics in Human Life)

There is a common saying that beauty lies in the eyes of the beholder. The best and most beautiful things in the world cannot be seen or even touched-they must be felt with the heart. It infers that beauty does not exist independently but has been created by observers. This famous quote can help us remember that a beholder is someone who sees or otherwise experiences things, becoming aware of them. Fuzzy Mathematics has made mathematics lovers, mainly and all learners, feel about the subject and attracted the admirer due to its usefulness. It has an aura of its own. No one can put it in the bounds of words. This chapter portrays different shades of fuzzy mathematics in the knowledge domain and some illustrative examples.

7.1 Introduction

Fuzzy relations have various applications in diverse areas such as Computer Science, Management Science, Banking and Finance, Social Science, Medical Science, Engineering Science, etc. Fuzzy set theory has already made an attractive imprint in Computer Science, particularly in dealing with various matters that require storing and manipulating knowledge and information in a way adaptable to human thinking. The application of fuzzy mathematics comprises fuzzy information retrieval systems, fuzzy expert systems, fuzzy databases, pattern recognition, image processing, and cluster analysis. The primary benefit of applying fuzzy sets in these computer-based systems is gaining greater generality, higher expressive power, and enhanced ability to represent and manipulate knowledge and information expressed in natural language. We get a methodology for exploiting the tolerance for vagueness or imprecision. This potentiality provides us—with a method that makes the systems more flexible and realistic and permits easy handling, lower solution cost, and robustness.

M. K. Singh, *Applied Fuzzy Mathematics*, Forum for Interdisciplinary Mathematics,
https://doi.org/10.1007/978-981-97-3257-9_7

7.2 Fuzzy Sets in Computer Science and Engineering

7.2.1 *Fuzzy Sets and Computer Science*

Plenty of matters show the applications of fuzzy sets in fuzzy information retrieval systems, fuzzy expert systems, and fuzzy databases. At the same time, FST has also established its utility in pattern recognition, image processing, and cluster analysis. Pattern recognition technology can be restricted into two broad types: unsupervised and supervised techniques. An unsupervised approach does not use a given set of unclassified data points, whereas, in supervised methods, we use a data set with known classifications. These two categories of techniques are complementary to each other. As far as clustering is concerned, one of the main issues is evaluating the clustering result of an algorithm. It is usually referred to as clustering validity. In a nutshell, the problem of clustering validity is to find an objective criterion for ascertaining how good a partition is; clustering algorithms generate that. Such type of yardstick is essential as it enables us to attain three goals:

(i) To differentiate the yield of alternating clustering algorithms for a data file.
(ii) Find the best number of clusters for the given (structured data collection).
(iii) To ascertain whether a structured data collection contains any structure. It is pertinent to mention that hard and soft clustering needs to address the matter of clustering validity by any method.

As far as image processing is concerned, it is associated with cluster analysis and pattern recognition because data are always provided in the form of digital images. Digital image processing has many advantages over analog image processing. It is a method in which we perform some kinds of operations on an image to derive useful information. After fuzzification, its impact is awe-inspiring for human interpretation.

7.2.2 *Fuzzy Sets in Science and Engineering*

The applications of FST are at the cutting edge in engineering compared to science. In engineering and industrial environment exploitation of the applications of FST, the most attractive potential is fuzzy controllers. Fuzzy controllers are nothing but knowledge-based controllers using AI techniques and fuzzy logic to compute an appropriate control action. These fuzzy knowledge-based controllers are integral to distributed control systems, including standard controllers in an extended range of industrial process control systems and consumer products. For Japanese industrialists, applications of fuzzy controllers have become a standard feature. In particular, for in-home appliances, the label "fuzzy controllers" pushes up positive associations in the sense of high quality, user-friendly, or modern. Indeed, it is a matter that fuzzy control has contributed significantly to the number of home appliances. For instance, the "one-touch–button" washing machine constitutes a landmark w.r.t. many control panels of the older machine. In the present scenario, this and some other fuzzy control applications have become a marketing argument. Due to this reason, most of the newly made home appliances are marked “fuzzy” or “neuro-fuzzy” on the product. Fuzzy control leads us to more automation for complex and ill-structured processes. The most crucial fact about fuzzy controllers is that they are more robust than conventional controllers. Researchers expect fuzzy controllers to be used in some industrial applications even though it is not required. It is to display their technological supremacy in the area of fuzzy controllers. Their point of view is that the marketing argument will have to focus also on user-friendliness, energy saving, reduction of freshwater consumption for washing machines, and environmental aspects despite labeling “fuzzy” or “neuro-fuzzy.”

The fuzzy boom in Japan resulted from the partnership between industries and universities in technology transfer. The *Panasonic Company* was the leader, who first applied fuzzy logic to a consumer product, called a showerhead for controlling the water temperature. *Matsushita Electric Industrial Company* named their washing machine based on a fuzzy controller as “Asai-go Day Fuzzy” (*beloved wife*) and launched a drive “fuzzy product.” A new meaning was introduced in Japan for “fuzzy” as “intelligence”. Several companies in Japan started following the approach of the Panasonic Company and introduced several consumer products such as fuzzy refrigerators, fuzzy rice cookers, and fuzzy vacuum cleaners, etc. This evolved into a fuzzy craze in Japan.

Apart from mathematics, researchers are paying attention to fuzzy biology, fuzzy chemistry, fuzzy physics, and fuzzy ecology are being developed substantially. Fuzzy sets and fuzzy measures have an effective impression on these areas. The usefulness of FST has an everlasting impact, and these were displayed in the areas of finite resolution limit by **Fridrich** in 1994, non-equilibrium thermodynamics by **Singer** in 1992, etc. It will restrict the disparity between theoretical and practical. FST has also been applied in studying earthquake and solid waste management and other disciplines of human knowledge.

7.3 Fuzzy Sets and Industrial Applications

Blue Circle Cement and *SIRA*, in Denmark, in 1976, were the first companies regarded as pioneers in the industrial application of fuzzy logic. The system was a cement kiln controller that incorporates the “know-how” of a skilled operator to magnify the capability of clinker with the help of smoother grinding. In 1987 **Seiji Yasunobu** and his colleagues presented a fuzzy logic-based automatic train operation control system in Sendai city’s subway. Later, *Fuji Electric* applied fuzzy logic and proposed a model for a water treatment system. Its development enabled *Fuji Electric* in 1985 to present the first Japanese general-purpose fuzzy logic controller in the name of **FRUITAX**. These two areas of applications brought a revolution in fuzzy logic technology in Japan in particular and among the world’s scientific community in general.

In the area of engineering design, FST plays an important character. Under FST, the designer can describe the sketched memento as roughly as desired at the beginning of the design activity. As acknowledged by skilled designers, this is the key, as early design decisions restrict the set of accessible alternatives design. If someone eliminates a few suitable substitutes by an earlier decision, one cannot recover them later in the design exercise. At the final stage of the design cycle, the ambiguity is effectively abolished. FST is exceedingly suited to robotize the process.

7.4 Product Based on Fuzzy Logic

1. The *Nissan* company developed a product, "**Anti-lock brakes**," which controls brakes in hazardous cases depending on wheel speed, car speed, and acceleration.
2. The *NOK/Nissan* Company developed a product, "**Auto transmission**," which checks the fuel injection and oxidization based on throttle setting, cooling water temperature, RPM, etc.
3. *Honda, Nissan* developed a product, "**Auto engine**," which selects gear based on engine load, driving style, and road conditions.
4. *Canon* developed a product, "**Copy machine**," used for adjusting drum voltage based on picture density, humidity, and temperature.
5. *Nissan, Isuzu*, and *Mitsubishi* developed a product **called** "**Cruise Control**" use it to adjust throttle settings to set car speed and acceleration
6. *Matsushita* developed a product, "**Dishwasher**," used for adjusting and cleaning cycles, strategies based on the number of dishes and the quantity of food served on the plates.
7. *Fujitec, Mitsubishi Electric, and Toshiba* developed an "**Elevator control product**." Use it to reduce waiting time based on passenger traffic.
8. *Maruman Golf* developed a "**Golf diagnostic system**" that selects golf clubs based on the golfer's swing and physique.
9. *Omron* Developed a Product, "**Fitness management**," based on Fuzzy rules implied by them to check the fitness of their employees.
10. *Mitsubishi Chemical* developed a product, "**Microwave oven**," sets lunes power, and cooking strategy.

7.4.1 Advantages of Fuzzy Logic System

- The structure of Fuzzy Logic Systems is understandable and straightforward.
- Fuzzy logic is broadly exercised for economic and practical reasons.
- Fuzzy logic in AI provides us to control machines and customer commodities.
- It may not offer accurate reasoning but sole admissible reasoning.
- Fuzzy logic in Data Mining helps us to deal with uncertainty in engineering.

- Fuzzy logic has proven well its broad potential in industrial automation.
- Fuzzy logic supervisory control for coal power plants is beneficial because the control problem is strongly non-linear and involves multiple measures and command variables.
- Mostly robust as no precise inputs are required.
- Fuzzy logic is a standard design technology in the automotive industry.
- It can be programmed in the situation when the feedback sensor end functioning.
- It can be modified to enhance or change system accomplishment.
- Cheap sensors can be used, which assists us in keeping the altogether system price and complexity economical.
- It provides a most fruitful solution to complex issues.

7.4.2 *Drawback of Fuzzy Logic Systems*

- Fuzzy logic is only sometimes accurate, so the results are based on supposition, which may not be broadly welcomed.
- Fuzzy systems do not have the potentiality of machine learning besides neural network type and pattern recognition.
- Validation and confirmation of a fuzzy knowledge-based system need substantial testing and machinery.
- Setting precise, rinse, and wash fuzzy rules and membership functions is difficult.
- Some fuzzy time logic needs to be clarified with probability theory and the terms.

7.5 Fuzzy Relation Equations and Neural Network

A *neural network* is a series of algorithms that venture to acknowledge fundamental relationships in a data set through a procedure that imitates the human brain's operation. In other words, neural networks refer to the system of artificial or organic neurons. Fuzzy relation equations can be expressed in terms of neural networks. Fuzzy relation equations play a rudimentary role in FST. It provides an ample structure within which several issues that appear from practical applications of the theory can be formulated. The high computational demands of the known solution methods foil this utility of fuzzy relation equations. To eliminate this trouble in solving fuzzy relation equations, neural networks have evolved and been explored.

In addition to neural networks and fuzzy logic, another area that has become popular is known as *evolutionary Computing*. It includes evolutionary strategies, genetic algorithms (GA), and evolutionary programming. The first two are optimization techniques. There are several amalgamations of fuzzy logic, GA, and neural networks. To differentiate them from traditional methodologies, **Lotfi Zadeh** proposed the notion of "soft computing." *Soft Computing* is an umbrella for computing techniques comprising GA, AI, and fuzzy logic. Here fuzzy logic acts as the building block of soft Computing. It elegantly deals with vague, imprecise, blurred, ambiguous, and uncertain information.

7.6 Application of Fuzzy Relational Equation in Evaluating Cardiovascular Diabetic Mellitus

Data Mining aims to discover knowledge from data and present it in an easily compressible form to humans. It is a process developed to examine large amounts of data routinely collected. Nowadays, fuzzy systems are an effective tool to solve various problems in different application domains, including the Genetic Algorithm for designing. Fuzzy Systems allows us to introduce learning and adaptation capabilities. The fuzzy set framework is in use in several different disease diagnosis processes. Fuzzy logic is a computational paradigm that equips us with a mathematical tool for dealing with the uncertainty and the imprecision typical of human reasoning. Based on the expert's medical knowledge, fuzzy relations between symptoms and risk factors for Diabetic have been considered related complications. Some common metabolic disorders may lead to vision loss, heart failure, stroke, foot ulcers, and nerve damage. Researchers have evaluated the fuzzy set A as symptoms observed in the patient. They have assumed fuzzy relation R as representing the medical knowledge that relates the symptoms in set S to the diseases in set D. By employing the compositional rule of inference, the fuzzy set B of the possible disorders of the patient can be obtained. In the present scenario, Neural Networks are efficiently used to learn membership functions, fuzzy inference rules, and other context-dependent patterns. The fuzzification of neural networks extends their capabilities' in applicability. The learning function can be supervised or unsupervised. In a supervised learning algorithm, learning is guided by specifying each training input pattern class to which the pattern belongs. The desired response of the network to each training input pattern and its comparison with the network's actual output is used in the learning algorithm for appropriate adjustments of the weights. First, expert detection is only based on the patient's articulation compared to medical knowledge, which may lead to various modifications, and the patient's rejection of specific symptoms may be inappropriate. The proposed detection system uses one committee of *Multilayer Perceptron Neural Networks (MLP)* for each one of the entities. Researchers use the back propagation algorithm, and the multilayer perceptron repeatedly works to remove errors in the network [IEEE, Vol.7, No.2, 2003].

7.7 Use of Fuzzy Sets in Life Insurance Underwriting When Insurer is Diabetic

In life insurance medicine, the mortality of the applicants within the period of insurance is required and is analyzed on the ground of present risk factors or diseases. Unlike hypertension or overweight, where risk assessment of the medical values is possible directly, the risk assessment of diabetes is a complex problem. If a person has diabetes, then finding out the mortality of insurers for life insurance medical underwriting is a complex problem due to many medical risk factors. For life insurance medical underwriting to handle this complex problem, the insurance companies need to have a reliable expert system that can help them evaluate the mortality of the applicants within the period of insurance. Here, researchers have considered a model based on a fuzzy expert system that will allow insurance companies to determine insurers' mortality in the existence of diabetes for life insurance underwriting (**Kumar and Pathak** 2010; **Gupta et al.** 1991).

7.8 Applications of Fuzzy Numbers

Fuzzy sets whose domain is the set of real numbers that is, fuzzy sets that can be characterized by a membership function having the form

$$A : \mathbb{R} \to [0, 1]$$

are of particular importance. Such membership functions have a quantitative meaning. Indeed they attract our intuitive perceptions of approximate intervals or numbers, such as "class of real numbers that are close to 5," "approximately equal to 2," and "around 3". Sometimes we say that the time is now "about 7 o'clock," today dinner time is "around eight-thirty." Under certain conditions, it may be viewed as fuzzy numbers or fuzzy intervals. Such notions are indispensable for characterizing linguistic variables and linguistic terms. Fuzzy numbers are significant in several applications, such as fuzzy control, statistics with imprecise probabilities, decision-making, approximate reasoning, and optimization. Let us consider a decision-making situation in which a stockbroker or analyst presumes that a particular share or stock reaches about Rs.15,000/; then, the fund manager decides to sell almost half of his available shares. Consider an example of an income tax table, which states that we owe Rs.30,000/ if our taxable income as an individual lies between the interval [Rs.500,000, Rs.1,000,000]. In decision-making, intervals are used for signifying acceptable or unacceptable values of pertinent numerical variables. In the same way, in engineering design, they are frequently used for illuminating acceptable variations of design parameters.

7.9 Fuzzy Set Theory and Psychology

Fuzzy Set Theory is intimately related to psychology. In the area of psychology, fuzzy set theory has a substantial prospect. The fascination of psychologists in the domain of FST has been flourishing rapidly. Psychology is not just a sphere in which it is feasible to foresee the solicitation or implementation of fuzzy set theory. Still, it is equally likely for the growth of FST oneself. The notion of context has an extraordinary relationship with FST. Psychologists' interest in fuzzy set theory came to the fore in the late 1980s. For the development of this field, it is essential to study the interconnection between the human use of linguistic terms in varied context and the fuzzy sets and their operations that represent these linguistic terms. Psychologists believe that humans use categories with vague or ambiguous edges and gradation of membership values. Due to this reason, psychologists are engrossed in psycholinguistics, FST, and fuzzy logic for studying the connection between the human uses of linguistic terms in a distinct state of affairs. It is desirable to acknowledge the interrelation for appropriately expressing the concept of a context in the conformity of FST. It deals with gradient thesis and measuring psychological concepts regarding linguistic variables and terms. Researchers have used linguistic variables and linguistic terms for questionnaires investigating the relationship between social stresses and psychological factors are being made very extensively (**Saitta and Torasso** 1981).

One of the crucial areas in psychology is the area of psychometric assessments. This test is used extensively to objectively measure the individual's personality traits, aptitude, intelligence, behavioral style, and abilities. It has widely been used in career guidance, personality for a suitable job, and employment to match a person's abilities. In some domains, measurements are associated with the property of data. Such as tall and long are concepts overlaid on the measured values of height and period. However, some models use fuzzy sets corresponding to notions unsupported by any estimated date. Such fuzzy sets often represent the subjective mapping between an idea and its degree of realization. Hence, such fuzzy sets use a psychometric scaling for their realm and are so-referred to as *psychometric fuzzy sets*. It comprises various notions, such as aggression, acceptable, aptitude, indicated, risk, substantially, and compatible. Here is the faint set representation of "risk." The substantive domain must conform to the coherency of the model. The wider the realm, the larger the degree of apparent detachment between the relative domain points. In the case of "Risk," with a bit of movement along with the membership function around the slope at 0.5, it would be difficult to ascertain that if the domain were [0, 20], it would be reasonably easy to see with the range [0, 100], and would give rise to a large displacement with the range [0, 500]. Simultaneously, the range of the psychometric scale must be related to the model's principles in settling any small or substantial changes along the axis of abscissa (see Fig. 7.1).

7.10 Fuzzy Set Theory and Economics

Present-day researchers opined that economic theories would become more faithful, realistic, and foretelling proficient if mathematics were based on fuzzy sets. Fuzzy set theory has attracted a sizeable group of world economists, as revealed in 1994 by a society named *"International Association for Fuzzy Set Management and Economics."* It is fascinating that the need for possibility theory in economics was perceived by the British economist **Shackle** (1961) much before it emerged from set theory or evidence theory given by **Shafer** (1976). The first book devoted to FST in economics was by **Ponsard and Fustier** in (1986). Apart from this book, **Ponsard's**

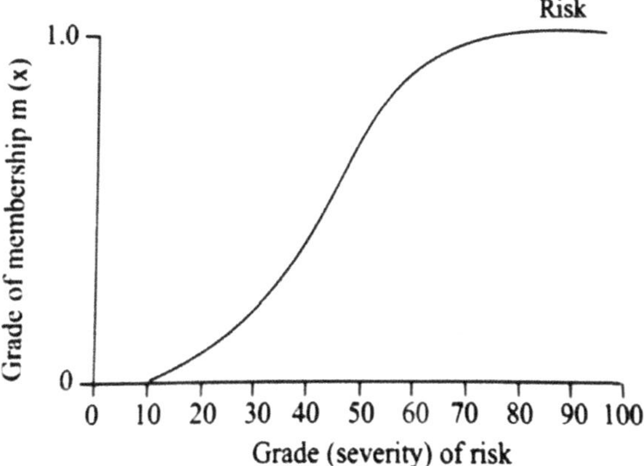

Fig. 7.1 The psychometric risk fuzzy set

two papers published in (1981 and 1988) show his significant contribution made by him in the realm of fuzzy economics. **Butnariu** (1978, 1985) proposed the notion of a vague game. Human life is full of shades of gray. Capturing these shades of gray has been problematic using classical mathematics. Mathematics of uncertainty looks to overcome some of these problems by directly integrating shades of gray and uncertainty into the model. Such models lead to emerging notions of fuzzy economics that deal with recent economic applications in both micro and macroeconomics.

7.11 Applications of Fuzzy Graphs

Azriel Rosenfeld 1975 propounded the fuzzy graph's notion. Since then, its applications have been excelling to new heights. Fuzzy graphs theory has extensive applications in diverse realms of engineering and science. The graphic models describe network traffic, rail network, telephone network, road network, social network, data hypothesis, broadcast communications, artificial reasoning, neural systems, and producing and arranging traffic light control. *Bang and Yeh* have introduced the various notion of connectedness in fuzzy graphs. For characterizing human behavior and mental activities, inaccurate information is being used. The particulars bank on a person's consciousness, and it is strenuous to process without prejudice. One can evaluate fuzzy details by exploiting a fuzzy graph. The fuzzy graph is nothing but an assertion of ambiguous relation, and therefore the fuzzy graph is continually expressed as a fuzzy matrix. The growth of fuzzy graph theory will be further accelerated by developing fuzzy software and hardware. We eagerly hope that the wide range of applications of graph theory and the beautiful and interdisciplinary nature of FST, if appropriately blended, will pave the way for the exponential growth of fuzzy graph theory.

7.12 Fuzzy Set Theory and Medical Science

Generally, the science of decision-making in the medical process is fuzzy. Linguistic concepts emerge when the physician decides the diagnosis and prognosis. The conversion from this murky nature into a crisp real-world result causes the accuracy to drop. Fuzzy logic is a potent way to provide the physician with the support he needs to handle linguistic concepts and eliminate the loss of accuracy. Fuzzy logic technologies apply to every domain of medicine, and they are prosperous. The literature review shows that the medical area has sizeable compatibility with fuzzy logic technology. Fuzzy medical image processing, fuzzy cognitive maps, fuzzy expert systems, incomplete applications in information retrieval from fuzzy medical data mining, medical databases, and hybrid fuzzy applications are the standard and well-known fuzzy logic exercise areas in the medical field.

Albin first exhibited the benefit of fuzzy set theory in medical diagnosis in 1975. So far, medical practitioners with open hearts have acknowledged the gift of fuzzy set theory in medicine. In the territory of treatment, the vagueness or blurriness initiated in the procedure of detection of ailment has been the center of attention of FST. The focal point has been on using fuzzy set theory in modeling the activity of detection of the disease. We know that ailments are usually distinguished or identified by natural language in terms of linguistic terms. Fuzzy set theory is not an astonishing fact but an indisputable fact. For instance, let us consider a statement.

The assertion identifies COVID-19:

> Reduction in Lymphocyte populations and the appearance of Atypical Monocytes, severe acute respiratory syndrome. The immunological changes associated with this infection are largely unknown.

The most familiar symptoms of COVID-19 are *slight fever*, *dry* cough, *tiredness*, *loss of taste or smell*, *slightly increased aches and pains*, *reddish eyes*, *sore throat*, *skin rash*, or *discoloration of fingers* or *toes*.

Here we observe that the linguistic terms marked in italics are intrinsically cloudy, hazy, indistinct, or unclear, i.e., fuzzy. The concept of a fuzzy set is required to derive meaningful information from these terms. Furthermore, fuzzy sets are used as a potent means for expressing the emergence of the symptoms in the sufferer and describing the connection between the ailments and the signs. Generally, doctors collect information about the patients from their history; after clinical tests, results such as ultrasonography, X-ray, lipid profile, biochemical analysis, pulmonary function, and glucose test are analyzed. The information collected by the above methods carries some uncertainty. The clinical reports from the pathological laboratory are unclear; think of the borderline cases. Thus, we can claim that the doctor can pronounce the ailments of the sufferer and will have a restricted degree of accuracy. To do this job better, researchers have attempted to model it with the help of FST. In recent years, the fuzzy set framework has been applied in varied ways by expert systems designed to help physicians diagnose ailments for some specified types of diseases. FST is also involved in dealing with the treatment of dangerous diseases like cancer. After all, medical experts have yet to endorse the usefulness of FST.

Example 7.1 An unquestionable kind of virus called COVID-19 struck human body cells worldwide in 2020–21. Clinical experts can visualize the infected cells by using a peculiar microscope. The microscope brings about digital images that can be investigated by expert doctors and will identify the infected cells. It has been detected that, due to the Covid-19 virus, the infected cells are bearing a black spot along with a dark gray region.

Then after, digital image processing can be enforced on the images. Two variables, say, U and V, are being generated by digital image processing. The initial variable, U, is associated with the aggregate of a black patch. The second variable, V, is related to the appearance of the black patch, whether elliptical or circular. Counting the number of black pixels or recognizing a whole circular bunch of pixels among the images is tough. Therefore, both of these variables must be evaluated as linguistic variables.

We assume that we have two fuzzy sets, U and V, where U is associated with the number of black pixels. Let it be of three kinds, say u_1, u_2, u_3, where u_1 represents having no black pixels; u_2 stands for a couple of black pixels, and u_3 represents a group of pixels. The fuzzy set V that describes the appearance/shape of the black pixel bundles is of two types, say v_1 and v_2. v_1 represents bundles of an ellipse, and v_2 represents bundles of a circle. Thus, we have

$$\mathrm{U} = \left\{ \frac{0.2}{u_1} + \frac{0.6}{u_2} + \frac{0.9}{u_3} \right\} \quad \text{and} \quad \mathrm{V} = \left\{ \frac{0.7}{v_1} + \frac{1.0}{v_2} \right\}$$

We wish to determine the relationship between the aggregate of black pixels in the virus and the shape of the black pixels that are centrally located. For this, it is required to determine the Cartesian product of U and V, which is given by

$$R(x, y) = \{U \times V\}(x, y) = \min\{U(x), V(y)\}$$

$$\mathrm{R} = U \times V = \begin{array}{c} \\ u_1 \\ u_2 \\ u_3 \end{array} \begin{array}{c} \begin{array}{cc} v_1 & v_2 \end{array} \\ \begin{bmatrix} 0.2 & 0.2 \\ 0.6 & 0.6 \\ 0.7 & 0.9 \end{bmatrix} \end{array}$$

Now, we consider the situation when another microscopic image has been taken in which the number of black pixels in the coronavirus is somewhat different. Let it be expressed by a fuzzy set W, where W is given by

$$W = \left\{ \frac{0.6}{u_1} + \frac{0.8}{u_2} + \frac{1.0}{u_3} \right\}$$

Let T be the fuzzy max–min composition of W and R. The max–min composition is defined by the set-theoretic and membership function-theoretic expressions. This process will provide another value for the fuzzy set of pixel bunch shapes related to the new black pixel quantity.

$$T = W \circ R$$

$$T(x, z) = \bigvee_{y \in Y} \{W(x, y) \wedge R(y, z)\}$$

$$\therefore \quad T = W \times R = [0.6 \ \ 0.8 \ \ 1.0] \circ \begin{bmatrix} 0.2 & 0.2 \\ 0.6 & 0.6 \\ 0.7 & 0.9 \end{bmatrix} = [0.7 \ \ 0.9]$$

7.13 Medical Diagnosis and the Law of Modus Ponens

There is a tautology known as modus ponens deduction. It is an essential plan of action, generally used in forward-chaining rule-based expert systems. It is an executive whose duty is to discover the actual value of a consequent in a production rule when the truth value of the antecedent in the rule is given. Under modus ponens deduction, we conclude that when two propositions p and $p \to q$ are given, and both are true, then the truth value of the proposition q can be drawn immediately. Each deductive rule of inference possibly is expressed in the form of a proposition that is a tautology. Modus Ponens, for instance, maybe demonstrated by the proposition.

$$\left[(p \Rightarrow q) \wedge p\right] \Rightarrow q$$

and is a tautology on a classical truth table. The aim of classical symbolic logic is only to justify complex reasoning by virtue of simple inference rules—like modus ponens and hypothetical syllogism—which represent forms of inference. Modus ponens, for instance, may be demonstrated by the proposition:

$$p \Rightarrow q$$
$$p$$
$$----$$
$$q$$

In our day-to-day activities, we always make use of common-sense reasoning. For this, let us consider the following inference:

Rule: If a building is spacious, then it is overpriced

Fact: Building B is quite spacious

Conclusion: Building B is quite spacious.

For two reasons, one cannot formulate such an inference in classical two-valued logic. The first one is that the notions such as spacious, quite spacious, and overpriced are fuzzy notions or concepts which cannot be formulated decently as two-valued propositions. The second cause is the antecedent of the if–then rule. In the case of the above inference, it is not the case. It follows that the classical modus ponens is not applicablc hcrc. Furthcrmore, thus, the above one is usually referred to as generalized modus ponens. In this section, we will illustrate by an example that modus ponens play a significant role in diagnosing the cardiovascular ailment.

The architecture of fuzzy mathematics evolves mathematical prototypes enforced or tested in dissimilar scientific and non-theoretical areas. They have also made outstanding philanthropy in the highly technical and beneficial domain for humanity, called medical diagnosis. In order to illustrate things more accurately, let us consider three non-fuzzy sets for making an appropriate diagnosis in one patient. We suppose that

The set of syndromes $= S = \{S_1, S_2, S_3, \ldots\ldots\ldots., S_n\}$

The set of prognoses $= P = \{P_1, P_2, P_3, \ldots\ldots\ldots., P_m\}$

The set of injured $= I = \{I_1\}$.

The syndromes occurring in set S are connected with the prognosis from set P. The syndromes $S_1, S_2, S_3, \ldots\ldots\ldots., S_n$ depicted in a set S are also represented as an ordered pair $(P_1, S_1), (P_1, S_2), (P_1, S_3), \ldots\ldots\ldots., (P_1, S_n)$. Let IS be any fuzzy relation displayed as a one-row matrix as follows.

$$\begin{array}{ccccc} S_1 & S_2 & S_3 & \ldots\ldots\ldots\ldots\ldots\ldots & S_n \end{array}$$
$$IS = \left[\delta_{IS}^{(P_1,S_1)}, \delta_{IS}^{(P_1,S_2)}, \delta_{IS}^{(P_1,S_3)}, \ldots\ldots\ldots\ldots\ldots., \delta_{IS}^{(P_1,S_n)}\right]$$

where $\delta_{IS}^{(P_1,S_i)}$, i = 1, 2, 3, ..., n gives us the values of the degree of membership, catering to us with an assessment of the strength of S_j in the injured I_1.

Another relation comprises the pairs $(S_1, P_1), (S_2, P_2), (S_3, P_3), \ldots\ldots\ldots., (S_n, P_m)$

$$SP = \begin{matrix} S_1 \\ S_2 \\ \vdots \\ S_n \end{matrix} \begin{bmatrix} \delta_{SP}^{(S_1,P_1)} & \delta_{SP}^{(S_1,P_2)} & \ldots\ldots\ldots & \delta_{SP}^{(S_1,P_m)} \\ \vdots & \vdots & \ldots\ldots\ldots & \vdots \\ \delta_{SP}^{(S_n,P_1)} & \delta_{SP}^{(S_n,P_2)} & \ldots\ldots\ldots & \delta_{SP}^{(S_n,P_m)} \end{bmatrix}$$

The fuzzy relation in which everyone's value of the degree of membership is chained up to the pair (S_i, P_j), where I = 1, 2, 3, …, n and j = 1, 2, 3, …, m and the injured and prognosis relation is a matrix given by the relation

$$\begin{array}{ccc} P_1 & P_2\ldots\ldots\ldots & P_m \end{array}$$
$$IP = \delta_{IS}^{(I_1,P_1)}, \delta_{IS}^{(I_1,P_2)}, \ldots\ldots\ldots, \delta_{IS}^{(I_1,P_m)}$$

In the end, the relation that permits us to predict the alliance between the injured with the prognosis is given by the max–min relation, that is

$$IP = \mathrm{IS} \circ \mathrm{SP}$$

Example 7.2 Let us consider an example to obtain a pertinent prognosis of the injured under the relation between injured to syndromes and syndromes to forecast. For doing this, suppose

The set of injured $I = \{I_1\}$

The set of syndromes $\mathrm{S} = \{S_1, S_2, S_3, S_4\}$

$$= \{\text{chest pain},\ \text{smoking},\ \text{increased bad cholestrol(low} - \text{density lipoprotein}\}$$

The set of prognoses $P = \{P_1, P_2, P_3\}$

The set of prognoses $P = \{\text{stroke},\ \text{cardiovascular disease},\ \text{coronary artery disease}\}$

$$= \{\text{smoking},\ \text{hypertension},\ \text{increased level of LDL - cholesterol},\ \text{pain in chest}\}$$

This investigation aims to uncover one of the prognoses in the injured I_1, which depicts the existence of syndromes at a definite level. Now, we obtain IS (that is formed by the help of physician questions to the injured person)

$$IS = \begin{bmatrix} 0.924\ 0.83\ 0.67\ 0.23 \end{bmatrix}$$

In the next step, we will consider the formation of syndromes-prognosis relation incorporating the help of "linguistic variables" and "Presence" (see Table 7.1).

We have

$$\text{Presence} = \left\{ \begin{array}{c} \text{never, almost never, very seldom, seldom, rather seldom,} \\ \text{moderately, rather often,} \\ \text{often, very often, almost always, always} \end{array} \right\}$$

Therefore, we can form the relation SP as

Table 7.1 Figurative elucidation of the fuzzy variables "Presence"

Fuzzy variables	T	$\delta_{\text{common}}^{(t)}$
Never	7.6	0
Almost never	17	0.017
Very seldom	23.6	0.72
Seldom	32	0.15
Rather seldom	38.5	0.27
Moderately	55	0.53
Rather often	67	0.83
Often	71	0.89
Very often	78.5	0.939
Almost always	87	0.987
Always	93.5	1

$$SP = \begin{bmatrix} 0.83 & 0.987 & 0.15 \\ 0.89 & 0.53 & 0.939 \\ 0.15 & 0.53 & 0.987 \\ 0.27 & 0.987 & 0.89 \end{bmatrix}$$

And the relation IP signifies the relation given by

$$IP = IS \circ SP$$

$$= \begin{bmatrix} 0.924 \ 0.83 \ 0.67 \ 0.23 \end{bmatrix} \circ \begin{bmatrix} 0.83 & 0.987 & 0.15 \\ 0.89 & 0.53 & 0.939 \\ 0.15 & 0.53 & 0.987 \\ 0.27 & 0.987 & 0.89 \end{bmatrix}$$

$$IP = \begin{matrix} P_1 & P_2 & P_3 \end{matrix} \\ \begin{bmatrix} 0.83 \ 0.924 \ 0.83 \end{bmatrix}$$, by using the max–min composition rule.

In conformity with the rule, the higher degree values signify the more prone prognosis the doctor can choose, i.e., P_2 will be the desired result.

Next, we will consider an example that can be used in **Civil engineering**.

Example 7.3 In India, Janakpur was initially called Mithilanchal, as this is the place where the Maithili culture originated. It is also said that Buddha and Mahavira spent some of their days in Janakpur. Janakpur is called the "city of ponds." These ponds stock overland flow from rainstorms and discharge the water downstream at a disciplined rate to bring down or eliminate flooding in downstream regions. In order to exemplify the situation, let us consider the level of water in the adjoin pond system based on 1-in-50-year storm volume capacity with nearest to three rain meters that measure the total rainfall.

Let P = pond system relative to depths based on 1-in-50-year capacity. Further, we assume that the capacities of the four ponds are c_1, c_2, c_3, and c_4, and all combine in one to form one exit to the trunk line sewer. Suppose that Q = total rainfalls for an event based on 1-in-50-year values from three distinct rain meter stations r_1, r_2, and r_3. Assume that we have the following distinct fuzzy sets P and Q given by

$$P = \left\{ \frac{0.3}{c_1} + \frac{0.7}{c_2} + \frac{0.6}{c_3} + \frac{1.0}{c_4} \right\} = \begin{bmatrix} 0.3 \\ 0.7 \\ 0.6 \\ 1.0 \end{bmatrix}$$

$$Q = \left\{ \frac{0.5}{r_1} + \frac{0.8}{r_2} + \frac{0.9}{r_3} \right\} = \begin{bmatrix} 0.5 \ 0.8 \ 0.9 \end{bmatrix}$$

We wish to illustrate a relation using the Cartesian product of two fuzzy sets, P and Q. Let the Cartesian product of P and Q be R. That is,

$$R = P \times Q = \begin{array}{c} \\ c_1 \\ c_2 \\ c_3 \\ c_4 \end{array} \begin{array}{c} \begin{array}{ccc} r_1 & r_2 & r_3 \end{array} \\ \begin{bmatrix} 0.3 & 0.3 & 0.3 \\ 0.5 & 0.7 & 0.7 \\ 0.5 & 0.6 & 0.6 \\ 0.5 & 0.8 & 0.9 \end{bmatrix} \end{array}$$

From this Cartesian product, we mean that the rain meter's prediction of sweeping storms to the genuine pond discharge during this event; the soaring values designate sketches and station particulars that could model and control the flooding sensibly. However, the lower values may designate a sketch problem or an abnormal meter location.

Now, we will exemplify the constitution for an identical situation. Let us seek to ascertain if the cloudburst is extensive or confined. For doing this, we will differentiate the results from a pond system deleting from the preceding locality during the same storm.

Let us assume that we have an association linking the magnitude of five additional ponds within a new pond structure $\{c_5, c_6, c_7, c_8, c_9\}$ and the downfall data from the initial downfall meter $\{r_1, r_2, r_3\}$. Let this correlation is given like

$$S = \begin{array}{c} \\ r_1 \\ r_2 \\ r_3 \end{array} \begin{array}{c} \begin{array}{ccccc} c_5 & c_6 & c_7 & c_8 & c_9 \end{array} \\ \begin{bmatrix} 0.4 & 0.7 & 0.6 & 0.3 & 0.2 \\ 0.5 & 0.8 & 0.6 & 0.4 & 0.4 \\ 0.3 & 0.7 & 0.9 & 1.0 & 0.9 \end{bmatrix} \end{array}$$

Suppose that T is the fuzzy max–min composition for the two ponding structures. The max–min composition is defined by the set-theoretic and membership function-theoretic expressions. Then we have

$$T = R \circ S$$

$$T(x, z) = \bigvee_{y \in Y} \{R(x, y) \wedge S(y, z)\}$$

$$= \begin{bmatrix} 0.3 & 0.3 & 0.3 \\ 0.5 & 0.7 & 0.7 \\ 0.5 & 0.6 & 0.6 \\ 0.5 & 0.8 & 0.9 \end{bmatrix} \circ \begin{bmatrix} 0.4 & 0.7 & 0.6 & 0.3 & 0.2 \\ 0.5 & 0.8 & 0.6 & 0.4 & 0.4 \\ 0.3 & 0.7 & 0.9 & 1.0 & 0.9 \end{bmatrix}$$

$$= \begin{array}{c} \\ c_1 \\ c_2 \\ c_3 \\ c_4 \end{array} \begin{array}{c} \begin{array}{ccccc} c_5 & c_6 & c_7 & c_8 & c_9 \end{array} \\ \begin{bmatrix} 0.3 & 0.3 & 0.3 & 0.3 & 0.3 \\ 0.5 & 0.7 & 0.7 & 0.7 & 0.7 \\ 0.5 & 0.6 & 0.6 & 0.6 & 0.6 \\ 0.5 & 0.8 & 0.9 & 0.9 & 0.9 \end{bmatrix} \end{array}$$

Indeed, this fresh relation describes the individuality of the rainstorm for the two topographically isolated pond structures. The former structure is from the four ponds, c_1, c_2, c_3, and c_4, and the latter is from the group of five structures c_5, c_6, c_7, c_8, and c_9. Here, we observe that if the entries in the matrix are substantial, then it signifies that the rainstorm is extensive. Otherwise, if the entries are adjacent to 0, it expresses that the rainstorm is confined, and the foremost rain meters are not a good forecaster for both structures.

7.14 Composition of Fuzzy Relations and Prototype for Forecasting Score of a Cricket Match

We can combine fuzzy relations in distinct product spaces through composition. There are distinct notions of the composition of relations, which vary in their results. At the same time, they have distinct mathematical attributes. One of the best-known and regularly accustomed compositions is the max–min composition. Though the so-called max-average or max-product composition yields more attractive and appealing results. Let us now describe an example of the use of the composition of fuzzy relations in forecasting cricket scores in a 20–20 or a 50-over cricket match played between two teams in a league match of the Asia Cup.

Let us now describe an example of applying fuzzy set theory in a highly different zone, the area of the most popular game in this sub-continent, Cricket. In this section, we will forecast scores using max-av composition, max-product composition, and max–min composition. For this, let us consider two fuzzy sets, B and P, where B represents the fuzzy set comprising "types of bowling," and P represents the kinds of pitches available. Thus, we have

B = Types of bowling and

P = Types of pitches available.

Let S represents the correlation between the type of bowling and the state of affairs of the pitch, and T represents the correlation between the state of affairs of the pitches and runs on the scoreboard.

$$
\begin{array}{c} \\ S = \end{array}
\begin{array}{r} \\ \text{bouncer} \\ \text{outswinger} \\ \text{inswinger} \end{array}
\begin{array}{c} \begin{array}{cccccc} g.p & f.t.p & d.p & w.p & du.p & de.p \end{array} \\ \begin{bmatrix} 0.7 & 0.6 & 0.5 & 0.2 & 1.0 & 0.6 \\ 0.9 & 0.7 & 1.0 & 0.3 & 0.2 & 0.7 \\ 0.8 & 0.9 & 0.7 & 0.8 & 0.2 & 0.3 \end{bmatrix} \end{array}
$$

$$\text{and}\quad T = \begin{array}{c} \\ g.p \\ f.t.p \\ d.p \\ w.p \\ du.p \\ de.p \end{array} \begin{array}{c} \begin{array}{ccc} \text{lo.rate} & \text{av.rate} & \text{hi.rate} \end{array} \\ \begin{bmatrix} 0.5 & 0.9 & 0.8 \\ 0.4 & 0.9 & 0.9 \\ 0.3 & 0.8 & 0.9 \\ 0.9 & 0.7 & 0.5 \\ 0.8 & 0.6 & 0.5 \\ 1.0 & 0.5 & 0.3 \end{bmatrix} \end{array}$$

Now, one by one will find out the relationship between types of bowling and runs on the scoreboard. Firstly, we will find the relation by ***max-av composition***.

$S \underset{av}{\circ} T$ = Relationship between the type of bowling and runs on the scoreboard.

That is

$$\left[S \underset{av}{\circ} T\right](x, z) = \left\{\left[(x, z),\ 1/2. \underset{z \in Z}{\max} \{S(x, y) + T(y, z)\}\right] : x \in X, y \in Y\right\}$$

$$\therefore \quad \left[S \underset{av}{\circ} T\right](x, z) = \begin{array}{c} \\ \text{bouncer} \\ \text{outswinger} \\ \text{inswinger} \end{array} \begin{array}{c} \begin{array}{ccc} \text{lo.rate} & \text{av.rate} & \text{hi. rate} \end{array} \\ \begin{bmatrix} 0.9 & 0.8 & 0.75 \\ 0.85 & 0.9 & 0.95 \\ 0.85 & 0.9 & 0.9 \end{bmatrix} \end{array} \tag{7.1}$$

In the next step, we will find out the relationship between types of bowling and runs on the

scoreboard by ***max-prod composition***. We have

$S \underset{\cdot}{\circ} T$ = Relationship between the type of bowling and runs on the scoreboard.

That is

$$\left[S \underset{\cdot}{\circ} T\right](x, z) = \left\{\left[(x, z),\ \underset{y \in Y}{\max} \{S(x, y) \cdot T(y, z)\}\right] : x \in X, y \in Y, z \in Z\right\}$$

$$\therefore \quad \left[S \underset{\cdot}{\circ} T\right](x, z) = \begin{array}{c} \\ \text{bouncer} \\ \text{outswinger} \\ \text{inswinger} \end{array} \begin{array}{c} \begin{array}{ccc} \text{lo.rate} & \text{av.rate} & \text{hi.rate} \end{array} \\ \begin{bmatrix} 0.8 & 0.63 & 0.56 \\ 0.7 & 0.81 & 0.9 \\ 0.72 & 0.72 & 0.81 \end{bmatrix} \end{array} \tag{7.2}$$

Finally, we will examine the relationship between types of bowling and runs on the scoreboard by the composition of relations; in this, we will follow the procedure named the ***max–min composition***.

$S \circ T$ = Relationship between the type of bowling and runs on the scoreboard.

That is,

$$[S \circ T](x,z) = \left\{ \left[(x,z), \ \max_{y} \{\min\{S(x,y), T(y,z)\}\} \right] : x \in X, y \in Y, z \in Z \right\}$$

$$\therefore \quad [S \circ T](x,z) = \begin{array}{c} \\ \text{bouncer} \\ \text{outswinger} \\ \text{inswinger} \end{array} \begin{array}{c} \text{lo.rate} \quad \text{av.rate} \quad \text{hi. rate} \\ \begin{bmatrix} 0.8 & 0.7 & 0.7 \\ 0.7 & 0.9 & 0.9 \\ 0.8 & 0.9 & 0.9 \end{bmatrix} \end{array} \qquad (7.3)$$

After obtaining results in (7.1), (7.2), and (7.3), when we analyze these results, the result obtained in (7.3) is the best one. In other words, (7.3) is trustworthy and consistent. Keeping all the related arguments, the captain or team management of the concerned team can choose their bowlers.

7.15 Theory of Fuzzy Sets and Decision-Making

Now, we wish to depict an illustration that will exhibit the application of FST in the domain of decision-making. This exemplification concerned the issue of determining which of the handful of alternatives of old cars a person can choose that is available on "sell my car" under the tag—Sell my car in a car store.

Suppose five old cars are accessible on the sell their motorcar in a car store. Here, we presume that we wish to purchase a car in good condition provided that its cost is admissible, it has traveled a reasonable number of kilometers, year of purchase of the car is a little. Representing or describing all of these yardsticks in suitable fuzzy sets is possible.

Let us now signify the five old cars by the first letter of their names, which are B, H, F, N, and T. These symbols stand for Maruti Baleno, Hyundai i20, Toyota Fortuner, Tata Nexon, Mahindra Thar, respectively. The prices of these cars, the distances they have traveled, and the year of the purchase are depicted in the noted table (see Table 7.2).

Let the fuzzy sets representing the admissible cost of the car be C; the car traveled a reasonable distance is D, and the year of purchase is Y, as presumed by the customer, shown in Table 7.3.

Table 7.2 Primary particulars regarding five designated cars for sale in a car store

Name of cars	Prices of the cars in lakh	Traveled a reasonable number of kilometers	Year of purchase of the car
Toyota Fortuner–F	20	30000	5
Hyundai i20–H	7	5000	20
Maruti Baleno–B	8	10000	25
Tata Nexon–N	10	25000	15
Mahindra Thar–T	15	20000	10

Table 7.3 Degree of appropriateness of the cars in conformity with the individual yardstick

Cars	Degree of appropriateness of the cars in conformity with the individual yardstick					Degree of gross appropriateness
X	C(x)	D(x)	Y(x)	A(x)	S(x)	
B	0.61	0.41	0.91	1	0.5	0.41
H	0.95	1	0.58	0.58	1	0.58
F	0.88	0.58	0.58	0.68	1	0.58
N	0.81	1	0.75	0.28	0.9	0.28
T	0.75	0.75	0.75	0.88	0.9	0.75

Grades of membership of the cars in every one of the five fuzzy sets indicate the degrees of the properness of the cars following the yardstick characterized by the fuzzy set. As the customer is engrossed in fulfilling all the yardsticks, the extent of gross acknowledgment of the individual cars is identical to their grades of membership in the intersection of the five fuzzy sets. Membership grades of these cars, resolved by the standard fuzzy intersection, are depicted in the hindmost column under the noted table. Thus, the car to purchase has a towering degree of gross appropriateness, which is nothing but the car T.

7.16 Application of Fuzzy Group

The applications of fuzzy abstract algebra mostly revolve around computer science and its domain. Vague groups are widely applicable in image processing and understanding the nature of the image and computer vision. Nowadays, fuzzy subgroup theory is recycled to test a device's effectiveness that decodes messages sent through a noisy channel. Abstract pattern recognition analysis has also benefitted from fuzzy group theory (**Mordeson and Nair** 1998) and coding theory problems (**Cheng and Mordeson** 1996). The present scenario has adopted vague groups in classifying pieces of knowledge (**Mashinchi and Mukaidono** 1994). The anti-fuzzy version of group theory is now engaged in organizing bits of knowledge.

7.17 Application of Fuzzy Topology

Indeed, fuzzy spatial objects (rivers, buildings, trees, and roads) dominate GIS. In GIS, topological relations are used between spatial objects regarding positional and informational attributes. For topological consistency, it has been observed that topological links are helpful for spatial analysis, spatial queries, and data quality control. Topological relations may be classical or fuzzy. It depends on the certainty and uncertainty of the spatial objects and the characters of their concerns. In modeling spatial objects in GIS, it has been assumed that the measurement of spatial objects is free from errors. However, in the real sense, illustrations of the spatial objects in GIS consist of uncertainties such as vagueness and fuzziness in interpreting the boundaries of nature. It is hard to describe the fuzziness between rural and urban or the border between the states by traditional classical GIS. Hence it requires modification in the existing GIS system by further dealing with uncertainties in spatial objects and topological relations between spatial objects.

In this way, classical set theory needs to be improved as it is based on crisp boundaries, whereas FST is quite handy. It provides a potent tool for handling the uncertainty of a single object in GIS. With the advent of fuzzy sets, it has been observed that the theory of fuzzy topology can be used vigorously in modeling fuzzy topological relations among spatial objects. There are two steps in modeling ambiguous topological ties between spatial objects.

(i) To define and describe spatial relations qualitatively.
(ii) To compute the fuzzy topological relations quantitatively.

For the (i) step of modeling fuzzy spatial relations, several models have been developed by **Winter** (2000), **Cohn and Gotts** (1996), **Smith** (1996), and **Tang** and **Kainz** (2002, 2003, 2004). It provides all required notions of uncertain topological relation between spatial objects based on the descriptions of the interior and exterior of the spatial things in GIS for the (ii) step of modeling vague topological relations, i.e., to compute the membership values of interior boundary and external spatial objects based on fuzzy membership function. Real-world data is required for flood prediction. Then the developed GIS method is applied to compute spatial objects' interior, closure, and boundary. After analyzing the model, researchers derive the topological relation between spatial objects and the other flood forecasting work.

The boundary of a fuzzy set A is defined by $\partial A = A \wedge A^c$. The other notions commonly used to define fuzzy topological space are fuzzy interior, closure, complement, and boundary (**Liu and Luo** 1997). We know that each interior operator corresponds to one fuzzy topology, and each closure operator corresponds to another topology (**Liu and Luo** 1997). In general, if we define two separate operators, interior and closure, they will define two fuzzy topologies. These two topologies may not cohere with each other. If we consider two operators, interior and closure, separately, they will define two fuzzy topologies. These two topologies may not cohere with each other. **Pascali and Ajmal** (1997) also defined two similar operators, interior, and closure operators. These two operators may not further define a coherent fuzzy topology in their definitions. They give a necessary and sufficient condition for these two operators to be coherent with each other. In ordinary topology, when we define a topology, the boundary of a set A is defined as the intersection of the closure of A with the closure of the complement of A. That is, $\partial A = \overline{A} \wedge \overline{(A^c)}$.

Conversely, it has an equivalent uniform definition of $\partial A = \overline{A} - A^{\circ}$. The latter is no longer valid in fuzzy topology (**Liu and Luo** 1997; **Tang and Kainz** 2002; **Wu**

and Zheng 1991). However, to be consistent with the previous studies, researchers have adopted the former as the definition of fuzzy boundaries.

For applying fuzzy topology in flood prediction, they have considered a case study of Assam. Researchers have carried out a case study to compute spatial objects' interior boundary and exterior. This case study aims to determine the flood-affected area of upper Assam. Each area affected by the flood is recorded. The level of any area affected by a flood is marked by a percentage level within the interval [0, 1]. They have obtained the interior, closure, and boundary of each area affected by the flood for a particular value of α. For the case of the interior, if the value is 1, the effect is high concerning the overall affected area. If the value is closure zero, the effect is low concerning the widespread affected areas. They wish to classify spatial objects, fuzzy interior, boundary, and exterior for the actual Geography Information System up to date of the area affected by flood using a new model. **They have made some assumptions:-**

(i) Each land photo is shown as a fuzzy space.
(ii) The size of each area affected by a flood is used to calculate the unclear value of the fuzzy sets.
(iii) The undefined value of each region affected by the flood is defined by a formula which is a well-defined mapping from the interval $[1, \alpha]$ to the interval [0, 1].
(iv) The fuzzy interior and boundary have been computed for different α values in [0, 1], respectively.

For different α values, they have calculated the fuzzy exterior, interior, and fuzzy boundary, which are different in their values. Based on the calculations, they predict that the relation between *interior* and the size of the closure is directly proportional to each other. At the same time, the relation between the exterior and the size of the interior is inversely proportional. This model will be comfortable simplifying the fuzzy interior, boundary, and exterior of fuzzy spatial object. Thus, we can say that they have shown the dynamic application in GIS of the fuzzy interior, boundary, and exterior of affected areas by flood in Assam.

7.18 Application of Fuzzy Sets in Smart City Project

The innovative city concept integrates sensor data and communication technologies into city infrastructure to collect sensor data for allocating resources and assets and improving quality of life and sustainability. The problems and inefficient solutions in recent trends increase the requirement for intelligent applications to address the current issues efficiently. For a waste management system to be sustainable, it is desired to be environmentally effective, economical, affordable, and socially acceptable with maximum public involvement. Solid waste collection system selection is a crucial stage of a successful waste management plan involving multi-criteria decision-making problems that should be considered environmental, social, economic, and technical aspects simultaneously. Collection and transportation are the most costly due to these processes' labor intensity and massive vehicles in these processes. Type-2 fuzzy sets with membership functions are also fuzzy and collaborated with multiple criteria methodology to evaluate and rank waste collection systems for a smart city. Membership functions constitute a theoretical base for techniques that can handle problems with vague components. It can also be merged with other decision-making approaches like MABAC, TOPSIS, and VIKOR using fuzzy mathematics. In applying the type-2 fuzzy TOPSIS approach, attributes like environmental sustainability, innovativeness, and setup cost advantages are also considered. Fuzzy sets enabled us to successfully convert the vagueness and uncertainty in the decisions of subordinate decision-makers and scientific experts.

7.19 Transshipment Problem Under Uncertainty

The transshipment problem was introduced by *Alex Orden* in 1956. A transportation problem is a peculiar optimization problem to reduce the transportation price while transiting the trafficking of articles from different origins to destinations. However, intermediate nodes can reach the final destinations in some situations instead of transporting the commodities directly from the source to the landing place. Hence a transshipment problem can be familiar with to ship the available articles to the respective destinations through some intermediate nodes at a reasonable cost.

Zadeh introduced fuzzy sets to deal with problems in uncertain environments. *Atanassov* introduced an intuitionistic fuzzy set to resolve the accuracy problem with imprecision information characterized by its membership and non-membership values. *Smarandche*, in 1998, introduced the Neutrosophic set. The higher level literature review reveals a significant gap within the study of transshipment problems under a crisp environment. The gaps thus found can be studied. The journey can be predicted with the crisp range to confirm its occurrence. The majority of research works are focused on crisp problems, but there are still many open questions that have to be addressed in an uncertain environment. Therefore, this motivates us to consider uncertainty to studying transshipment problems. Studying the recent trends and developments in transshipment problems under uncertainty is a good cause. Analyzing and understanding the implementation of various fuzzy methods like intuitionistic fuzzy and Neutrosophic fuzzy in handling uncertainty will take the lead soon. The following table will present the transshipment problem studied in an uncertain environment by different researchers (see Table 7.4).

Although many models have been introduced to deal with uncertainty, a few limitations were found.

(i) The application of fuzzy sets in many real-life scenarios and decisions must be revised for accuracy.
(ii) The indeterminacy associated with the unawareness of the transshipment problem and poor data also exists.

Table 7.4 Extension of existing literature on transshipment problem under uncertain environment

Authors	Year	Significance
H. Prade	1980	Algorithms for fuzzy data
Chanas and Kuchta	1995	To find the optimal solution to transportation problem with fuzzy coefficients
Li and Lai	2007	Fuzzy compromise programming approach to multi-objective transportation problems
Jana and Roy	2007	Presents a new intuitionistic fuzzy optimization approach
Hussain and Kumar	2012	To obtain the optimal solution in terms of triangular intuitionistic fuzzy numbers
Pramila and Uthra	2014	To find the least transportation cost of some commodities through a capacitated network in the intuitionistic environment
Thamaraiselvi and Santhi	2016	Initiated the mathematical representation of a transportation problem in the neutrosophic environment
Ahmad and Adhami	2018	Multi-objective non-linear transportation problem formulated with fuzzy parameters
Pratihar et al.	2020	Neutrosophic set to solve the transportation problem

(iii) The intuitionistic fuzzy set can handle incomplete information but cannot handle indeterminate and inconsistent information.

(iv) The application of Neutrosophic logic may have some indeterminacy in deciding unit, transportation cost, supply, and demands due to several reasons like road factors, traffic regulations, etc.

Although various models of fuzzy sets in transshipment problems are observed to be the best fit, it fails to consider indeterminate and inconsistent information associated with the transshipment problems. Although exhaustive research is done on the transshipment problem under uncertainty, we observed specific significant gaps in the existing methods. In the future, researchers should aim to develop a better model that resolves these limitations of transshipment problems under uncertainty and applies it to more complicated community issues (Table 7.5).

This closing section of this text aims to exemplify the applications of fuzzy set theory in the domain of OR. "Program Evaluation and Review Technique and Critical Path Method are two closely related operations research techniques that use networks to coordinate activities, develop schedules, and monitor the progress of projects." Principally, PERT and CPM exhibit identical techniques. PERT presumes that the duration (activity time) is a random variable for each activity, whereas CPM uses known fixed activity timings. In the Program evaluation and review technique (PERT)—a project management technique employs three-time estimates for each activity, and in the Critical path method (CPM), a project management technique uses only one-time factor per activity. The various operations research methods play a significant role in our daily problems and requirements. This segment will present only CPM/PERT problems and work on their related aspects. It is observed that, in the recent past, the application focused mainly on crisp climate while failing to

Table 7.5 Transshipment

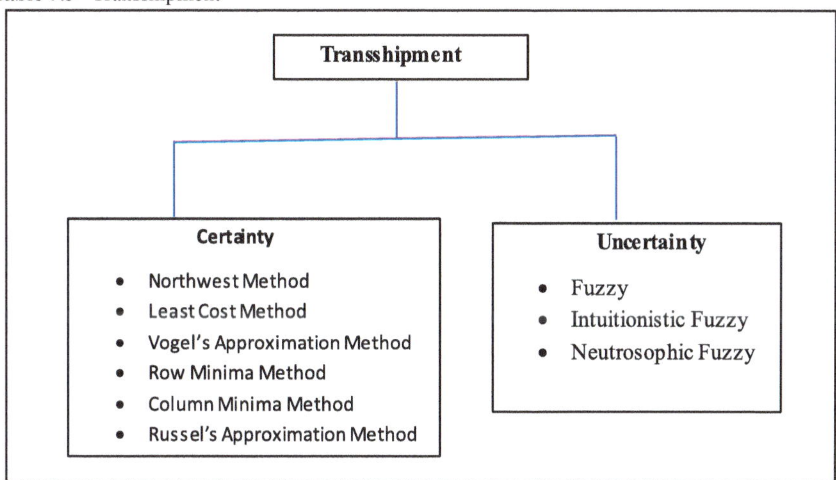

consider the uncertain environment, resulting in a gap. In 1975, **Zadeh** introduced the notion of fuzzy to handle uncertainty which various scientists later used to make an accurate prediction. On-going through the literature on CPM/PERT problems under uncertainty reveals substantial gaps in the existing models. As such, researchers may pay attention to these subsequent gaps.

Some investigations on CPM/PERT-related challenges still emerge in varied real-world applications. Although many researchers in this field have concentrated on crisp challenges, many unanswered questions still need to be answered. Few researchers are paying attention to CPM/PERT under various uncertain environments. The search for a solution to the problem of uncertain CPM/PERT remains a critical challenge. It requires a focus on approaches and modifications to existing methods. Therefore, this motivates researchers to consider practical uncertain analysis to solve CPM/PERT. Because of the above gaps, researchers' aim should be to analyze and review the content covering existing theories and research. Some in-depth analyzes of the methods are taken into consideration. As such, the objectives are to discuss the recent trends and improvements in CPM/PERT model. Identifying the current models' demerits under uncertain environments and providing insights on conflicts in previous studies and inconsistencies is required.

It is needed to analyze the implementation of various fuzzy methods, including intuitionistic fuzzy and Neutrosophic fuzzy, to handle the uncertainty discovered and the extent of it. In the under-noted table, the focus has been paid to the different methods of handling CPM/PERT in an uncertain environment (see Table 7.6).

Although the study shows the diverse methods to deal with uncertainty posed by the environment, a few limitations are observed.

- PERT/CPM, when not well-defined with required stability, can become ineffective for active project control.

Table 7.6 Extension of existing literature on CPM/PERT under uncertain environment

Authors	Year	Significance
H. Prade	1979	Developing an academic quarter schedule
Chanas and Kamburowski	1981	Apply fuzzy variables in activity time
Lorterapong	1994	Fuzzy resource-constrained project
Yogashanthi and Ganesan	2006	Solving networking problems in intuitionistic fuzzy environment
Jayagowri and Geetharamani	2014	Critical path problem by intuitionistic fuzzy set
Mousavi et al.	2015	Modeling production time
Mohamed et al.	2017	Using neutrosophic sets to obtain PERT estimates
Chakraborty	2020	Application in networking problem
Fernandez	2021	Computer system project management by using neutrosophic set statistics

CSMCS 22 Mar

- Although fuzzy PERT/CPM is observed to be a relevant fit to the nature of a problem compared to probabilistic models, it still fails to account for some subjective nature of parameters of an uncertain environment.
- While PERT applies to non-repetitive and new projects, fuzzy is inaccurate in certain instances. The results, heavily reliant on assumptions and human expertise, can be controversial.

These limits were then tackled by an intuitionistic fuzzy set [that considers the components dependent on membership and non-membership values]. Lack of data influences the uncertainty and indeterminacy of the situation, giving indeterminate and inconsistent information that limits this method. In 1995, **Smarandche** introduced the Neutrosophic set that best solves the current issues. However, the matter still has some indeterminacy in deciding various unpredictable factors and randomness.

7.19.1 Conclusion

The use and contribution of fuzzy set theory to problem-solving in production management research may be observed by comparing it to its application in operations research. In this section, we have investigated the previous and contemporary literature to understand the various Critical Path Method (CPM) and Project Evaluation Review Technique (PERT) problems under uncertainty. After doing an extensive literature survey and investigating the previous and contemporary literature to understand the various Critical Path Method (CPM) and Project Evaluation Review Technic

(PERT) problems under uncertainty, we conclude that the currently existing models have some demerits.

Although different methods of fuzzy set in PERT/CPM are observed to be a relevant fit to the nature of a problem when compared to probabilistic models, it still fails to account for some subjective nature of parameters, indeterminate and inconsistent information, and randomness of an uncertain environment. Even though extensive research has been conducted into the issues mentioned, there are still significant gaps in the future direction of present-day researchers. Their future aim should be to develop a new model that is better than the existing models in the literature by considering the uncertain environment and applying it to more complicated community issues.

7.20 Computing with Words: An Excellent Paradigm in Fuzzy Sets

In 1996 *Zadeh* published a paper that first equated Fuzzy Logic with Computing with Words (CWW or CW). *Zadeh* did not mean that computers would compute using words, a single word or phrase, rather than numbers. It was suggested that words could activate computers, which could then be converted into a mathematical model using fuzzy sets. These fuzzy sets would be mapped using a CWW engine into different fuzzy sets, which would be converted back into words. Computing with words means different things to different people. Humans have many remarkable capabilities; two stand out in importance. First is the ability to converse, communicate, reason, and in decision-making in an environment of imprecision, uncertainty, incompleteness, and partiality of truth.

Furthermore, second, the capability to perform various physical and mental tasks without any measurements and computations. These outstanding capabilities inspire Computing with words. In other words, Computing with Words is a methodology for decision-making, Computing, and reasoning with the information described in natural languages. It has been discussed in a paper entitled "A New Direction in AI—toward a Computational Theory of Perceptions" [AI Magazine, Springer 2001, pp.73–84]. The main point that was argued on the principal reasons for the AI's failure to address the problem of decision-making and reasoning with perception-based information can be the main reason for the slowness of the process. A key idea suggested in the paper was not dealing with perceptions but their descriptions in a natural language. Natural language is a system for the description of perceptions. At this point, when Computing with Words enters the picture. It is this methodology described in [IEEE Trans, on Circuits and Systems, vol. 45, no.1, pp. 105–119, 1999] that described "From Computing with Numbers to Computing with Words—from Manipulation of Measurements to Manipulation of Perceptions," that is needed for computation with natural language descriptions of perceptions. In many related works, Fuzzy logic is viewed in some publications as the machinery that will let "input" words provided by a human being transformed within the computer to "output" words, which are provided back to that human or other humans. Underlying Computing with Words are three principal rationales. First, most human knowledge is described in natural language.

Second, words are less precise than numbers—we use words when we need to know the numbers. Computing with Words can be a powerful tool for dealing with imprecise information. Lastly, precision carries a cost. Tolerance for imprecision can be handled through the use of words in place of numbers. The key idea that underlines the machinery of linguistic variables and fuzzy if–then rules is employed in all fuzzy logic applications, especially in consumer products and industrial systems. By adding AI, an effective tool for reasoning and Computing with the information used in natural language, Computing with Words opens opportunities for a wide-ranging enlargement of the role of natural languages in Science and Engineering. *Hani Hagras* extended the definition and commented, "Words are used when we do not know the numbers." Adding that when employing fuzzy sets, it will be better to employ higher order fuzzy sets like type-2 fuzzy sets rather than type-1 fuzzy sets where the higher order fuzzy sets will better model the uncertainty rather than type-1 sets which use precise and crisp membership function. *Sergio Guadarrama* points out that it is usually said that a natural language is a system for communicating information. However, we can communicate descriptions of perceptions, descriptions of actions, descriptions of intentions, and descriptions of feelings. *Lofti Zadeh* clarified a common misconception that CW and natural language processing are closely related. What is widely unrecognized at this juncture is that moving from computation with words has the potential to evolve into a fundamental paradigm shift that would open the door to a wide-range enlargement of natural languages' role in scientific theories. The connection between data and the fuzzy set model needs to be more present in the CWW literature. This connection should be made at the start of works about CWW because those works need to incorporate the uncertainties about words.

According to **Enric Trillas**, so far CWW is not a prevalent field of research; however it is one in course of development. So far it has not a clear-cut notion. By CWW, **Enric** means a challenge for those who are working in theoretic and the technology sides of fuzzy logic. In consequence, if this challenge is successful, it could give rise to CWW itself a new science, that will manage the concepts and tools of computer science and will interact with other disciplines. To work in CWW, it is good to have proficiency in fuzzy logic. Computing with words is a methodology for reasoning, computing and decision-making with information described in natural language.

It seems very interesting on the theoretical side that the design and realization of experiments inside natural language coupled with the meaning's depiction of complex linguistic expressions than those right now think about in fuzzy logic applications. Under experiments, one should go after two objectives:

(i) To get the present-day ceilings or limits of theoretical equipment's of fuzzy logic.

(ii) To develop and get new mathematical and algorithmic models, which made it to represent what cannot be possible employing it, for example, the algebra of fuzzy sets. Under such investigation, we should proceed to the conception of uncertainty arising from imprecision.

In the words of *Ronald Yager*, human beings can understand and express several kinds of knowledge by adopting language. According to him, one intention of Computing with Words is to empower the involvement of human-sourced knowledge in the formal computer-based decision-making models that are becoming attractive and potent. The translation process is the key to CWW. It involves considering the development of language's demonstrated knowledge and transforming it into a machine-manipulative pattern. The use of OWA furnishes an example of this transformation of processing rules.

To help the human perception of knowledge acquisition and details processing results. It requires a methodology of linguistic recap and interpretation. The OWA operator permits one to translate and implement aggregation rules expressed by linguistic quantifiers.

Here, it is pertinent to mention *Zadeh's* influential statement that Computing with words is a methodology in which the objects of computation are words and propositions drawn from a natural language. It is inspired by the remarkable human capability to perform various physical and mental tasks without any measurements or computations. This assertion is quite general, meaning that CWW is a comprehensive underlying methodology, making it affluent and prosperous as it is open to interpretations and incorporations.

As an engineer, **Mendel** asks, "How can I solve a specific problem(s) using CWW? He further says if I can do this, then CWW will be meaningful for me and have been defined/interpreted for this class of problems. Here, the word "solve" is foremost as it contains two main points-formulate (a solution) and implement (that solution). Both points are necessary for CWW to remain an imagination for solving the specific class of problems.

We believe "Computing with words" is an open-ended challenge and must remain so for its healthy growth. From time to time, many people have expressed their appreciation of CWW and its sublime rewards in fascinating words.

7.20.1 Fuzzy Logic = Computing with Words

The name signifies that Computing with words (CW) is a methodology in which words are used despite numbers for Computing and reasoning. Indeed fuzzy logic plays a significant role in computing with words and conversely. In brief, as an approximation, fuzzy logic may be equivalent to CW. Applications of fuzzy logic are growing day by day and in variety too. Its applications in basic sciences—particularly in mathematical and physical sciences—have become more apparent and tremendous. Even now, two questions are being raised repeatedly: (i) what do you mean by fuzzy logic? Moreover, (ii) What we can do with fuzzy logic that cannot be done by other programs such as probability theory, neural network theory, and predicate logic. Naturally, there is more to fuzzy logic than a technique for computing with words. Thus, equality in "Fuzzy Logic = Computing with Words" should be included, using parity to spotlight the importance of computing with words as an offshoot of fuzzy logic.

In its conventional perception, Computing requires handling/trickery of numbers and symbols. By divergence, individuals exercise mainly words in Computing and inference for reaching a conclusion exhibited as words from premises displayed in natural language or having the configuration of mental perceptions. Because when used by individuals, words have fuzzy implications. The same is true for the character performed by words in Computing with words.

The underpinning of Computing with words was set down some foretimes. Its emergence in terms of new methodology contemplates several developments in our interpretation of fuzzy logic and soft computing developments in the last few years. A crucial feature of Computing with words is that it requires a fusion of common languages and computations with fuzzy variables. This intermingling results in an evolution of Computing with terms into a fundamental procedure, having extensive consequences and implementation.

The point of escape in Computing with words is the notion of **granule**. In quintessence, a *granule* is a fuzzy set of points having the formulation of a clump of elements pinched together by similarity. The word w is a *label* of granule g and viceversa; g is the denotation of$w$. A word may be atomic (as in tall) or composite (as in not very tall). Unless otherwise stated, a word will always be assumed to be composite. The meaning of a word may be a higher order predicate. In computing with words, a granule$g$, the representation/meaning of a word w, is perceived as a fuzzy constraint on a variable. The central role in computing with words is played by fuzzy constraint propagation from premises to conclusion. One should remember that as a fundamental procedure, constraint propagation performs salient functions in several techniques, distinctly in mathematical programming, logic programming, and constraint programming.

To illustrate this fact, let us consider the proposition that Ananta is tall. In this case, tall is the level of a granule tall—the fuzzy set tall functions as the character of a fuzzy constraint on the height of Ananta.

As another example, we consider the propositions.

$$p_1 = \text{Brajesh lives near Ananta}$$

and

$$p_2 = \text{Ananta lives near Aniket}$$

In this case, "lives near" in p_1 and p_2 function as the character of fuzzy constraints on the distances between the homes of Brajesh and Ananta and Ananta and Aniket, respectively. If the inquiry is, "How far is Brajesh from Aniket?" an answer given by fuzzy constraint propagation may be expressed as p_3, where

$$p_3 = \text{Brajesh lives not far from Aniket}$$

The fundamental assumption in computing with words is that the details are delivered by limiting the values of variables. Moreover, information is supposed to contain a collection of propositions demonstrated in a common language or artificial language.

A general fundamental problem in computing with words is the following:

We have a given set of propositions demonstrated in a common language which compose the initial data set (IDS). We want to draw an answer to an inquiry shown in a common language from the IDS. The solution demonstrated in a natural language is called the terminal data set (TDS). The problem before us is obtaining TDS from IDS.

Exercise-7

1. Discuss, what are the various applications of fuzzy logic?
2. Explain the use of fuzzy logic in industrial applications.
3. Write a brief note on fuzzy logic applications in home appliances.
4. Discuss the applications of fuzzy topology in the area of geography.
5. Describe the meaning of CWW? Demonstrate it with the help of suitable examples.
6. Could you write a brief note on using fuzzy graphs in different domains of life?
7. Describe the uses of fuzzy logic in forecasting the score/ result of a cricket match.
8. Describe the applications of Fuzzy Sets in Science and Engineering.
9. Write a brief note on Fuzzy Sets and their uses in Computer Science.
10. Describe the applications of fuzzy sets and fuzzy logic in industries.
11. Explain in brief about the product based on fuzzy logic.
12. Discuss the advantage of fuzzy logic.
13. Describe **the** drawback of fuzzy logic systems.
14. Describe fuzzy relation equations and their uses in **a** neural network.
15. Illustrates the application of fuzzy relational equation in evaluating cardiovascular diabetic Mellitus.
16. Explain fuzzy sets in life insurance underwriting when **the** insurer is diabetic.
17. Write a brief note on fuzzy numbers in income tax filing.
18. Briefly describe the use of fuzzy set theory in psychology.
19. Analyze the application of fuzzy sets in the area of economics.
20. Describe the applications of fuzzy graphs.
21. Explain the use of intuitionistic fuzzy graphs.
22. Write a brief note on the uses of fuzzy set theory in medical sciences with the help of a suitable example.
23. Describe the relation between medical diagnosis and the law of modus ponens.
24. Describe the uses of the composition of fuzzy relations and prototypes for forecasting the score of a cricket match.

25. Explain the use of the fuzzy mathematical technique in decision-making.
26. Describe the uses of fuzzy groups in other areas.
27. Discuss the application of fuzzy sets in innovative city projects.
28. Write a brief note on the transshipment problem under uncertainty.
29. Why Computing with Words is becoming popular day by day? Explain its benefits.
30. Describe an example that shows the uses of fuzzy sets in Civil engineering.

Objective-Type Questions and Answers

1. Application of fuzzy mathematics in Computer Science provides us—with a method that makes the systems more

   (a) Flexible, realistic, and permits easy handling (b) lower solution cost,
   (c) Robustness (d) lower expressive power

2. The primary benefit of applying fuzzy sets in computer-based systems is gaining

   (a) Greater generality, (b) higher expressive power
   (c) enhanced ability to represent and manipulate knowledge and information expressed in natural language (d) none of these

3. Fuzzy set theory has also established its

   (a) Utility in pattern recognition (b) image processing (c) cluster analysis (d) soft computing

4. Pattern recognition technology can be restricted into

   (a) Unsupervised technique (b) supervised technique (c) complementary technique
   (d) none of these

5. Image processing is concerned with

   (a) cluster analysis (b) human knowledge (c) pattern recognition (d) none of these

6. A general fuzzy controller consists of some modules, such as

   (a) Fuzzy rule base (b) fuzzy inference engine (c) controlled process (d) fuzzification module

7. The most crucial fact about fuzzy controllers is that they are ……than conventional controllers

   (a) Less robust (b) more robust (c) much less robust (d) much more robust

8. Matsushita Electric Industrial Company named their washing machine based on a fuzzy controller as "Asai-go Day Fuzzy." What does it mean?

   (a) Beloved daughter (b) beloved son (c) beloved wife (d) beloved husband

9. Fuzzy relation equations can be expressed in terms of

(a) Fuzzy controller (b) defuzzification module (c) fuzzy inference engine (d) neural networks

10. Neural networks refer to the system of

    (a) Neurons, (b) artificial neurons
    (c) organic neurons (d) none of these

11. The area of evolutionary computing includes

    (a) Evolutionary strategies (b) genetic algorithms (GA)
    (c) fuzzy controller (d) evolutionary programming.

12. Genetic algorithms are a particular class of evolutionary algorithms that use techniques inspired by evolutionary biologies such as

    (a) Crossover (b) Inference rule (c) mutation (d) inheritance

13. Out of the following, which are included in soft Computing

    (a) Neural networks (b) linguistic hedges (c) fuzzy technology (d) fuzzy logic & neural networks

14. If we define a type-2 fuzzy set (a population of human beings) as "intelligent," membership grades assigned might be type-1 fuzzy sets such as

    (a) Superior (b) light (c) genius (d) heavy

15. In fuzzy clustering, a data point can have a membership in

    (a) One and only one cluster (b) multiple clusters (c) cluster analysis (d) none of these

16. The process of adding a new point to a cluster and then recomputing the center is called

    (a) Euclidean distance (b) entropy (c) iteration (d) cluster seeds

17. The data used to design a pattern recognition system are usually divided into some categories, such as

    (a) Design data (b) test data (c) clusters (d) none of these

18. The first applications of the fuzzy theory were primarily industrial, such as process control for

    (a) Water treatment plant (b) elevators (c) subway (d) cement kilns

19. Soft computing is an umbrella used for a set of computing techniques that consists of

    (a) Genetic algorithm (b) artificial intelligence (c) Fuzzy grammar (d) fuzzy logic

20. Fuzzy logic is a computational paradigm that equips us with a mathematical tool for dealing with the
    (a) Uncertainty, (b) imprecision
    (c) imperfect environment (d) with unsharp boundaries.

Answers

1. (d); 2. (a, b, & c); 3. (a, b, c, and d); 4. (a & b); 5. (a & c); 6. (a, b, & d); 7. (b); 8. (c); 9. (d); 10. (a, b, & c); 11. (a, b, & d); 12. (a, c, & d); 13. (a & d); 14. (a & c); 15. (b); 16. (c); 17. (a & b); 18. (d); 19. (a, b, & d); 20. (a, b, c, & d).

True/False Type Statements and Answers

1. Application of fuzzy mathematics comprises fuzzy information retrieval systems, fuzzy expert systems, fuzzy databases, pattern recognition, image processing, and cluster analysis
2. Bezdek developed a powerful classification method to accommodate fuzzy data in 1981, called -means clustering.
3. Pattern recognition technology can be restricted into two broad types: unsupervised and supervised techniques.
4. Clustering algorithms do not enable us to attain goals to differentiate the yield of alternating clustering algorithms for a data file.
5. Clustering algorithms enable us to ascertain whether a structured data collection contains any structure.
6. As far as image processing is concerned, it is not associated with cluster analysis and pattern recognition.
7. Data are always provided in the form of digital images.
8. Digital image processing has no advantages over analog image processing.
9. Fuzzy controllers are a class of knowledge-based controllers using AI techniques and fuzzy logic to compute an appropriate control action.
10. In-home appliances, the label "fuzzy controllers," does not push up positive associations in high-quality, user-friendly, or modern appliances.
11. Fuzzy control leads us to more automation for complex and ill-structured processes.
12. The most crucial fact about fuzzy controllers is that they are less robust than conventional controllers.
13. The word intelligence was introduced for the word fuzzy in Japan.
14. Fuzzy set theory has also been applied in studying earthquake and solid waste management.
15. Under FST, the designer can describe the sketched memento as roughly as desired at the beginning of the design activity.
16. A product, "Auto transmission," which works to check the fuel injection and oxidization based on throttle setting, and cooling water temperature, was developed by Nissan.
17. Fuzzy logic may not offer accurate reasoning but sole admissible reasoning.

18. We need more than fuzzy logic in data mining to help us deal with engineering uncertainty.
19. Fuzzy logic is the most robust, as no precise inputs are required.
20. Fuzzy logic cannot simply be modified to enhance or change system accomplishment.
21. Some fuzzy time logic needs to be clarified with probability theory and the terms.
22. Fuzzy relation equations cannot be expressed in terms of neural networks.
23. Neural networks refer to the system of neurons, either artificial or organic.
24. *Soft Computing* is an umbrella for computing techniques comprising GA, AI, and fuzzy logic.
25. Data Mining does not aim to discover knowledge from data and present it in an easily compressible form to humans.
26. Fuzzy Systems allows us to introduce learning and adaptation capabilities.
27. Fuzzy logic is a computational paradigm that equips us with a mathematical tool for dealing with the uncertainty and the imprecision typical of human reasoning.
28. The fuzzification of neural networks extends their capabilities in applicability.
29. Fuzzy numbers do not perform a significant role in several applications such as fuzzy control, statistics with imprecise probabilities, decision-making, approximate reasoning, and optimization.
30. In decision-making, intervals are used for signifying acceptable or unacceptable values of pertinent numerical variables.
31. Linguistic variables and linguistic terms for questionnaires investigating the relation between social stresses and psychological factors are being extensively made in psychology.
32. Psychometric assessments are used extensively to objectively measure the individual's personality traits, aptitude, intelligence, behavioral style, and abilities.
33. The fuzzy graphic models do not describe traffic networks, rail networks, telephone networks, road networks, social networks, broadcast communications, artificial reasoning, and neural systems.
34. The fuzzy graph is nothing but an assertion of ambiguous relation; therefore, the fuzzy graph is continually being expressed as a fuzzy matrix.
35. The science of decision-making in the medical process is fuzzy.
36. Fuzzy logic is not a potent way to provide the physician with the support he needs to handle linguistic concepts and eliminate the loss of accuracy.
37. In the realm of medical diagnosis, it was *Albin* who first exhibited the enjoyment of fuzzy set theory.
38. The theory of fuzzy topology can be used vigorously in modeling fuzzy topological relations among spatial objects.
39. In applying the type-2 fuzzy TOPSIS approach, attributes like environmental sustainability, innovativeness, and setup cost advantages do not apply to innovative city projects.
40. Computing with Words is a three principal rationales.

Answers

(T); 2. (T); 3. (T); 4. (F); 5. (T); 6. (F); 7. (T); 8. (F); 9. (T); 10. (F); 11. (T); 12. (F); 13. (T); 14. (T); 15. (T); 16. (T); 17. (F); 18. (F); 19. (T); 20. (F); 21. (T); 22. (F); 23. (T); 24. (T); 25. (F); 26. (T); 27. (T); 28. (T); 29. (F); 30. (T); 31(T); 32. (T); 33. (F); 34. (T); 35. (T); 36. (F); 37. (T); 38. (T); 39. (T); 40. (T).

References

1. A. Rosenfeld, Fuzzy groups. J. Math. Anal. Appl. **35**, 512–517 (1971)
2. A. Rosenfeld, *An Introduction to Algebraic Structures* (Holden-Day, San Francisco, 1968)
3. A. Rosenfeld, *Fuzzy graphs; fuzzy sets and their applications (Zadeh, Fu, Tanaka and Shimura)* (Academic Press, New York, 1975), pp. 77–96
4. A. Rosenfeld, How many are a few? Fuzzy sets, fuzzy numbers, and fuzzy mathematics. J. Math. Anal. Appl. **04**, 139–143 (1982)
5. A. Kaufman, *Introduction to the Theory of Fuzzy Subsets* (Academic Press, New York, 1975)
6. G.J. Klir, B. Yuan, *Fuzzy Sets and Fuzzy Logic; (Theory and Applications)* (PHI. PVT Limited, New Delhi, 2012)
7. G.J. Klir, U.H. St. Clair, B. Yuan, *Fuzzy Set Theory (Foundations and Applications)* (PHI. Prentice Hall PTR Upper Saddle River, NJ 07458, 1997)
8. G.J. Klir, T.A. Folger, *Fuzzy Sets, Uncertainty, and Information* (PHI Learning, New Delhi, 1988)
9. H.J. Zimmermann, *Fuzzy Set Theory and Its Applications* (Kluwer, Boston, 1985)
10. T.J. Ross, *Fuzzy Logic with Engineering Applications* (McGraw-Hill, New York, 1995)
11. D. Dubois, H. Prade, *Fuzzy Sets and Systems: Theory and Applications* (Academic Press, New York, 1980)
12. A. Kaufmann, M.M. Gupta, *Introduction to Fuzzy Arithmetic: Theory and Applications* (Van Nostrand Reinhold Company, New York, 1985)
13. C.L. Chang, Fuzzy topological spaces. J. Math. Anal. Appl. **24**, 182–190 (1968)
14. R. Lowen, Fuzzy topological spaces and fuzzy compactness. J. Math. Anal. Appl. **56**, 621–633 (1976)
15. J.A. Goguen, L-fuzzy sets. J. Math. Anal. Appl. **18**, 145–174 (1967)
16. J.A. Goguen, The fuzzy tychonoff theorem. J. Math. Anal. Appl. **43**, 734–742 (1973)
17. P.P. Ming, L.Y. Ming, Fuzzy topology I: neighbourhood structure of a fuzzy point moore smith convergence. J. Math. Anal. Appl. **76**, 761–764 (1980)
18. N.P. Mukherjee, Fuzzy normal subgroups and fuzzy cosets, information. Science **34**, 225–239 (1984)
19. P. Bhattacharya, Fuzzy subgroups: some characterizations I. J. Math. Anal. Appl. **128**, 241–252 (1987)
20. P. Bhattacharya, Fuzzy subgroups: some characterizations II. Inf. Sci. **38**, 293–297 (1986)
21. P.S. Das, Fuzzy groups and level subgroups. J. Math. Anal. Appl. **84**, 264–269 (1981)
22. J.M. Anthony, H. Sherwood, Fuzzy groups redefined. J. Math. Anal. Appl. **69**, 124–130 (1979)
23. V.N. Dixit, R. Kumar, N. Ajmal, Level subgroups and union of fuzzy subgroups. Fuzzy Sets Syst. **37**, 359–371 (1990)
24. F.I. Sidky, M. Atif Mishref, Fuzzy cosets and abelian fuzzy subgroups. Fuzzy Sets Syst. **43**, 243–250 (1991)

M. K. Singh, *Applied Fuzzy Mathematics*, Forum for Interdisciplinary Mathematics,
https://doi.org/10.1007/978-981-97-3257-9

25. C.K. Wong, Fuzzy topology: product and quotients theorems. J. Math. Anal. Appl. **45**, 512–521 (1974)
26. C.K. Wong, Fuzzy points and local properties of fuzzy topology. J. Math. Anal. Appl. **46**, 316–328 (1974)
27. C.K. Wong, Fuzzy topology: product and quotient theorems. J. Math. Anal. Appl. **45**, 512–521 (1974)
28. L.A. Zadeh, Fuzzy sets. Inf. Control. **8**, 338–353 (1965)
29. L.A. Zadeh, Similarity relations and fuzzy orderings. Inf. Sci. **3**, 177–200 (1971)
30. L.A. Zadeh, Fuzzy logic and approximate reasoning. Synthese **30**(1), 407–428 (1975)
31. L.A. Zadeh, The concept of linguistic variables and its application to approximate reasoning. Inf. Sci. **8 & 9**, 199–249 (1975)
32. J.G. Brown, A note on fuzzy sets. Inf. Control **18**, 32–39 (1965)
33. M. Mizumoto, K. Tanaka, Some properties of fuzzy sets of type 2. Inf. Control **31**, 312–340 (1976)
34. M. Mizumoto, Fuzzy sets and their operations. Inf. Control **48**, 30–48 (1981)
35. M. Akgul, Some properties of fuzzy groups. J. Math. Anal. Appl. **18**, 145–174 (1988)
36. M.K. Singh, *Theory of Fuzzy Structures and Applications (Fuzzifications of Mathematical Notions)* (Lambert Academic Publishing, Germany, 2010)
37. M.K. Singh, A Study in The Axiomatic Foundations of Set Theory; Ph. D. Thesis, M. U. Bodh-Gaya (1985)
38. M.K. Singh, On fuzzy subgroupoids. J.B.M.S. **23**, 109–113
39. D. Driankov, H. Hellendoorn, M. Reinfrank, *An Introduction to Fuzzy Control* (Narosa Publishing House, 2001)
40. J. Yen, R. Langari, *Fuzzy Logic, Pearson Education* (2003)
41. R.R. Yager, A characterization of the extension principle. Fuzzy Sets Syst. **18**, 205–217 (1986)
42. R.R. Yager, L. Hedge, Their relation to context and their experimental realization, 357–374 (1982)
43. H.M. Al Mutab, Fuzzy graphs. J. Adv. Math. **17**, 232–247 (2019)
44. A. Nagoor Gani, K. Radha, The degree of a vertex in some fuzzy graphs; I. J. Algorithms Comput. Math. **2**, 107–116 (2009)
45. A. Nagoor Gani, K. Radha, On regular fuzzy graphs. J. Phys. Sci. **12**, 33–40 (2008)
46. A. Nagoor Gani, S.R. Latha, On irregular fuzzy graphs. Appl. Math. Sci. **6**, 33–44 (2012)
47. A. Nagoor Gani, M. Basheer Ahmed, Order and size in a fuzzy graphs. Bull. Pure Appl. Sci. **22E**(1), 145–148 (2003)
48. K.T. Atanassov, Intuitionistic fuzzy set. Fuzzy Sets Syst. **20**, 87–96 (1983)
49. K.T. Atanassov, A. Shannon, A first step to a theory of the intuitionistic fuzzy graph, in *Proceedings of the First Workshop on Fuzzy Based Expert System*, D. Lakav, Sofia (1994), pp. 59–61
50. K.T. Atanassov, A. Shannon, On a generalization of intuitionistic fuzzy graph. NIFS **12**,24–29 (2006)
51. N. Gani, S.S. Begum, Degree, order and size in intuitionistic fuzzy graph, I. J. Algorithms Comput. Math. **3** (2010)
52. M.G. Karunambigai, R. Parvathi, Intuitionistic fuzzy graphs. Comput. Intell.-Theory Appl. **6**, 139–150 (2006)
53. F. Samarandache, Neutrosophic set-a generalization of the intuitionistic fuzzy set. Int. J. Pure Appl. Math. **24**, 38–42 (2005)
54. F. Samarandache, Plithogenic set. APS Meeting Abstracts (2018)
55. F. Samarandache, *Plithogenic Set an Extension of Crisp, Fuzzy, Intuitionistic Fuzzy, and Neutrosophic Sets-Revisted*. Infinite Study
56. D. Dubois, H. Prade, New results about properties and semantics of fuzzy set-theoretic operators, in *Fuzzy Sets; Theory and Applications to Policy Analysis and Information Systems* (Plenum Press, New York, 1980), pp. 59–76
57. W. Bandler, L.J. Kohout, On new types of homomorphisms and congruences for partial algebraic structures and n-ary relations. Int. J. Gen. Syst. **12**(2), 149–157 (1986)

58. W. Bandler, L.J. Kohout, On the general theory of relational morphisms. Int. J. Gen. Syst. **13**(2), 47–68 (1986)
59. J.J. Buckley, Solving fuzzy equations. Fuzzy Sets Syst. **50**(1), 1–14 (1992)
60. M. Sugeno, *Fuzzy Measures and Fuzzy Integrals; Fuzzy Automata and Decision Processes* (North Holland, 1977), pp.89–102
61. R.R. Yager, On a general class of fuzzy connectives. Fuzzy Sets Syst. **4**(3), 235–242
62. R.R. Yager, Measuring tranquillity and anxiety in decision making: an application of fuzzy sets. Int. J. Gen. Syst. **8**(3), 139–146 (1982)
63. H.T. Nguyen, A note on the extension principle for fuzzy sets. J. Math. Anal. Appl. **64**, 369–380 (1978)
64. P.R. Halmos, *Naïve Set Theory* (Van Nostrand, New York, 1960)
65. J. Lake, Sets, fuzzy sets, multi sets and functions. J. Lond. Math. Soc. **12**, 323–326 (1976)

GPSR Compliance

The European Union's (EU) General Product Safety Regulation (GPSR) is a set of rules that requires consumer products to be safe and our obligations to ensure this.

If you have any concerns about our products, you can contact us on ProductSafety@springernature.com

In case Publisher is established outside the EU, the EU authorized representative is:

Springer Nature Customer Service Center GmbH
Europaplatz 3
69115 Heidelberg, Germany

Batch number: 10370708

Printed by Printforce, the Netherlands